T0296631

The Physics of the Violin

The Physics of the Violin

Lothar Cremer
translated by John S. Allen

The MIT Press
Cambridge, Massachusetts
London, England

New material and English translation © 1984 by the Massachusetts Institute of Technology. Original edition published under the title *Physik der Geige* by S. Hirzel Verlag, Stuttgart. © 1981 by S. Hirzel Verlag.

This book was set in Times New Roman by Asco Trade Typesetting Ltd., Hong Kong

Library of Congress Cataloging in Publication Data
Cremer, Lothar, 1905–
  The physics of the violin.

  Translation of: Physik der Geige.
  Includes bibliographical references and index.
  1. Violin—Construction. 2. Music—Acoustics and physics. I. Title.
ML802.C7313   1983   787.1′2′0153   84-3920

ISBN 978-0-262-03102-8 (hc : alk. paper), 978-0-262-52707-1 (pb)

Contents

Preface to the American Edition

In the present translation I have made some improvements over the original German edition. It is self-evident that I have corrected all errata found during translation. I have also rewritten some text to make the derivation of ideas and formulas clearer (e.g., section 9.6 and section 10.2). I have taken advantage of the unavoidable delay between publication of the original and that of the translation to insert some text concerning essential progress of our scientific knowledge about the violin, as follows. The new section 5.6 concerns the influence of possible oscillations of the string perpendicular to the bow. The new section 8.8 includes the important question of transients, and extends the signal flow diagrams for computer simulation of a bowed string to transients. The extension of section 13.5 and the new section 13.6 discuss the general problem of whether the sound field of any body vibrating in air can be synthesized by means of spherical sound fields immediately outside its surfaces.

The improvements in this edition would not have been possible without the correspondence of colleagues, especially E. Jansson (Stockholm) M. E. McIntyre (Cambridge, England), T. J. Schultz (Cambridge, Massachusetts), and G. Weinreich (Ann Arbor). I am also indebted to The MIT Press and to the translator, J. Allen, for their readiness to include all these additions, which I hope will prove to be profitable to the reader.

Lothar Cremer
Miesbach, October 1983

Preface to the German Edition

The present monograph is the result of more than twenty years' involvement with the physics of the violin. As director of the Institute for Acoustical Engineering of the Technical University of Berlin, I had the opportunity to direct many students in their class and thesis work in this area, and later to assist colleagues in their dissertation work. Studies involving string instruments were among the many problems we dealt with in all areas of acoustics. Apparatus and personnel for these investigations were funded by the Deutsche Forschungsgemeinschaft (German Research Association), to which I once more express my thanks at this time.

After I retired from my directorship of the Institute, I reviewed this material, compiled it, and conducted additional investigations of my own. Because most of the research reports were unpublished, and the rest were widely dispersed through the literature, I decided to assemble them into a book. Naturally, I have included contributions by authors from all around the world, and from the classics to the present; yet it has been possible to review only the results of selected investigations, and these are cited as references.

An extremely large number of papers on the physics of string instruments has appeared over the past two decades. The *Catgut Acoustical Society Newsletter* alone has been published twice yearly since 1964, presenting a continuing series of interesting reports. Additionally, the cofounder and secretary of this society, Carleen M. Hutchins, has published two volumes of reprints of works on the instruments of the violin family as part of the series Benchmark Papers in Acoustics (Hutchins 1975, 1976). Of this store of work, too, only a part could be included and evaluated.

I have deliberately avoided certain controversial topics, such as the physical significance of the varnish used on string instruments, or the attributes that determine the quality of the violins made by the old master builders. I have generally avoided questions concerning the subjective evaluation of musical instruments; I consider research in this area not to have reached sufficient maturity. In the second edition of *Principles and Applications of Room Acoustics* (Cremer and Müller 1982), the psychological aspects were given an entire part of their own, amounting to several chapters. In architectural acoustics, however, it is now possible to compare different rooms by means of sound recordings made using dummy heads. These can be played back, one after another, to the subjects immediately, over a constant signal transmission path. The subjects do not know where the signals originated, and so the judgments are not influenced by such knowledge. Furthermore the judgments are made by many subjects, and so can be evaluated using multivariate methods. Until a corresponding procedure is applied to the subjective judgments of musical instruments, the reliability of the those judgments will remain a matter of our confidence in the ability of the judging personality. Though such confidence may be well justified in many cases, a natural scientist find it no substitute for the statistical results of psychometric investigations.

It goes without saying that the material presented here will be most easily accessible to acousticians; in fact, they will find much in this book that they already know well. The general equations for a string under tension, incorporating its torsional and bending stiffness, the equations for plates, including anisotropy and curvature, and especially the equations for sound fields in air are intentionally reviewed in order to make the book accessible to all physicists and, in fact, to all persons working in the natural sciences who are familiar with mathematical representations of such phenomena. People who read the text between the formulas will, however, also be able to draw on the book for many facts that need not be expressed mathematically. The book deals with theory, but it also explains the most recent experimental techniques.

Although I have great respect for the craftsmanlike achievements of the violin-maker's trade, and though believe that these achivements will remain indispensable in the future, I am of the opinion that the master violin maker, too, can derive useful benefit from the newer methods of measurement, in order to attain the goal of consistently producing first-rate instruments. I hope, therefore, that violin makers will also take an interest in this book.

Similarly, the book may also be of interest to players of string instruments. We acousticians do not expect our appreciation of musicians and their performance to be returned often, not so much because our education differs from that of musicians as because artists tend to be skeptical about scientific undertakings—particularly when these involve electronics and computers. This tendency has been reinforced by a widespread critical attitude toward the natural sciences, which has followed as a reaction to a period of overvaluation of scientific successes.

I render my sincerest thanks to the publishers, Hirzel Verlag of Stuttgart, for the excellent presentation and for their sensitivity to my wishes.

Finally, this monograph would not exist without the support of colleagues who furnished illustrations and who shared information with me by letter. I thank them with footnotes and citations at the relevant places in the text.

References

Cremer, L., and H. A. Müller, 1982. *Principles and Applications of Room Acoustics*. Tr. by T. J. Schultz. London: Applied Science.

Hutchins, C. M., ed., 1975. Benchmark Papers in Acoustics, vol. 5: *Musical Acoustics, Part 1*. Stroudsburg, Pa.: Dowden, Hutchinson & Ross.

Hutchins, C. M., ed., 1976. Benchmark Papers in Acoustics, vol. 6: *Musical Acoustics, Part 2*. Stroudsburg, Pa.: Dowden, Hutchinson & Ross.

Glossary of Symbols

The notation in this book is based upon that recommended by the AEF (Normen-Ausschuss für Einheiten und Formelzeichen: Commission for the Standardization of Units and Symbols). This notation follows International Standards Organization recommendations. We adopt the following conventions:

underlining	complex quantity
* to the right of symbol	conjugate complex value
⌢ above symbol	peak value
⁻ above symbol	arithmetical average
~ above symbol	rms value
′ after symbol	differentiation, or (sub)division with respect to length
″ after symbol	(sub)division with respect to area

The following lists give the meanings of mathematical symbols and abbreviations. Also given is the number of the formula where each symbol is first used.

Latin capital letters

A	Equivalent absorption area (16.14)
$\underline{A}$	Admittance (inverse of impedance $\underline{Z}$) (8.62)
B	Bending stiffness (7.2)
C	Capacitance (8.3a); clearness index (16.24)
D	Torsional stiffness (9.12)
E	Modulus of elasticity (2.38); electrical field strength of a light wave (9.7); energy density (14.15)
F	Force. F_x, tension on a string; F_y, force transverse to a string (2.2); F', force per unit length (4.9)
G	Shear modulus (6.4); electrical conductivity (8.4b); $\underline{G}(\varphi, \delta)$, directional Green's function (10.17)
He	Helmholtz number
I	Area moment of inertia. I', area moment of inertia per unit width (11.2)

J	Intensity (9.1)
K	Adiabatic volume stiffness of air (modulus of compression) (10.9)
L	Inductance (8.16). Length of a plate (11.8); L_p, sound pressure level (11.55)
M	Moment of a force. Torsional moment (6.3); bending moment (7.2); dipole moment (13.28)
N	Ordinal number (4.15), particularly the order of a natural resonance (11.17)
P	Power (4.42); static air pressure (10.9)
$Q'_{mn}(\vartheta, \varphi)$	Normalized distribution of normal velocity on any surface producing a spherical field (section 13.6)
R	Frictional force (1.1); ohmic resistance (8.2a); radius of curvature (11.28)
$R(r)$	Normalized function of the radial distance r in a spherical field (13.43)
S	Area (2.38)
T	Period (1.13); torsional stiffness (6.3); reverberation time (16.22a)
U	Circumference (10.36); velocity component normal to a surface (section 13.6)
V	Volume (10.10)
W	Energy (9.11); elements of the impedance matrix of the violin body (9.41)
Y	Elements of the admittance matrix of the violin body (9.44)
Z	Impedance; characteristic resistance of the string (2.21)

Latin lowercase letters

a	Radius of the string (6.2); half the distance between the feet of the bridge (9.2); element of an admittance matrix (9.5); width of a shell (11.53); radius of a sphere (13.3)
b	Length of a shell (11.53)

c	speed of propagation of a wave (2.1); distance in a model of a bridge (9.32c); c_L, speed of propagation of longitudinal waves (2.38); c_T, of torsional waves in a string (6.7); c_S, of transverse waves in a string (6.8); c_B, phase speed of a bending wave (11.13)
d	Distance in a model of a bridge (9.12)
e	Distance in a model of a bridge (9.23b)
f	Frequency of oscillation; with subscripts, natural frequencies; force divided by twice the characteristic resistance of the string (8.43)
$g(\varphi, \delta)$	Normalized direction function (14.3)
$g(t)$	Green's function (impulse response of the velocity of the string) (8.44)
g	Subscript for sliding ("gliding") friction (1.11); for propagation away from ("going from") the bow (8.5)
h	Thickness of a plate (11.1)
h	Subscript for sticking ("holding") friction (1.12)
i	Electric current (8.1b); radius of inertia (1.12)
k	Wave number (2.29)
$k(t)$	Green's function for excitation by periodic impulses (8.56)
k	Subscript for propagation toward the bow (kommend = coming) (8.5)
l	Length; length of the string (2.16); length of the opening of a Helmholtz resonator (10.5a)
m	Mass (1.1); ordinal number (3.26)
n	Ordinal number (2.23)
p	Sound pressure in air (11.55)
q	Coordinates of position (9.11); sound flux (10.3); volume flow (10.4)
r	Coefficient of friction (1.21); reflection factor (5.21)
s	Spring stiffness (1.1); spring stiffness of the bridge (5.18a)
t	Time (1.1); Transmission factor (6.45); t_T, transformation factor (6.46)

u Velocity relative to its lowest value (5.6); electrical voltage (8.1b); coordinate of position (11.47)

$\ddot{u}$ Force transfer factor (9.6)

v Velocity; v_B, v_b, bowing speed (1.10), (8.15)

w Resistance of the input of a bridge model (5.18a); angular velocity (6.2)

x Coordinate of position, usually parallel to the string (2.1)

y Coordinate of position, usually parallel to the bow (11.14); length of an arc across a curved plate (11.30)

z Coordinate, usually perpendicular the top and back plate (11.57b)

Greek capital letters

Γ Direction factor (14.4)

Δ A large increment in the following term (1.3), for example a step in velocity in Helmholtz motion (3.4)

Θ Normalized period (5.41); Θ', moment of inertia per unit length of the string (6.5); $\Theta_{mn}(\vartheta)$, normalized function of the declination angle ϑ in a spherical field (13.43)

$\Phi_n(\varphi)$ Normalized function of the azimuth angle φ in a spherical field (13.43)

Ω Normalized radian frequency (11.52); solid angle (13.4)

Greek lowercase letters

α Angle of inclination of the string (2.2); absorption coefficient (4.23); angle of incidence of a laser beam (9.10a)

β Parameter increasing with bowing pressure (5.4); exit angle of a laser beam (9.10a); frequency parameter (11.113)

γ Angle of inclination of the string for a lateral point load (7.18)

δ A small increment in the following term (1.20); decay coefficient (2.25)

ε Specific extension (2.39); loss parameter at the bridge (5.24)

ζ Displacement in the z-direction (perpendicular to the plate) (2.41b); torsion parameter of the string (5.21)

η Displacement in the y-direction (1.1); loss factor (10.53)

ϑ Normalized time (5.35); angle of incidence (9.7); angle of declination (13.26a); distinctness coefficient (16.23)

$\varkappa$ Bending-stiffness parameter of the string (7.50); specific heat capacity ratio (10.9); curvature parameter (11.52a)

λ Wavelength (2.30)

μ Coefficient of friction (3.32); Poisson's number for contraction (11.3)

ξ Displacement in the x-direction (11.56)

ϱ Density (11.1); $\varrho(v)$, normalized sliding friction as a function of relative velocity

σ Coefficient of increase (1.23)

τ Decay time (2.25)

φ Phase angle (figure 10.9); azimuth angle (14.13)

χ Angle of rotation. $\chi(t)$, reflection response to a unit step of incident angular velocity, χ_0 at the bridge, χ_1 at the nut or finger

ψ Reflection response to a unit step of incident velocity, ψ_0 at the bridge (5.33), ψ_1 at the nut or finger (8.6)

o Angular frequency of an antiresonance (10.48)

ω Angular frequency; radian frequency; with subscripts, natural frequencies

Other symbols

$\mathscr{R}_{rad}$ Radiation (flow) resistance (13.24b)

The Physics of the Violin

Introduction

Status and goal of physical research on string instruments

String instruments have been played for probably more than a thousand years. Performance on string instruments has reached a level today that cannot be improved upon, and the quality of instruments built nearly three centuries ago is maintained in the best instruments made today. The quality of both performance and instruments has been achieved through empirical methods; consequently, our application of the "exact" methods of the natural sciences will not result in any fundamental improvement.

Nonetheless, natural scientists—including some well known for first-class work in other fields of study—have again and again undertaken research into string instruments, and especially into the bowing process. They have done this not in the expectation that they might effect improvements, but rather with the hope that they might be able to explain the phenomena underlying the operation of string instruments in physical terms and to describe these phenomena quantitatively. String instruments pose "only" problems of classical mechanics and acoustics, fields which were well understood in Helmholtz's time; no new effects are today to be expected. And yet today, modern physics, especially through electronics, allows a continuing improvement in the measurement of subtle forces and motions. We now recognize to an even greater degree than our precursors that performance on string instruments is subject to many influences, not all of them yet understood. Neither can we judge which are significant and which are not.

Though we have by no means reached the mountain peak that would allow a comprehensive overview, we will still find it worthwhile to define the present status of our knowledge. In addition to describing processes, it is important also to show how they relate to the basic laws of acoustics, as long as we do not overextend the scope of our survey. A mathematical description of particular processes is an inescapable part of our exposition, not so much in an attempt to pin down the exact behavior of the system we are studying as to cast light on the principal phenomena. As is almost always the case in the natural sciences, we strive toward practical applications based on knowledge obtained for its own sake.

Some string-instrument builders are now beginning to take advantage of scientific methods of measurement and calculation (Hutchins 1962). Also, researchers are attempting to arrive at explanations of the phenomena experienced by musicians (Schelleng 1973, Meyer 1978). In the author's opinion, only one goal remains elusive: that of deriving credible, objectively measurable criteria for the evaluation of instruments. To achieve this goal it will first be necessary to gain a more extensive understanding of the process by which the sound of string instruments is evaluated subjectively and of the underlying phenomena of the physiology of hearing. (We understand, of course, that differences in the prices of string instruments, which range over more than four orders of magnitude, are not based entirely on acoustical criteria.) And finally, the scientific study of string instruments cannot help but deprive some myths of their justification.

Organization of the material

A string instrument is, from the physicist's point of view, a sound-radiating device that functions as follows: a nonperiodic energy source, the bow, is moved along at a particular speed by the right hand. A string is tensioned between two points of support. Frictional forces between the bow and the string alternate between sticking and sliding, setting the string into "self-sustained" oscillations. One of the points of support is on the bridge, which transmits these oscillations to the body of the instrument. All dimensions of the body are comparable to the wavelengths of audible sound. Consequently, the sound borne by the structure of the instrument leads to the radiation of sound in air. This sound reaches the ears of the performer and of listeners partly directly and partly after reflections from the boundaries of the room in which the instrument is played.

This causal sequence suggests the following organization of the material:

I. The bowing of the string
II. The body of the instrument
III. The radiated sound.

Later links in this chain of causality do, however, have an effect on earlier ones. Especially, the behavior of the elements of the system beyond the support points affects the results of bowing. The string of the monochord, supported on rigid bridges, exhibits a behavior different from and in some ways more complicated than that of the more highly damped string of a violin. Even on different violins, however, the behavior of the strings is different; if it were not, there would be no difference between instruments that "speak easily" and ones that speak with difficulty. The most obvious case of reaction of the instrument on the string is the so-called wolf tone, a phenomenon that occurs when the body of the instrument has a natural resonance equal or nearly equal to the fundamental frequency of the string's oscillation.

Despite reactive effects, it is possible to examine the string and the body separately. It is not necessary to understand every detail of the body in order to examine reactive effects. Rather, it is necessary only to know how the input of the bridge reacts; the rest of the instrument can be treated as a black box. It is possible, for example, to describe reactive effects by reference to the admittance of the input of the bridge, defined as the velocity of the input point of the bridge divided by the sinusoidally varying force applied to this point.

If we know this admittance, it is of no importance whether the velocity or the force at the input of the bridge is taken to be the term that excites the body (including the bridge). However, it makes better sense in physical terms to examine the force, because it takes more force to move the bridge than to move the string. The admittance of the bridge is much lower than that of the string. If this were not the case, the string could not be excited into oscillation at its well-known natural frequencies.

We can consider that giving the force on the bridge is suitable description of the output of the string and the input to the bridge. Similarly, giving the velocity perpendicular to the surface of the body is a suitable discription of the instrument's output and is the most appropriate manner in which to describe the way in which sound is excited in air. This consideration also applies to the difference in admittance between the body and air: It takes less force to move air than to move the body of a violin.

The admittance mismatch between the body and air is even greater than that between the string and the body. Consequently, the reaction of air on the body is less significant than that of the body on the string. Nonetheless, masses of air in the near field of the instrument, moving in phase with the body, can influence the body's natural resonances. Such masses of air are best regarded as part of the body. All oscillations in the f-holes and in the cavity of the instrument, for example, are those of sound in air; but it would create unnecessary methodological difficulties if these air masses were not regarded as parts of the body.

Even radiated sound can influence the output of the body, since it increases the damping of the body's natural resonances. However, even in a small room—and all the more in a large concert hall—the room in which the instrument is played has no significant effect on the pattern of velocity normal to the surface of the instrument. This pattern and the shape of the body of the instrument, along with that of the performer, determine how much and in which directions sound is radiated.

The effects of the shape of the room and of the materials that comprise its surfaces are problems of architectural acoustics. It will be necessary to refer to the relevant literature to examine these problems (e.g., Cremer and Müller 1982). In the final chapter of this book, a few problems especially important to the study of string instruments will be mentioned.

Most of the material in the following chapters refers to the violin, a fact that may have something to do with the author's familiarity, as an amateur musician, with this instrument. Moreover, most of the available literature relates to the violin, whether giving measurements or numerical calculations. The acoustical problems encountered in the study of all bowed string instruments are, however, essentially the same.

References

Cremer, L., and H. A. Müller, 1982. *Principles and Applications of Room Acoustics*. Tr. by T. J. Schultz. London: Applied Science.

Hutchins, C. M., 1962. *Scientific American* November, pp. 78–93.

Meyer, J., 1978. *Physikalsche Aspekte des Geigenspiels* [*Physical Aspects of Violin Playing*]. Siegburg: Verlag der Zeitschrift Instrumentenbau.

Schelleng, J., 1973. *J. Acoust. Soc. Amer.* **53**, 26.

I THE BOWING OF THE STRING

1 Self-Sustained Oscillation by Dry Friction

1.1 The simplest possible representation, in one degree of freedom

Physicists encounter many self-sustained oscillating systems: systems that periodically extract from an aperiodic source the energy necessary to replenish their losses. String and wind musical instruments are the oldest of such systems, and are still the most common. It was probably discovered by accident that strings could be bowed; they had been plucked and hammered earlier. Clearly, a long process of empirical development was necessary to bring the bow, an intrinsically aperiodic energy source, to its present form, with its shape, hair, and rosin serving to increase the static friction.

The generation of a pleasant string-instrument sound using such a bow is subject to the same physical principles as is the highly unpleasant squeal of automobile brakes. Oscillations such as these, self-sustained by dry friction between solid objects in relative motion across one another, are difficult to analyze because they are switching oscillations. The frictional force that excites the waves repeatedly alternates between static and dynamic (sticking and sliding).

Even a simple oscillator can easily be excited into periodic oscillation by an alternation of regimes of sticking and sliding friction. It is therefore appropriate to start our analysis with a simplified model obtained by replacing the stretched string—a one-dimensional continuum—with a simple oscillator (Cremer 1974). We assume the mass $m = m'l$ (where m' is the mass per unit length and l is the length of the string) to be concentrated at the point of contact with the bow (see figure 1.1).

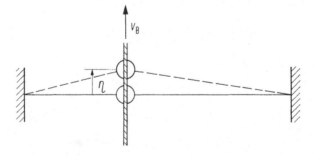

Figure 1.1
The simplest model, substituting a mass concentrated at the point of bowing for the mass of the string.

The rest of the string then functions as a reacting spring of stiffness s. Then, in the regime of sliding friction, the displacement of the string η, in the y-direction, can be represented by the inhomogeneous *equation of oscillation*

$$m\frac{d^2\eta}{dt^2} + s\eta = R_{\mathrm{g}}. \tag{1.1}$$

The sliding (or "gliding") friction R_{g} has been proven by experiment to be a function of the relative velocity v_{rel} between the bow and the string (see figure 1.2). It attains its highest value R_{max} when $v_{\mathrm{rel}} = 0$ and falls with increasing relative velocity.

As long as the system's losses, which we have neglected up to now, are not too great, this "falling characteristic" (which may be understood as the "negative resistance" of circuit analysis) allows the generation of an oscillation even without sticking friction. However, with string instruments, v_{rel} must reach zero from time to time; in other words, sliding friction alternates with sticking friction.

When sticking friction prevails, the characteristic becomes vertical; in other words, frictional force is replaced by a reaction force, and consequently the kinematic condition

$$\frac{d\eta}{dt} = v_{\mathrm{B}} \tag{1.2}$$

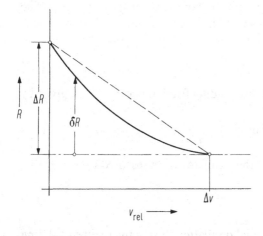

Figure 1.2
The dependence of frictional force on relative velocity.

holds, v_B being the speed of the bow. The sticking friction can give to the mass the speed of the bow as long as the sticking friction is below its maximum value, which may differ—if at all—only slightly from the maximum value of sliding friction that has already been described.

The alternation of sticking and sliding friction is so essential to self-sustained oscillations brought about by means of dry friction that there is relatively little need to consider the characteristic in any detail. It is possible, even in the present model, to assume a constant value for the sliding friction R_g. The falling characteristic is then reduced to a step

$$\Delta R = R_{max} - R_g \tag{1.3}$$

between the maximum sticking friction and the sliding friction.

In this case, however, the displacement plotted against time in the regime of sliding friction becomes an undamped sinusoid of average value

$$\bar{\eta} = \frac{R_g}{s}; \tag{1.4}$$

this equation comes into play when the reacting force due to the displacement reaches the value

$$\eta(0) = \frac{R_{max}}{s} = \bar{\eta} + \frac{\Delta R}{s}, \tag{1.5}$$

since inertia no longer plays a part. That is to say, in the regime of sticking friction the mass moves along with the bow at a constant speed v_B. The regime of sliding friction begins with the mass moving at this same speed:

$$\frac{d\eta}{dt}(0) = v_B. \tag{1.6}$$

The function of displacement after sliding friction begins is then given by

$$\eta = \frac{R_g}{s} + \left(\frac{\Delta R}{s} \cos \omega_0 t + \frac{v_B}{\omega_0} \sin \omega_0 t \right), \tag{1.7}$$

where the radian frequency of the oscillator is represented by

$$\omega_0 = \sqrt{\frac{s}{m}}. \tag{1.8}$$

The amplitude of the sinusoidal oscillation during the regime of sliding friction is also the peak value of the entire oscillation. This is given by

$$\hat{\eta} = \sqrt{\left(\frac{\Delta R}{s}\right)^2 + \left(\frac{v_B}{\omega_0}\right)^2}. \tag{1.9}$$

This value increases with the speed of the bow v_B and also with the force of the bow against the mass, F_z, since ΔR is proportional to this force. We will call this force the *bowing pressure*, as musicians do.

In the regime of sliding friction, the bow constantly tries to bring the mass up to its own speed, and then to carry it along by means of sticking friction. The following axiom may be stated: the bow succeeds in bringing the mass along at the moment when this is possible through the application of a frictional force no greater than the maximum value of sticking friction, R_{max}.

In the simple oscillator we are now examining, the frictional force works directly on the finite mass that is to be accelerated; thus the bow succeeds in bringing the mass up to its speed not before the second change in direction of that mass, when

$$\frac{d\eta}{dt} = v_B. \tag{1.10}$$

It is easy to understand this by noting that even if the speed of the mass differs only slightly from that of the bow, an infinitely high acceleration and an infinitely high frictional force are required to make the two speeds equal. The condition stated in (1.10), however, leads to symmetry of the motion about its average value in the regime of sliding friction. The same is also true, to be sure, in the regime of sticking friction, in which η increases linearly.

Figure 1.3a shows the periodic function of displacement against time, which is composed of segments in each of the two regimes.

Figure 1.3b shows the velocity function of the oscillating mass, obtained by differentiating the displacement function. We exhibit the velocity function here because it will be of more interest and easier to measure than the displacement function as we progress to the study of bowed strings. In an oscillation of a bowed string, the regime of sticking friction in which $d\eta/dt = v_B$ is much longer than in this example. But even our present example shows that the areas above and below the centerline must be equal, since the centerline of displacement cannot shift. Also, it is typical that the regimes of sliding and sticking friction can begin and end with corners (slope discontinuities).

The duration of the regime of sliding friction may be derived from the

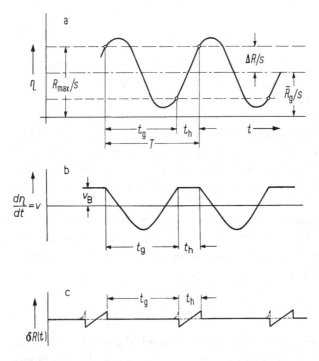

Figure 1.3
Displacement (a), velocity (b), and force (c) in a simple oscillator self-sustained by friction.

conditions for the initiation of sliding (release) and of sticking (capture), as follows:

$$t_g = \frac{1}{\omega_0}\left(\pi + 2\arctan\frac{v_B s}{\omega_0 \Delta R}\right). \tag{1.11}$$

In the adjacent time segment in the regime of sticking friction, η increases linearly by $2\Delta R/s$. The duration of this time segment is

$$t_h = \frac{2\Delta R}{s v_B}, \tag{1.12}$$

and the resulting duration of one period of the oscillation is

$$T = \frac{1}{\omega_0}\left(\pi + 2\arctan\frac{v_B s}{\omega_0 \Delta R}\right) + \frac{2\Delta R}{s v_B}. \tag{1.13}$$

All three of these durations depend on bowing pressure and bowing speed.

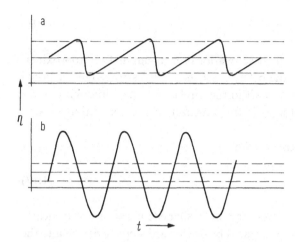

Figure 1.4
The limiting cases of displacement: a, switching oscillation; b, nearly free oscillation.

1.2 Nearly free oscillation and switching oscillation

Two limiting cases are particularly interesting. In one, sliding friction (figure 1.4, bottom) and in the other, sticking friction (figure 1.4, top) occurs over almost the entire duration of the cycle.

The first case occurs when

$$\frac{\Delta R}{s} \ll \frac{v_B}{\omega_0} \tag{1.14a}$$

or

$$\Delta R \ll v_B\sqrt{sm}; \tag{1.14b}$$

in other words, when the frictional forces are small compared with the spring forces and inertial forces. Barkhausen's insightful term for this limiting case may be translated as "nearly free oscillation" (Barkhausen 1907). In this case, the period T is almost the same as when the oscillation is not being excited through friction:

$$T = T_0, \tag{1.15}$$

and the amplitude of the displacement increases with the bowing speed:

$$\hat{\eta} = \frac{v_B}{\omega_0}. \tag{1.16}$$

We will encounter both of these relationships again as first approximations to a description of the behavior of the bowed string. The frictional forces R_{max} and R_g play no part in these relationships, although the difference between these forces is fundamental to the generation of self-sustained oscillations.

In the other limiting case, in which

$$\frac{\Delta R}{s} \gg \frac{v_B}{\omega_0}, \tag{1.17}$$

the oscillation becomes almost sawtooth-shaped; it consists very nearly of a series of linear ramps separated by downward steps. Friction and the spring constant determine the peak displacement

$$\hat{\eta} = \frac{\Delta R}{s}. \tag{1.18}$$

The same factors, along with the bowing speed, determine the period,

$$T \approx t_h = \frac{2\Delta R}{sv_B}. \tag{1.19}$$

Consequently, this limiting case can be categorized as a *relaxation oscillation*. (We may also call it a "switching oscillation," since it depends on a switching process.) The mass does not appear in the equation in this limiting case. Nonetheless, the mass is just as important to the generation of the oscillation in this case as is ΔR in the other case. Without the inertia of the mass, the displacement η would increase from $\eta(0)$ to $\bar{\eta}$ and remain at this value; only because of the inertia does the mass move beyond this point of equilibrium, to be carried back by the bow.

In the course of work on DIN 1311, sheet 2 (a publication of the German Standards Institute), the author became interested in developing the simplest possible model of a self-sustained oscillator which could demonstrate the behavior described by both limiting cases. Unlike the string model shown in figure 1.1 the oscillator illustrated in figure 1.5 is realizable. A pendulum lies in frictional contact with a tilted, rotating turntable. By changing the angle of the turntable, it is easy to change ΔR. Also,

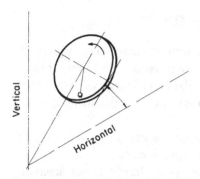

Figure 1.5
A simple oscillator self-sustained by friction.

the motion is slow enough with this model to make it possible to follow the alternations of sticking and sliding with the eye.

Using this model, it is especially easy to demonstrate the range of conditions between the two limiting cases. This range occurs in an electrical oscillator only if the oscillator has both a resonant circuit and a spark gap. The first investigations of oscillators combining resonance and switching were, in fact, of such electrical oscillators. Barkhausen (1907) and Wagner (1910) were important early investigators of this type of oscillator.

In connection with the problem of the string, let us take another look at the force function shown in figure 1.3. This gives only the difference in force

$$\delta R = R - R_{\mathrm{g}}, \tag{1.20}$$

which appears only during the regime of sticking friction, as a straight line rising from $-\Delta R$ to $+\Delta R$.

It is clear that release can occur only after $+\Delta R$ has been reached. No comparable phenomenon occurs at capture, as we have replaced the characteristic of sliding friction with one showing a constant average value. In the actual characteristic, however, there occurs a steady increase in ΔR as $d\eta/dt$ approaches v_{B} immediately before capture. If we take this into account, the function of η changes somewhat. Still, we can see that, just before capture, δR must rise to $+\Delta R$. Only after capture can δR fall to $-\Delta R$. (See the dashed line in figure 1.3c.)

1.3 Comparison to other self-sustained systems

In another respect as well, a falling characteristic, monotonically proceeding toward R_{max}, results in a more accurate physical description. If there is only sliding friction, the sinusoidal oscillation (1.7) applies to the entire period. This is the case when the maximum value of the string speed is lower than the bowing speed.

This is, however, not the case with a monotonically falling characteristic if, as earlier, we ignore additional losses of the system.

It is easiest to understand this fact if we replace the falling characteristic with a straight line:

$$R = R_{max} - r\left(v_B - \frac{d\eta}{dt}\right). \tag{1.21}$$

No part of this line may fall below zero; the characteristic must, in other words, must still be positive at the most negative value of $d\eta/dt$.

If (1.1) is taken to give the characteristic, then the linear differential equation

$$m\frac{d^2\eta}{dt^2} - r\frac{d\eta}{dt} + s\eta = R_{max} - rv_B \tag{1.22}$$

results. Its solution is composed of a shift in the rest value superimposed on an oscillation which increases exponentially, as described by

$$\eta = \bar{\eta} + e^{\sigma t}(\eta_C \cos \omega t + \eta_S \sin \omega t), \tag{1.23}$$

in which

$$\sigma = \frac{r}{2m} \tag{1.24a}$$

and

$$\omega = \sqrt{\omega_0^2 - \sigma^2}. \tag{1.24b}$$

In this case, the conditions for capture will certainly be reached. If the sign of $rd\eta/dt$ were not negative, the equation would describe the familiar damped sinusoidal oscillation.

The point at which the sign changes often delineates the threshold of the region in which an oscillator is stable. This is particularly true of electronic oscillators using feedback. In oscillators, the threshold is deliberately exceeded; in amplifying systems and control circuits, it is avoided.

If (1.23) precisely described the actual situation, oscillations would increase until they did mischief. Generally, however, as amplitudes increase, the characteristic lines shift from falling to rising. In almost all cases, the shift occurs gradually enough that it is still fruitful to analyze the oscillations in terms of sinusoids. This observation also applies to some wind instruments, such as organ pipes.

With string instruments, on the other hand, it is never possible to consider the fundamental alone, even as an approximation. This results not only from the nonlinear characteristic of sliding friction, but even more from the way capture and release change the frictional force from an impressed force to a reaction force, and vice versa. Since these step functions depend on the attainment of R_{max}, a condition describable only in the time domain, it is impossible to describe the behavior of the system for the entire period in the frequency domain. Analogies to problems for which a solutions in the frequency domain are possible can explain certain phenomena plausibly, but cannot lead to a quantitative descriptions. We have taken these introductory paragraphs to make this fact clear, though the simplest possible case of a self-sustained system is in many ways very different from the more difficult problem of a bowed string.

References

Barkhausen, H., 1907. Dissertation, University of Göttingen.

Cremer, L., 1974. *Acustica* **30**, 119.

Wagner, K. W., 1919. *Arch. Elektrotech.* **8**, 61.

2 The Plucked String

2.1 The wave equation

Before we turn to our study of the bowed string, we should first examine the characteristics of the stretched string and of free oscillations, that is, those oscillations which a string carries out when it is not in contact with a bow. Examples of free oscillation are those of a piano string after the hammer has moved away, or of a plucked violin string once the plucking finger is no longer in contact with it.

The string may be represented as a one-dimensional continuum in the x-direction, subjected to a longitudinal tension F_x and of mass per unit length m'. Consequently, its motions are not, as with the simple oscillator, functions only of time; they are functions of both space and time. The equation of oscillation (1.1) for $R_g = 0$ must be replaced by one valid for all of the quantities describing the wave field, including the displacement $\eta(x, t)$; this is the partial differential equation

$$c^2 \frac{\partial^2 \eta}{\partial x^2} = \frac{\partial^2 \eta}{\partial t^2}, \tag{2.1}$$

the one-dimensional *wave equation*. We assume that displacements and inclinations of the string from its rest position are small. The resulting force pulls nearly in the direction of the string, and so, since the angles are small, a perpendicular force results:

$$F_y = F_x \alpha = F_x \frac{\partial \eta}{\partial x}. \tag{2.2}$$

Now let us examine an element of a string whose length is dx, as illustrated in figure 2.1. We see that the force pulls the left end of the element toward the rest position, and pulls the right end away from the rest position by the same amount plus a differential. Applying Newton's law to an element of mass $m' dx$ leads to the equation

$$F_x \left(\alpha + \frac{\partial \alpha}{\partial x} - \alpha \right) = F_x \frac{\partial^2 \eta}{\partial x^2} dx = m' dx \frac{\partial^2 n}{\partial t^2}. \tag{2.3}$$

This is, however, precisely the wave equation (2.1), if we make a substitution for the constant c introduced in that equation:

$$c = \sqrt{F_x/m'}. \tag{2.4}$$

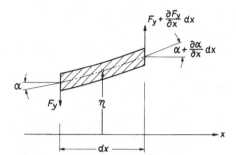

Figure 2.1
Displacements, angles of inclination, and forces on an element of the string.

2.2 D'Alembert's solution

In d'Alembert's solution to the wave equation, we can see how c represents the speed of a wave:

$$\eta(x, t) = \eta_+ (x - ct) + \eta_- (x + ct). \tag{2.5}$$

The argument of the function η_+ —which can represent any chosen waveform—remains constant if x increases at the same rate as ct. Consequently, the first term represents a wave propagating in the $+x$ direction without undergoing any changes. In the same way, the second term η_- represents a wave propagating in the $-x$ direction. The terms representing both waves satisfy the wave equation, as can be easily proven by carrying out the partial differentiations of (2.1).

That (2.5) represents the complete solution becomes clear physically from the fact that the two independent functions η_+ and η_- may represent any given initial conditions for two independent functions: that of displacement $\eta(x, 0)$ and that of velocity $\partial\eta/\partial t = v(x, 0)$. If we designate the derivatives of these two functions as η'_+ and η'_-, in accord with their arguments, then on the one hand

$$\frac{\partial\eta}{\partial x} = \eta'_+ + \eta'_-, \tag{2.6}$$

and on the other hand

$$\frac{1}{c}\frac{\partial\eta}{\partial t} = -\eta'_+ + \eta'_-, \tag{2.7}$$

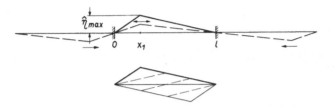

Figure 2.2
Oscillation of the plucked string seen in terms of D'Alembert's solution.

and, consequently,

$$\eta'_+ = \frac{1}{2}\left(\frac{\partial \eta}{\partial x} - \frac{1}{c}\frac{\partial \eta}{\partial t}\right),\tag{2.8}$$

$$\eta'_- = \frac{1}{2}\left(\frac{\partial \eta}{\partial x} + \frac{1}{c}\frac{\partial \eta}{\partial t}\right).\tag{2.9}$$

If, in the case of the plucked string, the wave process is initiated by displacement from the rest position, without initial motion,

$$\frac{\partial \eta}{\partial t} = v(x, 0) = 0,\tag{2.10}$$

then

$$\eta'_+ = \eta'_-.\tag{2.11}$$

Since the free oscillations contain only waves propagating without a shift of the rest position, so that there is no constant of integration, we have

$$\eta_+ = \eta_-.\tag{2.12}$$

In figure 2.2, top, the thick solid line shows the triangular pattern of the initial displacement, greatly exaggerated for clarity. Its peak is at $x_1 = l/4$.

To represent d'Alembert's solution graphically, it is necessary only to divide the amplitude of the displacement into two halves and to shift the two resulting half-height triangles to the right and the left, adding the resulting amplitudes at all points. The support points of the string, at $x = 0$ and $x = l$, generate inverted reflections; these are represented by the dashed lines continued beyond the support points.

The description can be simplified even further. The legs of the triangle are straight, and are in a state of rest. There is no force tending to move them away from their initial positions. Only the corner at the peak of the triangle is unstable; and note that, in contrast to the situation in figure 1.1, there is no mass at this point to accelerate. Since η_+ and η_- are equal, their second derivatives are also equal. The corner constitutes an ideal impulse of $\partial^2\eta/\partial x^2$. This divides into two equal parts which propagate to the right and to the left; they generate between them a straight line which moves at a constant rate (dashed lines in figure 2.2, bottom).

It is easy to account for the reflections as well. First at $x = 0$, then at $x = l$, the corner reaches the support points (bridge and nut or finger), which we will, for now, idealize as representing the boundary conditions

$$x = 0: \quad \eta(0, t) = \eta_+(0, t) + \eta_-(0, t) = 0, \tag{2.13}$$

$$x = l: \quad \eta(l, t) = \eta_+(l, t) + \eta_-(l, t) = 0. \tag{2.14}$$

The reflected waves which may be regarded as originating behind the boundaries must, then, be of opposite polarity:

$$\eta_+(0, t) = -\eta_-(0, t), \tag{2.15}$$

$$\eta_+(l, t) = -\eta_-(l, t), \tag{2.16}$$

and so must the corner after it is reflected. After half a period, a triangle of opposite polarity arises; its peak lies at $x_2 = l - x_1$, or, in figure 2.2, bottom, at $(3/4)l$. In the second half of the period, the process repeats, inverted, and the line returns to its original position.

The duration of the period is given by the time the corner takes to travel over the distance $2l$ at the speed c:

$$T_1 = \frac{2l}{c}. \tag{2.17a}$$

The inverse of this is the frequency of the fundamental (the first partial, hence the subscript 1) of the plucked string:

$$f_1 = \frac{c}{2l}. \tag{2.17b}$$

2.3 The requirement of pure fifths

Frequency is inversely proportional to the length of the string; control
of pitch by the positioning of the fingers of the left hand is based on this
principle. But also, the simpler the ratio of string lengths, the more
harmonious the resulting musical interval is perceived to be. This fact
must have caused considerable wonder in ancient times, when the fundamental frequencies corresponding to different string lengths could not
be measured. It is understandable that this wonder led to philosophical
speculations about the relationships between simple numerical ratios and
the sense of beauty. The psychological basis for perception of the harmony
of pure and complex tones was first discussed scientifically by Helmholtz
(1862), and is not yet a totally resolved issue. It should be mentioned that
today's science of psychological acoustics has uncovered exceptions:
consonances perceived in connection with frequency ratios that are not
simple.

Turning from psychology to physics, it is also not at all correct to
state that simple frequency ratios always occur in connection with simple
geometric ratios. Simple frequency ratios and simple geometric ratios
occur together only when the frequency-determining part of the system
is a one-dimensional homogeneous transmission line with ideal boundary
conditions. In instruments using columns of air, even cylindrical ones,
this is not the case. Aperture corrections must be made for the masses
of air oscillating in the vicinity of the ends of the air columns. In string
instruments as well, the simple relationship between frequency and string
length does not hold strictly if actual boundary conditions are taken into
account.

And, to an even greater degree, invisible variations in the composition
of the strings and in their diameters disturb the simple relationship. The
wave equation (2.1) has as a condition that c is constant. But according
to (2.4), the speed c of the waves depends on the mass per unit length,
m'. If m' is not constant, c varies from one point to another on the string.
We need not derive a wave equation for $c(x)$ to conclude that (2.17b)
can no longer hold if m' is somewhat different near the nut and near the
bridge. When notes are played one after another, the resulting alterations
of pitch are not as evident as when notes are played at the same time.
Especially, it is the fifths that give trouble, since, when two notes a fifth
apart are played at the same time on neighboring strings, the same finger

must hold both of them down. Consequently, strings whose m' is not constant are said to have "false fifths." Variations in m' can be a manufacturing problem. They can also result from playing the instrument. According to J. Meyer (1978, p. 71), "the moisture of the hand attacks the surface of steel strings and of metal-wound strings, and corrodes them. In this way, a string becomes somewhat thinner and lighter, especially the part played in the lower hand positions. . . . The situation is different with gut strings. . . . The reason is that the bow wears the string down (near the bridge), . . . and, apparently, also that the part of the string contacted by the left hand absorbs a certain amount of moisture, increasing the oscillating mass (near the nut)." Even these small changes can make strings unusable, due to the extraordinary sensitivity of the ear to the tuning of intervals. It is also for reasons of sensitivity to tuning that neighboring strings are preferably set in intervals of fifths.

2.4 Input force on the bridge

In our introduction, we noted that we would be interested in the "output" of the vibrations of the string—meaning the input of the body including the bridge, or the force on the bridge transverse to the string, in the direction of its oscillations, at $x = 0$. According to (2.2), this force is proportional to the constant longitudinal force F_x and to the variable angle α. From figure 2.2, bottom, we can easily see that, during one period of oscillation, the angle α jumps back and forth between two extreme values

$$\alpha_1 = \eta_{max}/x_1 \qquad\qquad\qquad\qquad\qquad\qquad (2.18a)$$

and

$$\alpha_2 = \eta_{max}/(l - x_1). \qquad\qquad\qquad\qquad\qquad (2.18b)$$

The function of force on the bridge is, then, rectangular. It can also be seen that the first value of the angle α at the bridge is maintained for the interval

$$t_1/2 = x_1/c \qquad\qquad\qquad\qquad\qquad\qquad\qquad (2.19a)$$

after the string departs from its original position, and for the same time interval immediately before it returns to this position. The total time per period during which the angle is α_1 is, then, $t = 2x_1/c$. The remainder of the period during which the angle is α_2 is

$$t_2 = T_1 - \frac{2x_1}{c} = \frac{2(l - x_1)}{c}. \tag{2.19b}$$

It follows that the areas above and below the t-axis in the function shown in figure 2.3 are equal. There is, then, no physical need for any balancing force.

We can see this even more clearly if we derive the transverse force from the velocity. In order to do this, we consider each of the terms in (2.5) separately. One of these is a wave propagating in the positive x-direction, the other, a wave propagating in the negative x-direction. If we differentiate η_+ with respect to time, we obtain the component of velocity corresponding to this wave:

$$v_+ = \frac{\partial \eta_+}{\partial t} = -c\eta_+'. \tag{2.20a}$$

We may also differentiate η_+ with respect to position and multiply it by the longitudinal force F_x. If we then consider that we are interested in the force which the segment of the string to the left of a given point exerts on that to the right, we obtain the result

$$F_+ = -F_x \frac{\partial \eta_+}{\partial x} = -F_x \eta_+'. \tag{2.20b}$$

We can therefore draw the important conclusion that force and velocity have the same dependence on the argument $(x - ct)$. The quotient of force and velocity is, therefore, a constant. This quotient is called the *characteristic resistance* of the string, and is represented by the symbol Z:

$$Z = \frac{F_x}{c} = m'c = \sqrt{F_x m'}. \tag{2.21}$$

Exactly the same result may be obtained using the wave that propagates in the negative x-direction, except that the sign is opposite.

In summary,

$$F = F_+ + F_- = Z(v_+ - v_-). \tag{2.22}$$

In order to make use of this general formula to derive the force at the bridge, we must return to d'Alembert's analytical solution as in figure 2.2, top. D'Alembert separated the initial displacement into two triangular components of peak value $\eta_{max}/2$. Differentiating one of these

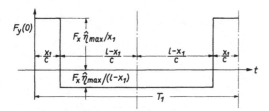

Figure 2.3
Force function of the plucked string.

components leads simply enough to a rectangular function in which the velocity of the arriving wave (appearing here as v_-) takes on the value $c\eta_{max}/2x_1$ during the first time interval x_1/c. For now, we idealize the bridge as rigid, and so velocity goes to zero at the bridge. Consequently, the arriving wave is reflected in opposite phase, and the velocity difference in (2.22) leads to a doubling of the transverse velocity of the arriving wave, to $c\eta_{max}/x_1$. Multiplying this value by the characteristic resistance of (2.21), we obtain exactly the same value as we did before from our simpler consideration of the angle α. The function over the remainder of the period is also exactly as in figure 2.3.

Later, we will compare the force on the bridge when the string is bowed. To prepare for this comparison, we will now switch briefly from the time domain to the frequency domain—from the time function of figure 2.3 to the corresponding line spectrum of figure 2.4, middle. It is well known that the envelope of the spectrum of a rectangular time function leads to the function

$$
\hat{F}_n = \frac{2}{T} \int_{-x_1/c}^{+x_1/c} \left[F_x\eta_{max} \left(\frac{l}{x_1} + \frac{l}{l - x_1} \right) \right] \cos\frac{2\pi nt}{T} dt
$$

$$
= \frac{4F_x\hat{\eta}_{max}}{(l - x_1)} \frac{\sin(\pi nx_1/l)}{\pi nx_1/l}, \tag{2.23}
$$

and that its position among the lines of the spectrum depends on the ratio defined by the position along the string at which it is plucked. For this reason, figure 2.4 shows not only the spectrum in the middle, which corresponds to the example in figure 2.3, with $x_1/l = 1/4$, but also spectra for $x_1/l = 1/2$ (top) and for $x_1/l = 1/8$ (bottom). The reference amplitude for these spectra is not equal maximum displacement, but equal static plucking force,

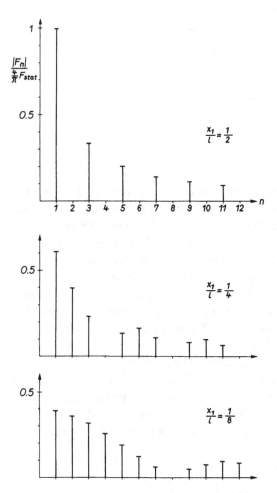

Figure 2.4
Line spectrum of force at the bridge for various plucking points.

$$F_{\text{stat}} = F_x \eta_{\max} \frac{l}{x_1(l - x_1)}. \qquad (2.24)$$

This makes the best sense in terms of violin-playing technique.

The case $x_1/l = 1/2$ is particularly interesting. We see that the amplitude of the partials corresponding to $1/n$ goes to zero for even values of n. The $(\sin n/n)$-envelope plays a part here only in that its zeroes correspond to even values of n. In any case, this result makes it clear that the point at which the string is plucked has a marked influence on the spectrum.

2.5 Taking damping into consideration

Any attempt to verify the differences in tone color shown in figure 2.4 is made difficult by the decay of the sound—and even more by accompanying changes in tone color over time.

Figure 2.5 shows the frequency dependence of decay times τ of a plucked string, measured by Reinicke (1973) using electrical filters to separate partials from one another.

The upper curve is for a string stretched between rigid bridges (the monochord). All damping is in the string itself. The lower curve is for an open a-string on an actual violin, with additional losses at the bridge and perhaps at the nut. A string stopped by a finger would certainly have much greater losses. Both of the cases shown make it clear that the upper partials have shorter decay times τ and larger decay coefficients

$$\delta = \frac{1}{\tau} \qquad (2.25)$$

than the lower partials.

The differences shown in figure 2.4 are valid only for the first few periods of vibration. During its decay, the tone continually loses brilliance. This means, too, that the corners visible in figure 2.2 become continually more rounded.

For actual strings, corners cannot in fact exist, since actual strings always have a bending stiffness, even if only a small one. Besides, sharp corners cannot be generated by a plucking finger. Here, however, as in the case of the bowed string which we will examine later, it is often possible to neglect these effects, since the damping, increasing with frequency, leads to rounded corners and to less-than-sharp transitions

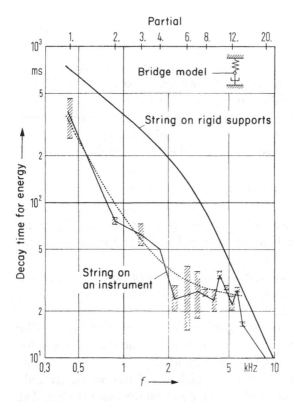

Figure 2.5
Measured decay times of a plucked string. Upper curve, for rigid supports; lower curve, on an instrument (after Reinicke). Dashed line, calculated according to the model of the bridge at the top right, as described in section 5.4.

in the velocity function. After a few periods, the triangle of figure 2.2 deforms to correspond to a realizable displacement.

2.6 Bernoulli's solution

Since the damping of a plucked string depends on frequency, it is appropriate to construct a mathematical description of the string's decaying oscillation by synthesis, using damped sinusoids. The synthesis is based on Bernoulli's solution for the wave equation and its boundary conditions.

This solution assumes that time dependence and spatial dependence can be factored; without damping, the time dependence may then be represented by sinusoidal oscillations, with any phase.

If the plucked string is initially pulled aside, but given no acceleration, all of its time functions can be represented as cosine functions. Nonetheless, we base our analysis on the general problem of initial conditions without phase constraints and choose a representation using rotating vectors and their projection onto the real axis.[1] We set

$$\eta(x, t) = \text{Re}\{\hat{\underline{\eta}}(x)e^{j\omega t}\}. \tag{2.26}$$

We now substitute this formula into the wave equation (2.1), obtaining, for the spatial dependence of a homogenous string, an ordinary second-order differential equation of the same structure as the equation of oscillation (1.1) with $R_g = 0$:

$$\frac{d^2\underline{\eta}}{dx^2} + \left(\frac{\omega}{c}\right)^2 \underline{\eta} = 0. \tag{2.27}$$

The general solution of this equation is composed of a cosinusoidal component and a sinusoidal component:

$$\underline{\eta} = \underline{\eta}_c \cos kx + \underline{\eta}_s \sin kx. \tag{2.28}$$

In this equation, we have set the quotients

$$\frac{\omega}{c} = \frac{2\pi}{cT} = \frac{2\pi}{\lambda} = k. \tag{2.29}$$

The parameter k is called the *wave number*; the wavelength

$$\lambda = cT = c/f \tag{2.30}$$

appears in the denominator.

In order to arrive at the correct spatial form for the natural oscillatory modes (Eigenformen) of stretched strings, we must also consider spatial boundary conditions. First, we see that

$$x = 0: \quad \underline{\eta} = 0, \quad \underline{\eta}_c = 0. \tag{2.31}$$

The oscillations are, therefore, of sinusoidal form, and begin at one end of the string. However, because of the second boundary condition

$$x = l: \quad \underline{\eta} = 0, \tag{2.32}$$

---

[1] Readers unfamiliar with this representation are referred to the relevant literature, e.q., Cremer and Müller (1982, vol. 2, sec. 1.7).

they must also go to zero at the other end of the string:

$$\sin kl = 0. \tag{2.33}$$

The length of the string can encompass a half-period, a full period, or any greater rational multiple of a half-period, since it follows from (2.33) that

$$kl = \frac{2\pi l}{\lambda} = n\pi; \quad l = n\frac{\lambda}{2}. \tag{2.34}$$

The corresponding natural frequencies, obtained by substituting (2.34) into (2.30), are

$$f_n = n\frac{c}{2l}. \tag{2.35}$$

We have already dealt with the case in which $n = 1$, in terms of d'Alembert's solution, as shown in (2.17b). Even in our earlier discussion, we were able to state that the ongoing periodic function can be divided arbitrarily into periodic segments.

In any case, we now see that the frequencies of the overtones are integral multiples of the frequency of the fundamental. This is by no means the case for all sound generators; in fact, it is true for the stretched string only if we disregard detuning due to nonideal boundary conditions.

Given ideal boundary conditions, Bernoulli's solution for the plucked string takes the form

$$\eta(x, t) = \sum_{n=1}^{\infty} \text{Re}\left\{\underline{\eta}_n \sin\frac{n\pi x}{l} e^{jn\omega_1 t}\right\}. \tag{2.36}$$

This is a complete solution, like the previous one. The parameter $\hat{\eta}_n$, still an independent variable, representing the spatial and temporal maximum value of each natural oscillation and its zero phase angle, allows any possible initial condition to be fulfilled. This allows us to proceed to a Fourier analysis of the initial displacement and of the initial velocity distribution. Note that the fundamental spatial period of this Fourier analysis is $\lambda_1 = 2l$.

If there is no initial velocity distribution, all of the natural oscillations are of the same (or opposite) phase and $\hat{\eta}_n$ is real. Then, simply,

$$\hat{\eta}_n = \frac{1}{l}\int_0^{2l} \eta(0) \sin\frac{n\pi x}{l} dx. \tag{2.37a}$$

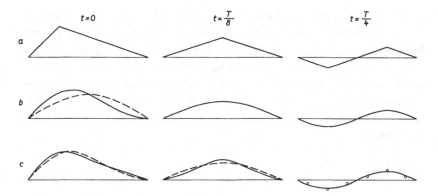

Figure 2.6
Comparison of the D'Alembert representation (a) with Bernoulli's accounting for the
first and second partials (b), and with the third as well (c), for $t = 0$, $T/8$, and $T/4$.

Calculation of $\hat{\eta}_n$ for the plucked string, with its rising and falling
straight-line segments, can be simplified if it is noted that spatial differentiation results in a rectangular function, as shown in (2.33). We
also can see that, after integration, a factor of n^2 appears in the denominator instead of n; the series converges rapidly. For any given value
of x_1/l, we have

$$\hat{\eta}_n = \frac{2\eta_{max}}{\pi^2 n^2} \frac{l^2}{x_1(l - x_1)} \sin\frac{\pi n x_1}{l}. \qquad (2.37b)$$

For the case dealt with in figure 2.2, in which $x_1/l = l/4$ we then have

$$\hat{\eta}_n = \frac{32}{e\pi^2}\eta_{max}\frac{\sin(n\pi/4)}{n^2}. \qquad (2.37c)$$

For this particular case, figure 2.6 gives a comparison of d'Alembert's
solution, discussed earlier, against a step-by-step synthesis using Bernoulli's solution. The points in time $t = 0$, $t = T_1/8$, and $t = T_1/4$, represented earlier in figure 2.2, bottom, are shown again here side by side.
The pattern shown in the figure repeats itself after this quarter-period.
The upper row shows the strict d'Alembert solution. Below this the
solutions for $n = 1$ only (dashed line) and for $n = 1$ and 2 (solid line)
are shown. Even these combinations show some of the basic features of
the strict solution: an asymmetrical beginning; symmetry at $t = T_1/8$,
where the $n = 2$ component goes to zero; and a division into two similar

displaced segments, one facing upward and the other downward, at
$t = T_1/4$, where the fundamental disappears and the $n = 2$ component
has changed its sign. The combination of $n = 1$ and $n = 2$ components
is shown again in the third row as dashed curves, and the $n = 3$ component
is added in the solid curve. The closer approximation to the triangular
function can be seen at $t = 0$ and $t = T_1/8$. On the other hand, it is disappointing that there is no change when $t = T_1/4$. There is none because
the value of

$$\sin \frac{n\omega_1 T_1}{4} = \sin \frac{n\pi}{2}$$

goes to zero for all odd values of n. But also, nothing is changed by
adding $n = 4$ for any point in time. This natural frequency and all of
its multiples are not excited, as we know from the spectrum shown in
figure 2.4. In the case shown, the next partial to make a contribution
corresponds to $n = 6$. In figure 2.6, the points of maximum displacement
of $n = 6$ are shown as small circles, in order to distinguish them from the
dashed lines. These points, too, can be seen to improve the approximation
to the strict solution.

2.7   Range of validity of the linear wave equation for a string under tension

Torsional waves, which we will examine later; longitudinal waves in a
string; and even electromagnetic waves in a homogeneous transmission
line are described by the wave equation (2.1) and by all of the analytical
methods based on this equation.

The wave equation was in fact first derived as it applies to transverse
waves of a string under tension. Still, its range of validity in this case is
quite limited, more than in the other examples just given. Tension F_x
must be high, and displacement η must be small. There is also one very
significant difference in the physical behavior of the string, compared with
these other examples, even when displacement is small and tension is
high. In the other types of natural oscillations listed above, kinetic and
potential energy both shift from one point to another, with maxima of
one corresponding to minima of the other; but for transverse waves of
a string under tension, only kinetic energy moves from one point to
another. Potential energy, in the form of slight increases in elongation

and tension, is always distributed over the entire length of the string. All differences that might be expected as a result of deformation propagate as longitudinal waves, with the speed

$$c_L = \sqrt{\frac{ES}{m'}}, \tag{2.38}$$

where E is the modulus of elasticity and S is the cross-sectional area. Longitudinal waves travel much faster than transverse waves in a string; the time necessary for elongation and tension to equalize themselves throughout the string can generally be ignored.

Granted that there is an equal increase in tension over the entire length of the string, a nonlinear effect occurs, as first described by Kirchhoff (1897, 29th lecture, sec. 7). The displacement of the string increases the tension F_x, which we previously have regarded as constant. From its rest value, the tension increases in proportion to the average increase in extension necessary to achieve the displacements,

$$\bar{\varepsilon} = \frac{\Delta l}{l} \tag{2.39}$$

to

$$F_x = F_{x0} + E\bar{\varepsilon}. \tag{2.40}$$

For a given distribution of displacement $\eta(x)$, however, the average extension is described by the integral

$$\bar{\varepsilon} = \frac{1}{2l} \int_0^l \left(\frac{\partial \eta}{\partial x}\right)^2 dx. \tag{2.41a}$$

This integral accounts for all overtones. Pure superposition, however, which has been an important feature of Bernoulli's and d'Alembert's solutions that we have discussed up to this point, cannot be extracted from this integral. We can now make only one general statement: as the peak value of the function of $\eta(x)$ increases, $\sqrt{F_x}$, c, and all natural frequencies rise if displacement is entirely—or even primarily—to one side at any one point in time. A strongly plucked string may, then, at first exhibit an elevation of its fundamental frequency. Because the strings have deliberately been put under high tension, this effect is generally small enough that it is not disturbing. It is, however, possible to make this effect audible: for example, by reducing the tension of an e-string to lower its pitch by an octave. (The distinctive sound of the Beckmesser harp is probably partly due to this effect.)

In a plucked string, this nonlinearity is also manifested in another way. When the string is vibrating freely, it does not remain in its original plane of displacement; an imperceptibly small impulse in the perpendicular plane will after a while cause the string to oscillate in that plane. Minor inhomogeneities in the string's material, can cause this behavior. However, it is beyond question that the displacements in the y- and z-directions, η and ζ, have a coupled influence on the tension of the string; when there are displacements in both directions, it is possible to extend (2.41a) to

$$\bar{\varepsilon} = \frac{1}{2l} \int_0^l \left[\left(\frac{\partial \eta}{\partial x} \right)^2 + \left(\frac{\partial \zeta}{\partial x} \right)^2 \right] dx. \tag{2.41b}$$

Suppose now that we place a metallic string in a magnetic field perpendicular to the string and, as will often be described in this book, we sense the motion of the string by amplifying the electrical voltage induced in the string. In this way or in other ways, we might register the motion of the string only in the plane in which it is plucked. In this case, the existence of two degrees of freedom and of their partly linear, partly nonlinear coupling is expressed as beats. Figure 2.7 gives an example of the fundamental, isolated by filtering, of a plucked string with rigid supports. Striking oscilloscope traces and a theoretical treatment of the nonlinear coupling have been published by Gough (1983).

It should be mentioned here that a string stretched over the bridge of a violin may exhibit coupling of motion between the direction of excitation (y) and perpendicular to it (z) even when the string's behavior is linear. The phenomenon was observed by Hancock (1975) using a microscope, with an electrodynamically excited string. Gough (1981) later discussed this in more detail. The cause of the coupling is that the feet of the bridge move differently where they are placed on the body; the notch in the bridge which supports the string is consequently forced to move in both the y- and z-direction. (See also section 5.6 and chapter 9.)

Nonlinearity due to changes in tension is displayed most clearly when a lightly tensioned string is excited with a sinusoidally alternating force; for example, once again by placing a metallic string in a magnetic field, but this time passing an electric current through the string. In this case, there occur "overhanging resonance curves" with instabilities, and subharmonics may be generated (Cremer 1941, Anand 1966, Raghunanden

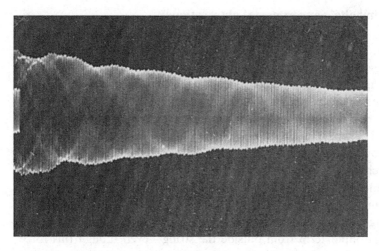

Figure 2.7
Recording of the displacement of a plucked string (filtered fundamental of η).

and Anand 1978). In the playing of stringed instruments, however, such forced oscillations never occur.

References

Anand, G. V., 1966. *J. Acoust. Soc. Amer.* **40**, 1517.

Cremer, L., 1941. *Forsch. Ing.-Wes.* **12**, 226.

Cremer, L., and H. A. Müller, 1982. *Principles and Applications of Room Acoustics.* Tr. by T. J. Schultz. London: Applied Science.

Gough, C. E., 1981. *Acustica* **49**, 124.

Gough, C. E., 1983. *J. Acoust. Soc. Amer.* (forthcoming).

Hancock, M., 1975. *Catgut Acoust. Soc. Newsletter* no. 23 (May).

Helmholtz, H. v., 1862. *Lehre von den Tonempfindungen* [*On the Sensations of Tone*]. Braunschweig. Vieweg.

Kirchhoff, G., 1897. *Vorlesungen über Mechanik* [*Lectures on Mechanics*]. Leipzig: Teubner.

.Meyer, J., 1978. *Physikalische Aspekte des Geigenspiels* [*Physical Aspects of Violin Playing*]. Siegburg: Verlag der Zeitsehrift Instrumentenbau.

Raghunandan, C. R., and G. V. Anand, 1978. *J. Acoust. Soc. Amer.* **64**, 1093.

Reinicke, W., 1973. Dissertation, Institute for Technical Acoustics, Technical University of Berlin.

3 The Bowed String Considered as a Free Oscillation

3.1 Helmholtz's experimental observations

Our next task is to combine the two areas of knowledge we examined in the preceding chapters. Chapter 2 was devoted to the possible motions of a string under tension, and in chapter 1 we examined the mechanism of self-sustained excitation by means of friction.

The labels we will give to different parts of the system are shown in figure 3.1. There is no doubt that regimes of sticking and sliding friction will follow one another in this system, just as before. Here, as before, the switching from one regime to the other will lead to a change in the significance of the frictional force, and to a change in the character of the system.

During the regime of sliding friction, the force at the point of bowing acts as an input term from outside the string—even though this force depends on the relative velocity of the bow and the string. The system upon which this force acts is the entire string, with its natural frequencies as determined by its total length l.

In the regime of sticking friction, however, the velocity at x_B is predetermined. Consequently the string separates into two independent segments, whose lengths are x_B and $l - x_B$, and each of which has its own natural frequencies.

If the system is to be periodic, the distribution of displacement and velocity must be the same at the moments of capture and of release. This complicated constraint reduces the likelihood of a solution based purely on theory. At best, a theoretical interpretation might be obtained by deduction from an experimentally derived understanding of the motion.

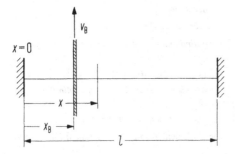

Figure 3.1
Choice of symbols for the study of the bowed string.

Figure 3.2
Lissajous figure of a bowed string observed by Helmholtz.

In this way, through experimentation, Helmholtz (1862) attacked the problem. It was first necessary for him to construct an appropriate experimental apparatus, which he called the vibration microscope. This consisted of an eyepiece attached to a tuning fork; this oscillated sinusoidally parallel to the string, driven by a magnetic field induced by means of an alternating electric current. The eyepiece was focused on a bright-colored grain of starch pasted onto the string. When Helmholtz bowed the string, he saw a so-called *Lissajous figure*; the displacement of the string was displayed as a function of a sinusoidally varying abscissa (not, as is more usual, as a function of an abcissa proportional to time). The figure stood still when the frequency of the tuning fork was an integral fraction of the frequency of the string, and could be compared to the projection of the function of time onto a cylinder, observed from a considerable distance; or, even better, the cylinder appeared to rotate slowly if the frequencies of string and fork were slightly different. If the frequency of the string was approximately four times that of the tuning fork, as in the example in figure 3.2, the time function at the middle of the pattern appeared more or less undistorted, allowing Helmholtz to sketch the function $\eta(t)$ by hand.

The observed displacement was surprisingly simple (see figure 3.3), considering the theoretical difficulty of the problem. At whatever point Helmholtz observed the string, its displacement followed a triangular pattern; the velocity, in other words, stayed at a constant positive value v_+ during the time interval t_+, and at a constant negative value v_- during the time interval t_-. Twice the peak value of displacement of both time intervals, $2\hat{\eta}$, is then

$$2\hat{\eta} = v_+ t_+ = -v_- t_-. \tag{3.1}$$

There exists, then, a proportionality of

$$t_+ : t_- = (-v_-) : v_+. \tag{3.2}$$

By introducing the period

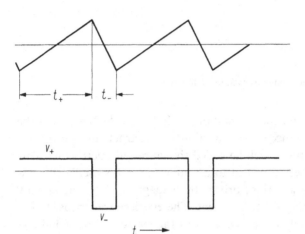

Figure 3.3
Displacement and velocity functions at any position along the string, according to Helmholtz's observations.

$$T = t_+ + t_- = t_+ \left(1 + \frac{v_+}{(-v_-)} \right) \qquad (3.3)$$

and the jump in velocity

$$\Delta v = v_+ + (-v_-) \qquad (3.4)$$

into the equation, it can be extended to

$$t_+ : t_- : T = (-v_-) : v_+ : \Delta v. \qquad (3.5)$$

Today, electromagnetic sensing of a metallic string (known, by the way, even in Helmholtz's time), sufficiently inertia-free oscillographs, and above all, electronic amplifiers, make it very much easier to observe the velocity of a string—and, after electronic integration, the displacement as well. In figure 3.4, top, the principle of electrodynamic sensing is illustrated;[1] at the bottom are shown two functions of $\eta(t)$ recorded by Müller (1962), corresponding to different points of observation along

[1] The author cannot be sure when this method was first employed using modern measuring equipment. It is to be found in Bladier (1961) and, independently, was also used by Müller (1962). Besides using a magnetic field, Bladier also developed a method using an electrostatic field; he did not, however, put this to use.

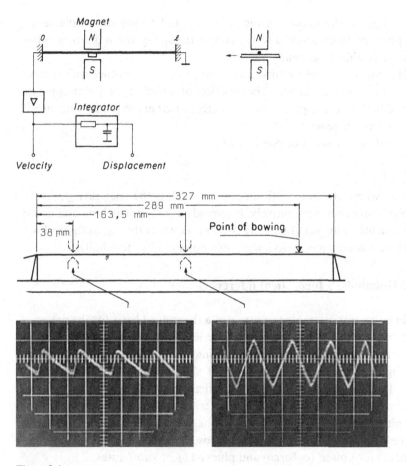

Figure 3.4
Top: Principle of electrodynamic sensing. Bottom: Examples of the displacement over
time (after Müller).

the string. At the middle of the string, t_+ and t_- are equal; the closer the point of observation is to the end of the string, the more they differ and the smaller $\hat{\eta}$ becomes.

The intervals t_+ and t_- are mirror images of one another with respect to the middle of the string. The function observed at the point opposite that of bowing corresponds, then, exactly to that at the point of bowing if the time axis is reversed.

Exactly at the point of bowing, where

$$v_+ = v_B,\tag{3.6}$$

that is, where sticking friction comes into play, the triangular function is easy to understand intuitively. It reminds us of the switching oscillation with the long interval of sticking friction shown at the top in figure 1.4.

These spatial dependences were also observed by Helmholtz.

3.2 Helmholtz's theoretical inferences

Helmholtz next attempted to arrive at a theoretical basis for the presence of a triangular function over the entire length of the string. In doing so, he considered the motion of the string as a free oscillation, without distinguishing between the regimes of sliding and of sticking friction; he did not explore in any detail the difficulties posed by such a simplified treatment. The simplification was certainly justified at least insofar as the pitch of the string while bowed is the same as that after it is plucked. If the situation were otherwise, composers would have to use different symbols for bowed (*col arco*) and plucked (*pizzicato*) notes.

By an extension of the analysis in the first chapter, especially that which is illustrated in figure 1.4, it is certainly justified to base theoretical inferences on the model of an undamped natural oscillation. This model has the advantage that the fundamental frequency is constant. The bow, however, tends toward the longest possible interval of sticking friction. The possible motions of the string in connection with the type of switching oscillation shown in figure 1.4 allow the bow to achieve this.

Consequently, Helmholtz made use of Bernoulli's solution, as applied to the undivided string [see (2.36)]:

$$\eta(x, t) = \sum_1^\infty \mathrm{Re}\left\{\underline{\hat{\eta}}_n \sin\frac{n\pi x}{l}e^{j2\pi nt/T}\right\}.\tag{3.7}$$

With this solution, the time function at any one point x_1 along the string allows us to determine the time functions for all other points:

$$\eta(x_1, t) = \sum_1^\infty \text{Re} \{\underline{\hat{\eta}}_{1n} e^{j2\pi nt/T}\}. \tag{3.8}$$

Comparing (3.7) and (3.8), we obtain the amplitudes of the partials:

$$\underline{\hat{\eta}}_n = \frac{\hat{\eta}_{n1}}{\sin(n\pi x_1/l)}. \tag{3.9}$$

In using this formula, we must make sure that the denominator is not zero; we must choose a value of x_1 that is not a node of the nth partial. This difficulty can indeed be avoided, in principle.

In section 2.6, we have already analyzed the triangular spatial function shown again in figure 3.3; the context of our earlier analysis was the initial spatial function of the plucked string. In order to reapply the earlier analysis, we need only place the unrestricted value of $t = 0$ on one of the zero crossings. With forethought, we choose the positive-to-negative zero crossing. Now $t_-/2 < T/4$, corresponding to the earlier value of x_1, as long as x_1, which here too corresponds to the distance from the bridge, is less than $l/2$. The oscillation is then built up of components of the form $-\sin(2\pi nt/T)$, in other words, the vector $\underline{\eta}_{1n}$ introduced in (3.8) takes on a positive imaginary value. In analogy to (2.37b), we obtain

$$\underline{\hat{\eta}}_{1n} = j\frac{2\hat{\eta}}{\pi^2 n^2} \frac{T^2}{t_+ t_-} \sin\frac{\pi nt_-}{T}, \tag{3.10}$$

and consequently

$$\underline{\hat{\eta}}_n = j\frac{2\hat{\eta}}{\pi^2 n^2} \frac{T^2}{t_+ t_-} \frac{\sin(\pi nt_-/T)}{\sin(n\pi x_1/l)}. \tag{3.11a}$$

The displacement $\hat{\eta}_n$ applies to the entire string, and so it cannot be dependent on x_1; if t_- and x_1 are restricted to the first quarter-period of the particular sinusoidal function, this is only then the case under the condition

$$\underline{\hat{\eta}}_n = j\frac{2\hat{\eta}}{\pi^2 n^2} \frac{T^2}{t_+ t_-} \tag{3.11b}$$

when

$$\frac{t_-}{T} = \frac{x_1}{l}.$$ (3.12a)[2]

Since T is composed of t_+ and t_-, we can extend this proportionality, just as in (3.5), to

$$t_+ : t_- : T = (l - x_1) : x_1 : l.$$ (3.12b)

Additionally, based on the relationship between the period and twice the length of the string,

$$T = \frac{2l}{c},$$ (3.13)

we can obtain

$$t_+ = \frac{2(l - x_1)}{c}$$ (3.14a)

and

$$t_- = \frac{2x_1}{c}.$$ (3.14b)

These equations indicate that the "corners" of the displacement, and the corresponding steps of velocity, propagate at the wave speed c.

Specifically, at the point of bowing $x_1 = x_B$, t_+ describes the interval of sticking friction

$$t_h = t_+(x_B) = \frac{2(l - x_B)}{c}$$ (3.15a)

and t_- describes the interval of sliding friction

$$t_g = t_-(x_B) = \frac{2x_B}{c}.$$ (3.15b)

These time intervals are, therefore, determined by the lengths of time for wave propagation between the bow and the nut (or finger) and between the bow and the bridge.

We could have reached the same conclusion based on our examination of the plucked string, without making use of Bernoulli's solution. It must

----

[2] Helmholtz (1862) obtained this same result by considering the ratio η_{21}/η_{11}.

be emphasized, however, that in the case of the plucked string two corners (i.e., steps of velocity) propagate in opposite directions; here, only one transition exists, and it is reflected at the ends of the string with a change of sign. Besides, the absolute values of η, v_+, and v_- differ from one point to another.

Consequently, we obtain a complete model of the bowed string only when we include the dependence $\hat{\eta}(x)$; in other words, when we determine the envelopes of the traveling peak which occurs at the corner. To accomplish this, we combine the proportions contained in (3.5) and (3.12a), obtaining

$$t_+ : t_- : T = -v_- : v_+ : \Delta v = (l - x) : x : l. \tag{3.16a}$$

From this, it follows that $|v_+|$ and $|v_-|$ increase linearly with the distances $(l - x)$ and x from the ends of the string:

$$v_+ = \Delta v \frac{x}{l}, \tag{3.16b}$$

$$v_- = -\Delta v \frac{l - x}{l}. \tag{3.16c}$$

Then the segments of the string between the ends and the propagating corner must behave as if they were rigid sticks, rotating in opposite directions around the end points with the angular velocity $\Delta v/l$.

Moreover, whether we cast it in terms of $\frac{1}{2}v_+ t_+$ or of $-\frac{1}{2}v_- t_-$, we can rewrite the relationship (3.1), which describes the displacement $\hat{\eta}$, as

$$\hat{\eta}(x) = \frac{\Delta v T}{2} \frac{(l - x)x}{l^2}. \tag{3.17}$$

The envelope, therefore, consists of parabolas, whose maxima are at the middle of the string:

$$\hat{\eta}_{\max} = \Delta v T/8. \tag{3.18}$$

Since the eye cannot follow the oscillations of the string, an observer sees only the parabolic envelope when looking at a bowed string, and might conclude that the string maintains its parabolic form throughout each period. With the aid of a stroboscope, it is possible—most easily in the case of the longer and deeper-pitched strings of the cello—to see the propagating corner and its change of sign at the ends.

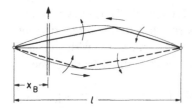

Figure 3.5
Helmholtz displacement along the string.

3.3   Representations of Helmholtz motion

The motion described at the end of the previous section and illustrated in figure 3.5 is known as *Helmholtz motion*. Figure 3.5 includes representations of the position of the string at two moments in time which are $T/2$ away from one another.

Helmholtz himself pointed out other possible motions; but he dealt thoroughly only with the motion shown in figure 3.5. Helmholtz was of the opinion that this motion corresponded to the best tone.

In figure 3.5 arrows indicate the directions of motion of the bow, of the corner, and of the straight segments of the string. These directions are interrelated; they are always to be interpreted such that the interval of sticking friction is longer than the interval of sliding friction. As already shown in chapter 1, the bow always tries to make the interval of sticking friction as long as possible; when there are two possibilities, the bow always forces the one with the longer interval of sticking friction. When there is a reversal of the direction of bowing—called a "bowing change" in performance technique—all of the arrows are reversed.

Helmholtz motion may described in term of other properties. When displacement is small, the parabolas may be replaced by arcs of circles; Then the angle $(180° - \Delta\alpha)$, complementary to $\Delta\alpha$, is the peripheral angle between the two straight-line segments of the string under an arc of a circle. This peripheral angle is, however, the same wherever along the string it is located. The motion can therefore be represented as the propagation of a peripheral angle. If $\Delta\alpha$ is constant, then so is the step in transverse force. And, if we consider the change in velocity Δv of the element of mass $m'cdt$ in the time interval dt, using the impulse relation,

$$F_x\Delta\alpha dt = \Delta vm'cdt, \tag{3.19}$$

Figure 3.6
Helmholtz distribution of velocity along the string.

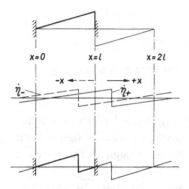

Figure 3.7
Generation of the velocity function by means of two waves propagating in opposite directions.

then the step in velocity Δv is also constant. We could already have concluded this from (3.5).

This constancy of Δv, and the linear changes of v_+ and v_- (3.16a), lead to Raman's representation of Helmholtz motion, shown in figure 3.6 (Raman 1918, 1927). This representation has certain advantages when motions are under examination in which there is more than one Δv per period.

The propagating corners of displacement and steps of velocity in figures 3.5 and 3.6 make it tempting to consider d'Alembert's representation as better suited to the problem than Bernoulli's. Yet Helmholtz used Bernoulli's representation; in fact, d'Alembert's representation is not as appropriate to this problem, as the triangle continually changes shape in this representation.

D'Alembert's representation is simple and clear only as it aplies to the velocity distribution. As shown in figure 3.6 and with a change of sign in the adjacent mirror-image regions, the velocity distribution follows an ideal sawtooth function, with a $2\Delta v$ step (see figure 3.7, top). By

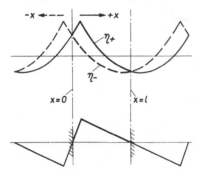

Figure 3.8
Generation of the displacement function by means of two waves propagating in opposite directions.

separating this step into two halves, allowing each half to propagate, and adding the resulting functions, exactly the function in figure 3.6 is obtained. This function repeats with a change of sign in the adjacent mirror-image regions (figure 3.7, bottom).

This function is composed of straight lines and steps. In describing displacement, however, D'Alembert's solution requires integration over time and also over length. The resulting functions are parabolic arcs with corners. The conversion from the velocity distribution to that of displacement requires a change of sign, and the opposing curvatures of the parabolas makes it very difficult to visualize the result. It is easier to grasp the result if we keep the two curves facing the same way and take their difference. Then, in the initial case in which displacement is zero, the two curves fall exactly on top of one another. As the two parabolas propagate, their difference (figure 3.8, bottom) is a triangular function along the length of the string and also in the mirror-image regions. This description based on d'Alembert's solution—a description introduced by Lippich (1914)—can hardly be considered more suitable than a description using Helmholtz motion.

3.4 Transverse force on the bridge

We return now to an examination of the transverse force on the bridge, which force results from Helmholtz motion. This force is of special interest to us: the sound pressure that excites our ears increases along with this force, since all sound transmission elements between the bridge and the listener are linear. Also, we may compare the behavior of the bowed string with that of the plucked string as described in section 2.4, in order to draw conclusions about differences in tone color.

In the case of the plucked string, the time function of transverse force is rectangular; in the case of the bowed string, however, it takes on an ideal sawtooth shape composed of ramps and steps.

This conclusion results from an examination of the two similar propagating waves $v_+(ct - x)$ and $v_-(ct + x)$ of d'Alembert's solution.[3] As shown in figure 3.7, these have the peak value $\Delta v/2$. Where the two waves are reflected, they add, producing the peak value Δv. Consequently, $F(0)$ is described in terms of (2.22) as

$$\hat{F}(0) = Z\Delta v. \tag{3.20a}$$

But also, in connection with figure 3.6, the representation

$$\hat{F}(0) = F_x \hat{\alpha}(0) \tag{3.20b}$$

leads to the same result. The segment of the string beginning at $x = 0$ rotates with the constant angular velocity $\Delta v/l$. Consequently, α increases by

$$2\Delta\alpha = \frac{\Delta vT}{l} = \frac{2\Delta v}{c} \tag{3.21}$$

during one period T. In this equation, as in (3.19), $\Delta\alpha$ represents the angle at the propagating corner of Δv. At the point where it is reflected, this angle jumps from $\Delta\alpha$ to $-\Delta\alpha$; in other words, it changes by $-2\Delta\alpha$.

In addition, we may substitute the bowing speed for Δv, as in (3.16a), and we may express Z in terms of F_x and m'. Then we may also express the peak value of the transverse force at the bridge as

$$\hat{F}(0) = \sqrt{F_x m'}\, \frac{1}{x_B} v_B. \tag{3.22}$$

This equation is formulated in terms of easily measured variables drawn from the structure of the system and on the technique of performance.

Tension and mass are the variables that relate specifically to the string.

---

[3] The subscripts + and − here indicate the directions of propagation, whereas in section 3.1 they were the signs of the velocity of a particle of the string.

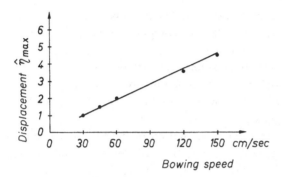

Figure 3.9
The increase in displacement with increasing bowing speed (after Müller).

In the course of the evolution of the violin, the tension of the string increased considerably; the construction of the body was changed to accommodate the higher tension. As is well known, these developments led to an increase in loudness. m', too, was increased with the introduction of metal-wound strings. In both cases, these changes reached a limit which, as was discovered empirically, could not be safely exceeded.

Equation (3.22) also describes ways in which performance technique can affect the input force at the bridge. Players continually make use of these. We note first that $\hat{F}(0)$ increases along with v_B. In the context of the purely linear model given in the present chapter, this relationship is obvious. If all of the kinematic variables are doubled, then the requisite sticking at the point of bowing corresponds to doubled bowing speed, and all forces are also doubled.

Nonetheless, it is important to determine within what range the assumptions underlying Helmholtz motion are valid. With this goal in mind, Müller (1962) used the method illustrated in figure 3.4 to measure displacement in the middle of the string. In accord with (3.18) and (3.20a), this displacement stands in the fixed relationship

$$\hat{F}(0) = 8Z\hat{\eta}_{max}/T \tag{3.23}$$

to $\hat{\eta}_{max}$.

The result, shown in figure 3.9, indicates that this linear relationship is valid over a 1 : 5 range of bowing speed; this range should be sufficient for normal performance.

An experiment such as Müller's is possible only through the use of an

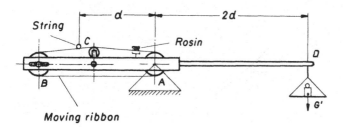

Figure 3.10
A mechanical bowing device.

apparatus which holds v_B constant and allows it to be measured accurately. Figure 3.10 shows the apparatus used by Müller. It presses a moving belt against the underside of the string. The string is mounted on rigid supports. The apparatus is balanced like a set of scales, allowing the bowing pressure to be changed or, as in the present case, to be held constant so as to remove its effect from the measurement. Bladier (1961) used a rotating hard-rubber disc with a rosined edge to perform similar measurements.

The importance of constant bowing speed and bowing pressure, achievable by such means, leads to the question whether results are the same as with a human performer, who can control these variables by ear. The question whether the feedback loop of the ear, brain, and hand influences bowing was explored at the Institute for Technical Acoustics of the Technical University of Berlin in the following way: the "tone" produced by the instrument was recorded on magnetic tape, and played back to the performer over headphones, sometimes with a delay of between 64 and 128 ms. The performer noticed the delay when changing the direction of the bow, but did not experience any unusual disturbances during the playing of prolonged notes. Results were, to be sure, different when the performer was asked to play a melody consisting of brief notes. In a further experiment, all reference to the tone produced by the instrument was removed from the performer. Very loud noise was played over the headphones, masking the tone. Even in this case, the bowing was hardly affected.[4] Such experiments suggest that it is not inherently inappropriate to use a mechanical bowing apparatus instead of a human

---

[4] The author owes special thanks to the violist Liselotte Schönewald and the cellist Brunhilde Schönewald for their assistance in these experiments.

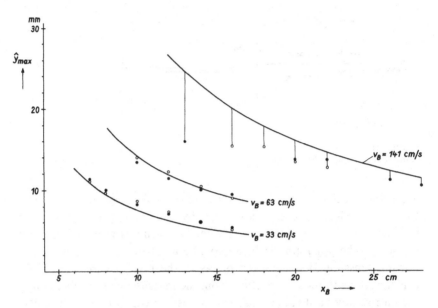

Figure 3.11
Measured displacements in the middle of the string, as they depend on the point of
bowing, for various bowing speeds (the circles drawn solid and in outline correspond to
two series of measurements) (after Völker).

performer, so that the bowing speed and pressure can be held constant
and measured.

Every violin player knows that it is possible to make the tone loud by
bowing faster. The relationship $v_{B\,max}/v_{B\,min}$, once the player comprehends
it, clearly becomes a working attribute of the performer.

In (3.22) there is a second relationship which affects loudness; namely,
$\hat{F}$ is inversely proportional to the distance x_B between the bow and the
bridge. The tone becomes stronger the closer the bow is brought to the
bridge, as long as it is possible to avoid making scratching noises instead.
This ability, too, distinguishes the good performer from the less accomplished one.

This dependence as well can be explored using the apparatus shown
in figure 3.10, by measuring $\hat{\eta}_{max}$. Figure 3.11 shows values of $\hat{\eta}_{max}$ measured by Völker (1961) for a 1 m long string with a bowing "pressure"
of 0.3 kg and bowing speeds of 33, 63, and 141 cm/s. Völker measured
these values using two screw-type micrometers whose zero settings were

calibrated to account for the thickness of the string. The micrometers were retracted until the string vibrated freely, and then closed until contact with the string made itself evident through impact noise. This method allowed for measurement not only of the maximum displacement of the string to each side, $\hat{\eta}_{+\max}$ and $\hat{\eta}_{-\max}$, but also of the superimposed static displacement resulting from frictional force. We will examine this static displacement in section 3.6. If the displacement is not too great, Völker's results for the given bowing pressure, and for other bowing pressures as well, corresponds to calculated values for Helmholtz motion; in other words, they correspond well if the bowing speed is not too great and the bow is not too close to the bridge. The calculated values for Helmholtz motion are shown as solid lines in figure 3.11.

Every violin player knows that it is necessary to press harder on the bow when it is near the bridge than when it is near the fingerboard. This chapter explores only differences that can be explained in terms of free oscillations; consequently, no external forces enter into the calculations. The forces of sticking and sliding friction increase along with the bowing pressure, but this effect is disregarded in pure Helmholtz motion.

Also, the following observations about tone color, based on the amplitude spectrum of the bowed string, are valid only insofar as they are independent of the bowing pressure. In this context, they represent only a good first approximation. The amplitudes of the partial tones corresponding to a sawtooth function of period T and peak value $\hat{F}$ are given by

$$|\hat{F}_n| = \frac{2\hat{F}}{\pi n}. \tag{3.24}$$

In contrast to the much more complicated spectrum of the plucked string, this spectrum, illustrated in figure 3.12, is independent of the point of excitation. We should recall here that it was necessary to exclude bowing at a node of displacement of a partial tone from our derivation. A significant feature of the spectra of plucked strings illustrated in figure 2.4 was the absence of a partial and all of its multiples in the analysis when plucking was at a node of that partial. The same is, however, also true here for the analysis of the spectrum as it corresponds to the point of bowing. Since Helmholtz motion can be generated only by excitation at a point of bowing, it is the case here, as with the plucked string, that the corresponding partial tones are absent.

Very generally, the theory of forced oscillations (in our example, oscilla-

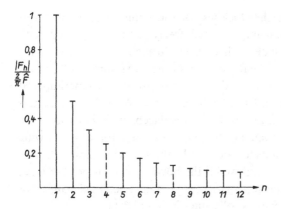

Figure 3.12
Line spectrum of force at the bridge for the bowed string.

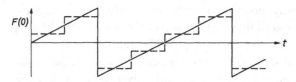

Figure 3.13
Force over time for bowing at $x_B = l/4$.

tions in which a given external force is exerted on a string) indicates that at such nodes of displacement—and hence of velocity—no power can be transferred into the string. In order for power to be transferred, a corresponding component of velocity must be present (see section 4.1).

If, however, we eliminate the partial tones of orders n, $2n$, $3n$, etc., then naturally the time function, as well as the spectrum, must change.

We take once more as an example the case in which $n = 4$. We have shown the spectral lines corresponding to this case with dashed lines in figure 3.12. These spectral lines correspond to a sawtooth curve in the time domain, just as does the overall spectrum. The period of the sawtooth curve corresponding to these lines is $T/4$ and its peak value is $\hat{F}/4$. If we subtract this from the sawtooth corresponding to the full spectrum, we obtain the stepped function shown as a dashed line in figure 3.13.

The function $\eta(x, t)$ is also altered; not at $x_1/l = 1/4$, $1/2$, $3/4$, ..., corresponding to the point of bowing, but in between these nodes, reach-

Figure 3.14
Displacement over time for bowing at $x_B = l/4$, observed at $x = l/8$.

ing a maximum at the antinodes. In figure 3.14, the idealized Helmholtz motion is shown by the solid line, and the motion corresponding to the corrected spectrum is shown by the dashed line. The maximum deviations here are $\hat{\eta}_{max}/n^2$, and so are smaller than the deviations in F and v. Helmholtz himself discovered these deviations with his vibration microscope and explained them in terms of the absence of the partial tones. He described the deviations as "ripples," in the displacement functions of the string.

In reality, bowing at $n = 4$ occurs only when the player uses fingering positions high on the fingerboard of the instrument. Generally, the string is bowed in the range from 2 to 4 cm from the bridge. In the case of the open string, this means that the lowest partials at whose nodes bowing can occur are the 8th through 16th. To place the bow in one of these nodes, great precision in positioning would be necessary; this is hardly possible and does not seem to be worth the trouble. Also, the bow has a finite width, making bowing at a node even more difficult. Even if bowing at a node does occur, the steps in the function F are only 1/8 to 1/16 the height of the overall step. In addition, as already demonstrated in the case of the plucked string, the increasing damping at higher frequencies evens out all of the steps and corners. Helmholtz's ripples are, consequently, generally of little importance.

It is also possible to reach this conclusion by noting that no musical notation includes the instruction to place the bow at a particular node. The only instructions of this type are to place the bow near the bridge (*sul ponticello*) and at (or somewhat overlapping) the end of the fingerboard (*sul tasto*). These instructions, however, would seem to have much to do with the involuntary change in bowing pressure which accompanies the change in position. We will discuss bowing pressure later.

3.5 Superposition of several Helmholtz motions

We have just discussed modifications of Helmholtz motion that depend on the position of the bow; but these are by no means the only complicated free motions which can occur. In principle, any superposition of Helmholtz motions is possible, including superpositions in which there are multiple alternations of sticking and sliding.

Helmholtz observed these, too, with his vibration microscope. He particularly noted that light and fast bowing near the bridge sometimes leads to a halving of the period of oscillation. In addition, he observed irregular alternation of sticking and sliding. This, however, always occurred in combination with a scratching sound.

The largest number of possible forms of oscillation of the bowed string were described by Helmholtz's pupil Krigar-Menzel and Krigar-Menzel's co-worker Raps (Krigar-Menzel and Raps 1891). Their technique, advanced for its time, was to register the motion of a well-lighted point on a string, isolated by an optical slit, using light-sensitive paper fixed to the outside of a drum which rotated at a constant speed. Their string was only 0.1 mm thick, and exhibited zigzag curves with surprisingly sharp corners. Krigar-Menzel and Raps confirmed that Helmholtz motion occurs only when the point of bowing is neither too near the bridge ($x_B/l > 1/15$), nor too far from it ($x_B/l < 1/7$).

They were also able to exhibit the ripples described above, which occur in every case in which bowing is at a node of a partial and the string is observed at a point away from a node of the same partial. But their results also includes superpositions of several Helmholtz motions which have equal steps of velocity; Raman (1918) dealt with such superpositions thoroughly in his monograph.

Only a few such superpositions are of interest in connection with the violin, especially as the important relationships shown in (3.15a, b) can be upset when Helmholtz motions are superimposed. The interval of sliding friction can, for example, be shorter than that corresponding to the propagation of waves from the bow to the bridge and back,

$$\frac{t_g}{T} < \frac{x_B}{l}. \tag{3.25}$$

The discrepancy cannot be explained in terms of a minor correction; it

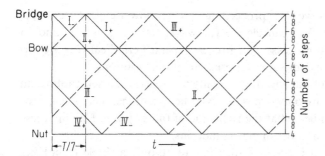

Figure 3.15
Space-time diagram for four Helmholtz motions that cancel each other three times at the point of bowing.

amounts to a factor of 1/3 or even 1/2. As Raman showed, this occurs only when

$$\frac{x_B}{l} = \frac{m}{n},$$

(3.26)

where n must be an odd number and $n > m > 1$.

Lippich (1914) showed that this behavior can be explained in terms of Bernoulli's representation of the motion, as described in its basic form by Helmholtz. The pure kinematics of this behavior still fall within the category of free oscillations.

The discussion becomes clearer and simpler through the use of graphic timetables of the individual corners or steps. Figure 3.15 shows such a timetable for the case in which $x_B/l = 2/7$ and $t_g/T = 1/7$. Positions along the string are represented vertically, one above the other, with the bridge at the top. Time advances from left to right. The scales of the two coordinates are chosen so that the straight lines of $x(t)$ lie at angles of 45°. Lines representing waves proceeding toward the bridge are dashed; at the bow, they represent steps $-\Delta v$ at release. Solid lines represent waves proceeding away from the bridge; at the bow, these lines represent steps $+\Delta v$ at capture. We begin with the moment of release. If the step-shaped wave I_- were the only one, then it would put an end to sliding friction after $\frac{2}{7}T$ when it returned to the bow as I_+. The interval of sliding friction ends at the moment of capture, after only $\frac{1}{7}T$, however, due to an earlier wave II_+ proceeding from the direction of the bridge. When the I_+ wave arrives, it is cancelled by the III_- wave, and when this in turn arrives as

III$_+$, yet another wave IV$_-$, proceeding from the lower end, is necessary. This, however, as IV$_+$, cancels the II$_+$ wave, returning as II$_-$, and so prevents a premature release. The process repeats periodically. It is composed, as we see, of four superimposed Helmholtz motions.

At the position of the bow, there is only one release and one capture; in other words there are only two steps. At other positions, there are 4, 6, or 8 transitions, as indicated at the right side of the diagram. Halving these numbers gives the number of zigzag motions.

Since, however, it is possible for there to be only a single Helmholtz motion, there arises a question which our purely kinematic analysis cannot answer: namely, when does the single Helmholtz motion with $t_g = 2/7$ occur, and when does the more complicated motion with the shorter interval of sliding friction $t_g = 1/7$ occur? The frictional mechanism tends toward the shorter interval of sliding friction. However, as Lazarus (1972, fig. 14) was able to show, four times the bowing pressure is necessary in this case.

In his investigations, Lawergen (1978) undertook to reproduce Krigar-Menzel's and Raps's oscillograms using present-day sensing techniques and while measuring bowing pressures precisely. Lawergen showed that the occurrence of complicated oscillations depends not only on the point of bowing but also on the bowing pressure and on increased frictional forces. These influences and their effects are still not understood well enough to permit anything more than a kinematic description of them.

For this reason, we restrict our discussion of the subject to this one example. There is another reason as well: Raman (1918), who brought about great progress in emplaining Krigar-Menzel's and Raps's results, came to the conclusion that "in the musical application of the subject we have to deal only with the first type of vibration [Helmholtz motion] and its modifications [the ripples]."

3.6 Superposition of static displacement resulting from constant friction

The motions of the bowed string can be explained quite well in terms of free oscillation; we have justified this approximation by pointing out that the pitch of the string when bowed is empirically found to be nearly the same as when it is plucked. Consequently, the energy input and output of

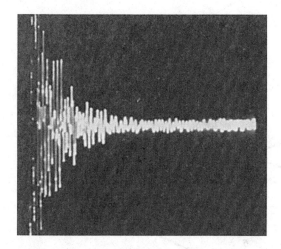

Figure 3.16
Decay of oscillations of a string plucked and sensed in the middle, as close as possible to
the bow, which is resting lightly on the string.

the string must be small in comparison to the oscillating energy. This is the
case with most self-sustained oscillators.

There is a very simple experiment, however, which can be performed on
a string instrument, and which casts this conclusion into doubt. When an
open string is plucked, its oscillation has a long decay time. However, when
the bow is laid on the string as nearly as possible to the point of plucking,
and so lightly that it does not act as a rigid support forming a node, the
decay time is short. Figure 3.16 shows an oscillogram corresponding to
this latter case. In comparing this decay to that shown in figure 2.7, it must
be noted that the time scale is compressed fivefold in figure 2.7.

If the bow did not damp the string so rapidly, *staccato* and *martelé*
playing would not be possible.

When the bow and the point of observation are at the middle of the
string, it is easy to analyze the damping process graphically (see figure
3.17).

From the description of the motion of the plucked string illustrated in
figure 2.2, we know that the point at the middle of the string moves transversely at a constant velocity until it reaches a position mirroring its
original one. Sliding friction must, then, be constant, and so a static dis-

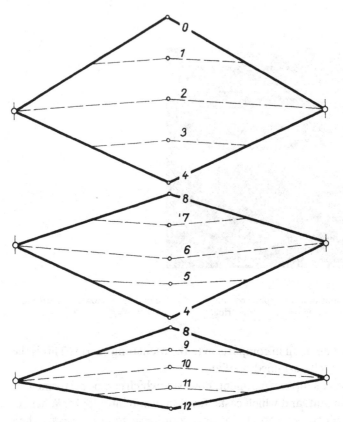

Figure 3.17
Displacement for the case illustrated in figure 3.16.

placement of constant form is superimposed (line 2 in figure 3.17). This static displacement causes the resulting displacement $\hat{\eta}_1$ reached after a half-period to differ from the original by twice the average,

$$\bar{\eta} = \frac{R_g}{F_x} \frac{l}{4}, \tag{3.27}$$

that is, the half-period displacement is

$$\hat{\eta}_1 = \hat{\eta}_0 - 2\bar{\eta}. \tag{3.28a}$$

When the string is traveling in the opposite direction, the frictional force is reversed (figure 3.17, middle); consequently, after a full period,

$$\hat{\eta}_2 = \hat{\eta}_0 - 4\bar{\eta}, \tag{3.28b}$$

as long as, for simplicity's sake, we ignore the dependence of R_g on the relative velocity. This sort of linear decrease of displacement is also well known in connection with simple oscillators damped by dry friction.

Unfortunately, this ideal case can be approximated only imperfectly in practice, since it is not possible to pluck the string at the same point at which the bow is placed. If the string is plucked only slightly away from its middle, partial tones are generated for which the bow lies at a node; these partial tones quickly overwhelm the behavior to be illustrated, since the bow has very little frictional effect on them.

Our experiment shows that when the string is bowed, large losses must also occur during the intervals of sliding friction. There is, however, an important difference: in this case, sliding friction always bears on the string in the same direction—the direction of bowing. Though the sliding friction may vary over time, it nonetheless has an average value $\bar{R}$ such that exactly as much energy is put into the string during the interval of sticking friction as is withdrawn from it during the interval of sliding friction. In order, then, to determine the energy input and output of the string, it is necessary only to consider this average value. The average value amounts to a superimposed static displacement, always in the same direction. This effects no change in the free oscillations.

The static displacement is triangular, with its peak at the point of bowing:

$$\bar{\eta}_{max} = \frac{R_g}{F_x} \frac{(l - x_B)x_B}{l}. \tag{3.29}$$

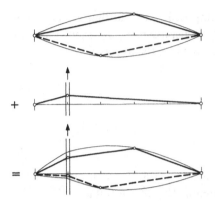

Figure 3.18
Helmholtz motion with a superimposed displacement due to constant frictional force.

The envelope in figure 3.5 is, therefore, transformed into one with two corners (see figure 3.18). The change of direction of the superimposed, static deformation can be seen with the naked eye, particularly on the violin's lower strings and on the cello.

Electrodynamic transducers are sensitive primarily to velocity, so they do not respond to $\bar{\eta}(x)$. Völker (1961), however, was able to measure the static displacement by determining the difference between positive and negative displacement, $\hat{\eta}_+$ and $\hat{\eta}_-$, using screw-type micrometers as described in section 3.4:

$$\bar{\eta} = \frac{1}{2}(\hat{\eta}_+ - \hat{\eta}_-). \tag{3.30}$$

In turn, he was able to determine the average frictional force, which is always proportional to the bowing pressure F_z and is equal to the average sliding friction,

$$\bar{R}_g = \bar{\mu}_g F_z. \tag{3.31}$$

From many understandably scattered measurements, he was able to arrive at an average value for the coefficient of sliding friction,

$$\bar{\mu}_g = 0.24. \tag{3.32}$$

Additionally, he was able to compare the energy loss per period,

$$W_{\text{loss}} = 2\bar{R}_g \hat{\eta}(x_B) = \bar{R}_g \frac{(l - x_B)x_B}{l^2} T \Delta v, \tag{3.33}$$

to the peak value of oscillating energy. This value is easiest to determine at the instant when the string passes through its rest position. At this instant, the energy is purely kinetic:

$$W_{\text{max,kin}} = \frac{1}{2}m'(\Delta v)^2 \int_0^l \left(\frac{x}{l}\right)^2 dx = \frac{m'l}{6}(\Delta v)^2. \tag{3.34}$$

Völker determined the ratio $W_{\text{loss}}/W_{\text{max,kin}}$ to be about $1/10$ when bowing pressure is low and about $4/10$ when it is high. Nonetheless, even in this case, a simple Helmholtz motion was superimposed on the static displacement.

Since only the alternating forces at the input of the bridge "make music," it is common not to mention the superimposed static deformation, which Lippich (1914) and Raman (1918) already have discussed. The effect of bowing pressure on the static deformation is of no importance as regards the generation of the tone; the alternating frictional forces, which also increase with bowing pressure, are of primary importance, and these have nothing to do with the large losses at the bow. Rather, these frictional forces are related to the much smaller losses at the bridge, at the nut, at the finger, and in the string. This fact, however, does not make the analysis any simpler.

References

Bladier, B., 1961. *Acustica* **11**, 373.

Helmholtz, H. v., 1862. *Lehre von den Tonempfindungen* [*On the Sensations of Tone*]. Braunschweig: Vieweg.

Krigar-Menzel, O., and Raps., A., 1891. *Berl. Ber.* **32**, 613.

Lazarus, H., 1972. Dissertation, Technical University of Berlin.

Lawergen, B., 1978. *Catgut Acoust. Soc. Newsletter* no. 30, 1.

Lippich, F., 1914. *Wiener Ber.* **125**, 1071.

Müller, H. A., 1962. *Untersuchungen zur Physik der Geige* [*Investigations of the Physics of the Violin*] Unpublished ms. of work carried out at the violin-builders' school in Mittenwald.

Raman, C. V., 1918. *Indian Assoc. Sci. Bull.* **15**.

Raman, C. V., 1927. In Geiger and Scheel, eds., *Handbuch der Physik* [*Handbook of Physics*], vol. 8, p. 370. Berlin: Springer.

Völker, E. J., 1961. Thesis, Institute for Technical Acoustics, Technical University of Berlin.

4 The Bowed String as a Forced Oscillation

4.1 Incorporating distributed viscous friction

The first researcher to investigate energy losses at points other than the point of bowing was Raman, who later was awarded a Nobel Prize for his work in atomic physics. His 158-page monograph (Raman 1918) on the physics of the violin attained fame, though it did not, in fact, cover all of the subjects listed toward the end of its table of contents. The primary aim of Raman's monograph was to present the simplest possible kinematic description of the observations by Krigar-Menzel and Raps described in section 3.5 of the present work. Raman's description made use of functions alternating between constant velocity with sticking friction and constant velocity with sliding friction at the position of the bow. Constant velocity is clearly a prerequisite for sticking friction; Raman justifies the assumption of constant-velocity sliding friction by stating that variations in frictional force must be small compared with its average value. This assumption is, however, certainly not valid at the moments of capture and release; at these moments, sliding friction must equal the highest values of sticking friction.

In order to represent mathematically the small fluctuations δR of friction that he assumes, Raman, like Helmholtz, takes the stepwise changes of velocity at the point of bowing as given, and then asks what forces are necessary to produce such stepwise changes at this point. He thus treats the problem of the self-sustained string as that of a forced oscillation, though he reverses the usual causal sequence: here the motion is given and the force is to be determined.

The problem of a string excited at a point can be solved by means of a strict mathematical procedure (Rayleigh 1877). This procedure describes the string as being divided into two segments, each of which is to be treated separately, with a corner at the point of bowing. But this procedure results in a treatment less easily understood than does an analysis in terms of natural oscillations. To be sure, Raman makes use of the latter analysis, although he presents in detail the less easily understood strict solution.

In a treatment incorporating these approaches, there is a way in which losses can be included in the analysis, yet the natural oscillations can remain the same as those of a lossless string with rigid supports; we have already encountered this model, though without losses, in section 2.6, in connection with Bernoulli's solution. Now losses are assumed to result from friction between the string and the surrounding air; Raman also mentioned this method of treating the losses. To the difference in transverse forces

$$F_x \frac{\partial^2 \eta}{\partial x^2} dx$$

exerted on an element $m'dx$ of the string, there is added a frictional force

$$-r' \frac{\partial \eta}{\partial t} dx$$

proportional to velocity, where r' represents viscous impedance per unit length. The differential equation for the freely oscillating string then is altered from (2.3) to

$$m' \frac{\partial^2 \eta}{\partial t^2} + r' \frac{\partial \eta}{\partial t} - F_x \frac{\partial^2 \eta}{\partial x^2} = 0. \tag{4.1}$$

The boundary conditions

$$x = 0; \quad \eta = 0$$

$$x = l; \quad \eta = 0 \tag{4.2}$$

remain the same as before.

Both the differential equation and the boundary conditions are satisfied by the functions

$$\eta(x, t) = \sum \eta_n(t) \sin \frac{n\pi x}{l}. \tag{4.3}$$

Substituting this into (3.1) results in a time function which is the well-known equation of damped oscillation

$$m' \frac{d^2 \eta_n}{dt^2} + r' \frac{d\eta_n}{dt} + F_x \left(\frac{n\pi}{l}\right)^2 \eta_n = 0, \tag{4.4}$$

whose general solution is

$$\eta_n(t) = \text{Re}\{\hat{\eta}_n e^{-\delta t + j\omega_n t}\}. \tag{4.5}$$

The coefficient of decay is

$$\delta = \frac{r'}{2m'} \tag{4.6}$$

and the natural frequencies, described as radian frequencies, are

$$\omega_n = \sqrt{\left(\frac{n\pi}{l}\right)^2 c^2 - \delta^2}. \tag{4.7}$$

Since it is assumed that $\delta^2 \ll (n\pi/l)^2 c^2$, these are practically the same as in the case of the lossless string:

$$\omega_n \approx \left(\frac{n\pi}{l}\right) c. \tag{4.8}$$

4.2 Temporal and spatial Fourier analysis of the excitation

We now turn to the problem of forced oscillation, extending (4.1) by at first proposing a general, unrestricted distribution of excitation force per unit length, $F_y'(x, t)$:

$$m'\frac{\partial^2 \eta}{\partial t^2} + r'\frac{\partial \eta}{\partial t} - F_x\frac{\partial^2 \eta}{\partial t^2} = F_y'(x, t). \tag{4.9}$$

Using this approach, we can synthesize the time dependence of F_y' in terms of sinusoidal components without restricting the generality of the solution. We are interested in periodic solutions:

$$F_y'(x, t) = \sum_{N=1}^{\infty} \mathrm{Re}\{\hat{F}_{yN}'(x)e^{jN\Omega_1 t}\}. \tag{4.10}$$

If this formula is substituted into (4.9), the result is the ordinary differential equation

$$[-(N\Omega_1)^2 m' + jN\Omega_1 r']\underline{\eta}_N - F_x\frac{\partial^2 \eta_N}{\partial x^2} = \underline{F}_{yN}'. \tag{4.11}$$

This, however, permits the synthesis of $\eta_N(x)$ and $\underline{F}_{yN}'(x)$ out of the same natural functions which also describe free oscillation. The boundary conditions, as well, are fulfilled:

$$\underline{\eta}_N(x) = \sum_{n=1}^{\infty} \underline{\eta}_{Nn} \sin\frac{n\pi x}{l}, \tag{4.12a}$$

$$\underline{F}_{yN}'(x) = \sum_{n=1}^{\infty} \underline{F}_{yNn}' \sin\frac{n\pi x}{l}. \tag{4.12b}$$

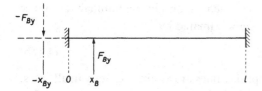

Figure 4.1
Excitation of the string by a point force.

In this case, as well, the values of F'_{yNn} are found through Fourier analysis, though to be sure this is a special case:

$$\underline{F}'_{yNn} = \frac{2}{l} \int_0^l F'_{yN} \sin \frac{n\pi x}{l} dx. \tag{4.13}$$

In order to carry out the integration over the spatial period $2l$, this formulation requires that the distribution of forces be $-F'_{yN}(-x)$ in the mirror image region $-l < x < 0$.

We are interested in the special case in which force is exerted at the point $x = x_B$; as illustrated in figure 4.1, we must therefore assume that an opposite force is exerted at $x = -x_B$. The result of the analysis in this special case is

$$\underline{F}'_{yNn} = \frac{2F_{yNB}}{l} \sin \frac{n\pi x_B}{l}. \tag{4.14}$$

This formula shows, as was already mentioned in the discussion of the ripples observed by Helmholtz, that a partial and its overtones are absent if excitation is at one of the partial's nodes, $x_B = ml/n$.

4.3 Calculation of the force spectrum from the velocity spectrum

If the type of synthesis shown in (4.12a, b) is also used to determine η_N, then the displacement vector η_{Nn} and the force density vector $\underline{F}'_{yNn}$ are related as

$$\left[F_x \left(\frac{n\pi}{l} \right)^2 - (N\Omega_1)^2 m' + jN\Omega_1 r' \right] \underline{\eta}_{Nn} = \underline{F}'_{yNn}. \tag{4.15a}$$

This relationship is familiar from the study of the forced oscillation of a simple oscillator. If it is assumed, as usual, that force is given and that the displacement is to be determined, then we obtain the familiar resonance curves, whose peak values are approximated by

$$jN\Omega_1 r' \underline{\eta}_{Nn} = \underline{F}'_{yNn}. \tag{4.15b}$$

In the present problem, these peak values are attained exactly in all cases.

The period of the function of force is the same as the period of the first partial of the string:

$$\Omega_1 = \frac{\pi c}{l}. \tag{4.16}$$

Consequently, (4.15a) assumes the form

$$\underline{\eta}_{Nn} = \frac{\underline{F}'_{yNn}}{F_x\left[\left(\dfrac{n\pi}{l}\right)^2 - \left(\dfrac{N\pi}{l}\right)^2\right] + jN\left(\dfrac{\pi c}{l}\right)r'}. \tag{4.17}$$

If we take n as a given and allow N to assume the values of all integers, then the peaks will always coincide with

$$N = n. \tag{4.18}$$

Nonetheless, the result is still, at least for now, a double sum. Terms corresponding to $N \neq n$, namely those in which $N = n \pm 1$, make a contribution.

If, however, r' is very small [if, more precisely, as we have already assumed in making the transition from (4.17) to (4.18), $\delta^2 \ll (n\pi/l)^2 c^2$], then as an approximation we can neglect the contribution of the values which do not correspond to the resonance peaks. Then (4.17) simplifies to

$$\underline{\eta}_n = \frac{\underline{F}'_{yn}}{jn\left(\dfrac{\pi c}{l}\right)r'}. \tag{4.19}$$

When we examine velocity, this leads to the relationship

$$\underline{v}_n = \frac{\underline{F}'_{yn}}{r'}, \tag{4.20}$$

which is independent of n, or, substituting from (4.14), to the relationship

$$F_{ynB} = \frac{r'l}{2 \sin \frac{n\pi x}{l}} v_n,$$

(4.21a)

or the more interesting inverse relationship

$$v_n = \frac{2F_{ynB}}{r'l} \sin \frac{n\pi x_B}{l}.$$

(4.21b)

(In these relationships, we can replace N by n, and also do without the double subscripts.)

4.4 Extending the model to include other losses

If, in (4.21a), we express r' using $\tau = 1/\delta$ as we did in (4.6), the result is

$$F_{ynB} = \frac{m'l}{\tau \sin \frac{n\pi x}{l}} v_n.$$

(4.22)

This way of writing the formula allows us to use it to approximate types of energy losses that are physically more significant in the case of the oscillating string than is air friction. As shown by the values of τ measured on the monochord and illustrated in figure 2.5, there certainly do exist losses distributed along the string. But these are for the most part "inner" losses, for which τ is a function of frequency. Since we have paid attention only to the peaks of the resonance curve in our derivation, there is no objection to substituting the value $\tau(n\Omega_1)$, valid only for the frequencies at the peaks, into our formula.

Equation (4.22), however, can also be extended to embrace the case in which losses occur at the ends of the string by the absorption of a certain fraction α of the power that arrives there. In room acoustics, the symbol α designates the absorption coefficient (see section 16.2). When the absorption at each end of the string, α_0 and α_1, is small, then τ may be found using the formula

$$e^{-2t/\tau} = (1 - \alpha_0)^{ct/2l}(1 - \alpha_1)^{ct/2l}$$

$$\approx e^{-(\alpha_0+\alpha_1)ct/2l},$$

(4.23)

which reduces to

$$\tau = \frac{4l}{c(\alpha_0 + \alpha_1)}.$$ (4.24)

In this case too, α, and consequently τ as well, can depend on the frequency. Also, losses in the string and at the boundaries can be combined additively.

Losses at the boundaries are bound up with a certain amount of change in the natural functions. Since these must now transfer power, there is a phase shift along the string (Cremer and Müller 1982, secs. 4.4ff. and 12.8).

These discrepancies, however, also are small enough that (4.22) can still be taken as an approximation. This holds true as well for the steps of phase at the ends of the string, which we will discuss in section 5.4.

4.5 Raman's model

As already mentioned, Raman (1918) assumes that the only changes in the velocity of the string at the point $x = x_B$ are steps between v_B and $v_B - \Delta v$. The value of Δv follows, then, from the requirement that velocity, integrated over an entire period, must disappear; there can be no endlessly increasing displacement of the string in the direction of bowing.

In what follows, we wish to exclude cases in which more than one capture and one release occur in each period; as Helmholtz observed, the tone is always scratchy in such cases. We therefore restrict ourselves to the case in which there is only one interval of sticking friction and one interval of sliding friction. The spectrum corresponding to this case is given by

$$\underline{v}_{nB} = \frac{2\Delta v}{\pi n} \sin \frac{\pi n t_h}{T},$$ (4.25)

if we set $t = 0$ at the middle of the interval of sticking friction. From this formula and from (3.11a), it follows that $\underline{v}_n$ is

$$\underline{v}_n = \frac{2\Delta v}{\pi n} \frac{\sin \dfrac{\pi n t_h}{T}}{\sin \dfrac{\pi n x_B}{l}}.$$ (4.26)

We must, however, recall that if x_B lies at a node of a natural oscillatory mode, the values of $\underline{v}_n$ for which the numerator disappears are absent from the spectrum.

This condition is more pronounced if, as in (4.22), we substitute the resulting values of $\underline{F}_{ynB}$; if we do this, an additional factor of $\sin(\pi n x_B/l)$ appears in the numerator:

$$\underline{F}_{ynB} = \frac{2\Delta v m' l}{\tau \pi n} \cdot \frac{\sin \dfrac{\pi n t_h}{T}}{\sin^2 \dfrac{\pi n x_B}{l}}. \qquad (4.27)$$

Also, for mathematical reasons having to do with the convergence of his Fourier series, Raman restricts his evaluations to cases in which

$$\frac{t_h}{T} = 1 - \frac{t_g}{T} = 1 - \frac{1}{n}; \qquad \frac{x_B}{l} = \frac{m}{n} \qquad (n = 1, 2, 3, \ldots; m = 2, 3, \ldots). \qquad (4.28)$$

Since the sinusoidal function in the denominator goes to zero at the same points as that in the numerator, the requirement to remove the corresponding values of F_n is the same as in the previous discussion of free oscillations.

Raman adds another condition that is more physically relevant: he assumes τ to be independent of frequency. As we have shown in section 4.1, this condition holds not only when losses occur through viscous friction between the string and the surrounding medium, but also when frequency-independent losses occur at the ends of the string. This condition, in turn, holds when the terminating impedances at the ends, representing the relationship of force and velocity, are real:

$$\frac{\underline{F}_0}{\underline{v}_0} = Z_0, \qquad \frac{\underline{F}_l}{\underline{v}_l} = Z_l. \qquad (4.29)$$

The energy loss is in this case given by $\dfrac{1}{2}\dfrac{\hat{F}_0^2}{Z_0}$; the arriving energy is given by $\dfrac{1}{2}\dfrac{F_a^2}{Z}$. In these formulas, Z represents the characteristic impedance of the string, as described in (2.21). Since the force at the end of the string is nearly doubled,

$$\hat{F}_0 \approx 2\hat{F}_a, \qquad (4.30)$$

the coefficient of absorption which determines τ is then

$$\alpha_0 = \frac{\frac{1}{2}\hat{F}_0^2/Z_0}{\frac{1}{2}\hat{F}_a^2/Z} = \frac{4Z}{Z_0} = \frac{4m'c}{Z_0}, \qquad (4.31a)$$

and correspondingly

$$\alpha_l = \frac{4Z}{Z_l} = \frac{4m'c}{Z_l}. \tag{4.31b}$$

In what follows, it makes no difference whether such losses are considered to occur at only one end of the string or at both ends. Raman considered losses at only one end, noting that losses at the bridge are necessary if sound is to be produced. However, using these simple formulas, it is also possible to model losses at the other end—losses at the finger would be the most significant example of these. It is, however, necessary to pay attention to what is meant in the literature: particularly when the terminating impedance is taken to be the same at both ends of the string. When losses occur at only one end, then, in accord with (4.24) and (4.31a), we have

$$\tau_1 = \frac{Z_0 l}{m' c^2}, \tag{4.32a}$$

and, on the other hand, when losses occur at both ends, we have

$$\tau_2 = \frac{Z_0 l}{2m' c^2}. \tag{4.32b}$$

Raman's assumption that the decay time is independent of frequency appears first as an aid in solving the problem into which the simplest possible representation of losses is incorporated. Not only is the assumption a simplification: as we will see in the next chapter, nothing else would be consistent with Raman's further assumption that the only changes in v_B are steps.

For this reason, it is appropriate to speak of a "Raman model." This model is not to be considered as an approximation to actual behavior, but rather as a physically unrealizable limiting case, which serves only to help make certain features of the actual behavior of the string understandable. There is yet another failing of the Raman model: it does not lead to any definition of the rest position of the string. We will discuss this failing in section 8.4.

Because of these problems, we will look at only a few of the many cases that Raman examined. Raman, whose work points to an extraordinary aptitude for seeking out the most appropriate transformations of circular (i.e., sine and cosine) functions, recognized that the time function corresponding to the spectrum shown in (4.27) must consist of series of steps.

This result can also be deduced from the physics of the system. The velocity function is composed of steps whose magnitude is diminished either as they propagate or when they are reflected, but without alterations of their form. For this reason, the propagating waves of force necessary to determine the velocity function must also consist only of steps (Lazarus 1970, Schumacher 1979). Since every periodic step of δR has a corresponding line spectrum $\delta R/j\pi n$, then $\underline{F}_{ynB}$ in (4.27) must also be analyzable as a series

$$\underline{F}_{ynB} = \frac{\delta R_1}{j\pi n} e^{-j\vartheta_1} + \frac{\delta R_2}{j\pi n} e^{-j\vartheta_2} + \cdots . \tag{4.33}$$

The restriction (4.28) indicates that these steps can occur only at intervals of mt_g/T. For each step, there must also occur an opposite one at a symmetric point on the time axis, and so the general term in (4.33) assumes the Raman form

$$\frac{\delta R_1}{j\pi n}(e^{j\pi t_g/T} - e^{-j\pi t_g/T}) + \frac{\delta R_2}{j\pi n}(e^{j2\pi t_g/T} - e^{-j2\pi t_g/T}) + \cdots$$

$$= \frac{2\delta R_1}{\pi n}\sin\frac{\pi t_g}{T} + \frac{2\delta R_2}{\pi n}\sin\frac{2\pi t_g}{T} + \cdots = \frac{2\Delta vm'l}{\tau\pi n}\frac{\sin\dfrac{\pi n t_h}{T}}{\sin^2\dfrac{\pi n x_B}{l}}. \tag{4.34}$$

Raman determined the values of δR by substituting $n = 1, 2, 3, \ldots$ until the equations began to repeat themselves.

We shall now work through this process for the case $t_g/T = x_B/l = 1/4$, which we have already examined in the context of Helmholtz motion. We obtain:

for $n = 1$,

$$\delta R_1/\sqrt{2} + \delta R_2 + \delta R_3/\sqrt{2} = \frac{\Delta vm'l\sqrt{2}}{\tau}; \tag{4.35a}$$

for $n = 2$,

$$\delta R_1 \quad + 0 \quad - \delta R_3 \quad = -\frac{\Delta vm'l}{\tau}; \tag{4.35b}$$

for $n = 3$,

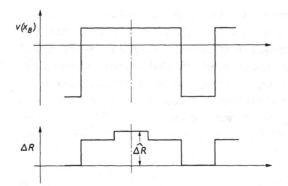

Figure 4.2
Velocity over time (top) and the required frictional force over time (bottom) at the point of bowing $x_B = l/4$, if the problem is regarded as a forced oscillation with frequency-independent damping (after Raman).

$$\delta R_1/\sqrt{2} - \delta R_2 + \delta R_3/\sqrt{2} = \frac{\Delta v m' l \sqrt{2}}{\tau}. \tag{4.35c}$$

The case $n = 4$ is absent from the spectrum, and, beginning with $n = 5$, the same equations repeat cyclically. If the last equation is subtracted from the first one, the immediate result is

$$\delta R_2 = 0 \tag{4.36a}$$

and, if this is substituted into the first two equations,

$$\delta R_1 = \frac{1}{2}\frac{\Delta v m' l}{\tau} \tag{4.36b}$$

and

$$\delta R_3 = \frac{3}{2}\frac{\Delta v m' l}{\tau}. \tag{4.36c}$$

Figure 4.2 shows the corresponding function.

4.6 Necessity of a minimum bowing pressure

From figure 4.2 it may also be deduced that a variation in frictional force

$$\Delta R = \delta R_1 + \delta R_2 = \frac{2\Delta v m' l}{\tau} \tag{4.37}$$

is necessary. This variation is possible, however, only if the maximum sticking friction $R_{h\,max}$ is not exceeded. Since ΔR is superimposed on the value of sliding friction $R_g(v_B - \Delta v)$, there results the condition

$$\Delta R < R_{h\,max} - R_g. \tag{4.38}$$

But, on the other hand, the difference $R_{h\,max} - R_g$ increases along with the bowing pressure F_{zB}:

$$R_{h\,max} - R_g = F_{zB}(\mu_{h\,max} - \mu_g), \tag{4.39}$$

and the coefficients of friction in this formula for the maximum sticking friction $\mu_{h\,max}$ and for the sliding friction are independent of the bowing pressure. Therefore (4.38) determines a minimum bowing pressure necessary for the generation of the type of motion under consideration:

$$F_{z\,min} = \frac{\Delta R}{\mu_{h\,max} - \mu_g}. \tag{4.40a}$$

In the present example, this minimum bowing pressure is

$$F_{z\,min} = \frac{2\Delta v m' l}{(\mu_{h\,max} - \mu_g)\tau}. \tag{4.40b}$$

If we replace Δv with the bowing speed, it becomes clear that the minimum bowing pressure increases along with the bowing speed:

$$F_{z\,min} = \frac{2m' l^2}{(\mu_{h\,max} - \mu_g)\tau x_B} v_B. \tag{4.40c}^{[1]}$$

This way of writing the formula, however, has the disadvantage that it suggests a further, false constraint: that $F_{z\,min}$ be proportional to x_B^{-1}; the present author once accepted this constraint (Cremer and Lazarus 1968, Cremer 1971).

Equation (4.40b) is valid only if $x_B/l = 1/4$. But the numerical factor in (4.40b) also increases approximately as (l/x_B).

In figure 4.3, the illustrations from Raman's 1918 monograph for $t_g/T = 1/n$ (labeled at the right) and $x_B = l/n$ (labeled at the left) are shown together. Since the two sets of curves are taken from different sets of

---

[1] Cremer and Lazarus (1968). In that published version of the lecture which the first author gave at the ICA in Tokyo, 1968, the formula for ΔR is incorrect. The correct formula is (4.37) of the present work.

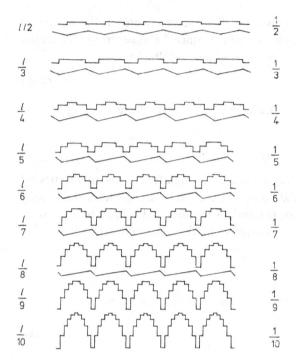

Figure 4.3
Alteration of force over time as the point of bowing $x_B = l/n$ is brought closer to the bridge (after Raman).

illustrations, the durations of the periods are represented differently. According to Raman's description, however, both sets correspond to the same bowing speed, as can be seen from the rising segment of the curve of displacement where it is included. The curves of force show a rise in ΔR approximately as $(l/x_B)^2$, and can be described by the formula

$$\Delta R = v_B \frac{Z^2}{Z_0}\left(\frac{l}{x_B}\right)^2 = v_B \frac{m'l}{2\tau}\left(\frac{l}{x_B}\right)^2, \qquad (4.41)$$

derived by Schumacher (1979) from considerations of behavior in the time domain. The formula is for a string at both ends of which the impedance is Z_0. The formula gives an exact result when l/x_B is even, and a result with a slight discrepancy when l/x_B is odd. Schumacher, however, distinctly emphasizes that the formula is valid only for integral n and not for arbitrarily chosen positions on the string.

74 The Bowed String as a Forced Oscillation

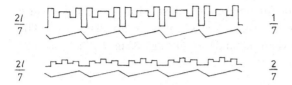

Figure 4.4
Force over time at the point of bowing $x_B = \frac{2}{7}l$, with intervals of sliding friction of $\frac{1}{7}T$ (top) and $\frac{2}{7}T$ (bottom), at the same bowing speed (after Raman).

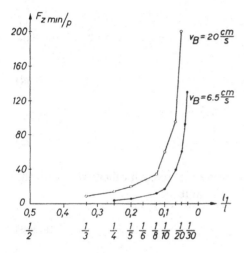

Figure 4.5
Measured minimum bowing pressures for a cello plotted against the distance of the bow from the bridge ($l_1/l \equiv x_B/l$) (increasing toward the left), for two different bowing speeds (after Lazarus).

The function for $x_B = 2l/7$ and $t_g/T = 2/7$ shown in figure 4.4, bottom, reveals another structure. This structure is even more evident in the case shown in figure 4.4, top, in which x_B is also $2l/7$ but $t_g/T = 1/7$. Comparing the two cases, we see clearly that a greater value of ΔR, and therefore also a greater minimum bowing pressure, occurs when the interval of sliding friction is half as long. We have already mentioned this phenomenon in section 3.5.

Measurements by Lazarus (1972) confirm that the bowing pressure necessary to produce Helmholtz motion decreases when the bow is farther from the bridge, and that minimum bowing pressure increases with bowing speed. Figure 4.5 shows the results of Lazarus's experiments.

It is appropriate to look for a more general basis for this behavior, since Lazarus obtained consistent results even when x_B was not equal to l/n. The relationship between power input and output suggests itself as a basis. The power input to the string at the bow is

$$P_1 = v_B \overline{\delta R(t)}. \tag{4.42}$$

If this amount of power is delivered by the string to the bridge, it is given there by

$$P_2 = \overline{F_y(0, t)^2}/Z_0. \tag{4.43}$$

The function $F_g(0, t)$ is, to a first approximation, a sawtooth, whose peak value is

$$\hat{F}_g(0) = Z\Delta v. \tag{4.44}$$

Consequently,

$$P_2 = \frac{1}{3}\frac{Z^2\Delta v^2}{Z_0} = \frac{1}{3}\frac{Z^2}{Z_0}\left(\frac{l}{x_B}\right)^2 v_B^2. \tag{4.45}$$

Setting the two values of the power equal to each other, we obtain the average value of the additional frictional force:

$$\overline{\delta R} = \frac{1}{3}\frac{Z^2}{Z_0}\left(\frac{l}{x_B}\right)^2 v_B. \tag{4.46a}$$

If the string is terminated in Z_0 at both ends, $\overline{\delta R}$ rises to:

$$\overline{\delta R} = \frac{2}{3}\frac{Z^2}{Z_0}\left(\frac{l}{x_B}\right)^2 v_B. \tag{4.46b}$$

But the minimum bowing pressure depends on the peak value $\widehat{\delta R}$, and the ratio $\widehat{\delta R}/\overline{\delta R}$ can, as Raman's calculations show, sometimes be different. Insofar as we may assume that this ratio depends only slightly on x_B/l or other conditions, then

$$\widehat{\delta R} = \Delta R \propto \left(\frac{l}{x_B}\right)^2 v_B. \tag{4.47}$$

In the functions shown in figure 4.3, the form factor $(\overline{\delta R}/\widehat{\delta R})$ approximates the factor 2/3 in (4.46b), and so approximates Schumacher's formula with increasing n. It must be noted, however, that stepwise—not sawtooth-shaped—functions of force at the bridge occur in connection with these form factors, in which excitation is at nodes. Consequently the power loss is smaller, though this difference decreases as n increases. When $n = 4$, the form factor is only $5/8 = 15/24$ instead of $16/24$. When $n = 10$, the form factor has already risen to 0.660.

Once a minimum bowing force is attained, the Raman model and Raman's analytical technique lead to no further change in the form of the oscillation with further increase in the bowing force, as Raman already emphasized. Only the static displacement, the corresponding sliding friction R_g, and the maximum sticking friction $R_{h\,max}$ increase.

References

Cremer, L., 1971. *Nachr. Akad. Wiss. Götingen* II, Math. Phys. K1., **12**, 223.

Cremer, L., and H. Lazarus., 1968. ICA Tokyo, 1968, N-2-3.

Cremer, L., and H. A. Müller, 1982. Principles and Applications of Room Acoustics. Tr. by T. J. Schultz. London: Applied Science.

Lazarus, H., 1970. *Technischer Bericht des Heinrich-Hertz Institutes* no. 117. Berlin-Charlottenburg.

Lazarus, H., 1972. Dissertation, Technical University of Berlin.

Raman, C. V., 1918. *Indian Assoc. Sci. Bull.* **15**.

Rayleigh, Lord, 1877. *Theory of Sound.* Reprinted 1929, London: Macmillan.

Schumacher, R. T., 1979. *Acustica* **43**, 109.

5 Extension of Helmholtz Motion Using Corrective Waves

5.1 Review of previous chapters

Up to this point, we have discussed the problem of excitation by dry friction only in chapter 1, in the context of self-sustained oscillations of simple oscillators. In that discussion, it was above all clear that the alternation between sticking and sliding friction always requires the attainment of the maximum sticking friction determined by the bowing pressure. (As elsewhere throughout this book, we use the musician's term "bowing pressure" when speaking of the bowing force.) This condition has not yet come up in our discussion of the bowed string. The alternation of sticking and sliding has up to now been modeled simply, based on kinematic considerations.

Nonetheless, the discussion of the bowed string in chapter 3, based on the work of Helmholtz, led to the examination of two important relationships which are qualitatively familiar to players of string instruments: the input force at the bridge—and with it the sound pressure at the ear—grows with the bowing speed; also, the input force at the bridge is inversely proportional to the distance x_B between the point of bowing and the bridge.

Raman's treatment of the system as a forced oscillation with a frequency-independent impedance at the bridge, discussed in chapter 4, showed, in addition, that there is a minimum bowing pressure. This increases with the bowing speed. Also, as the point of bowing is brought closer to the bridge, the minimum bowing pressure increases as $(l/x_B)^2$. This relationship, too, agrees qualitatively with the experience of the players.

There is no qualitative change in behavior dependent on increases of bowing pressure above the minimum value, if we analyze the system as a forced oscillation. This assertion follows from the assumption that velocity at the point of bowing can be modeled as a Helmholtz motion, composed of steps Δv.

This constancy of behavior also agrees, however, with the electrodynamic recordings of Müller (1962), which we have already illustrated in figure 3.9 in connection with our discussion of the dependence of displacement on bowing speed. Müller also confirmed that the displacement, which in this case is always determined by the peak value of the non-sinusoidal oscillation, does not change even if the bowing pressure is quintupled (see figure 5.1).

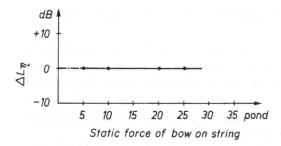

Figure 5.1
The measured maximum displacement level vs. bowing pressure (pond = gram force) (after Müller).

This assertion, however, would be perplexing to any string player. But the peak value alone does not determine the perceived loudness, and it certainly does not determine the tone color. In his book, still very much worth reading today, the physiologist Trendelenburg (1925) wrote:

As already noted, the strength of the tone does not correspond to the displacement of the string. A basic experiment is, for example, to bow a cello string with low bowing pressure and relatively high bowing speed: it is easy in this way to achieve a displacement so great that the string nearly hits the adjacent ones. The tone is dull and not very loud. Now, if the bow is moved more slowly and with somewhat more pressure, the tone becomes brighter, and somewhat louder, and so it carries to greater distances. In this case, however, the displacement is considerably smaller than before.

In addition to the bowing position and bowing speed, there is, then, a third degree of freedom, that of bowing pressure, which has not been examined up to this point in the discussion.

Finally, our objection to Raman's method of attempting to determine the influence of bowing pressure is not only that the function of velocity over time is given *a priori*; but also that the assumption of a frequency-independent decay time does not correspond to the observed behavior of the system. The increase in damping with frequency has been determined experimentally, as discussed in section 2.5. This increase in damping corresponds much better to what is to be expected physically, because it assumes neither points of inflection in the displacement function nor steps in the velocity function. The former are, in fact, physically impossible, as the string always has some bending stiffness.

5.2   Rounded corners and their relationship to bowing pressure

It is easy to conclude intuitively that the assumption of an *a priori* function of velocity over time is independent of the assumption that decay times are independent of frequency. However, a simple observation suffices to show that these assumptions belong together, and also that rounding of the corners is sufficient to account for the consistent influence of bowing pressure (Cremer and Lazarus 1968, Cremer 1974).

Now let us replace the sharp corners of Helmholtz motion with rounded corners of finite length and constant radius, proceeding from the nut toward the bow (figure 5.2a–c). Also, let us assume no alteration in bowing friction δR, and consequently a very low bowing pressure. Then, instead of dropping suddenly, velocity must fall at a constant rate from v_B to $v_B - \Delta v$. Similarly, when a rounded corner arrives from the bridge (figure 5.2d–f, in order from bottom to top corresponding to the direction of motion), there occurs a linear increase in velocity. The ends of the interval of sliding friction are defined no longer by steps, but instead by slopes whose rate of increase or decrease is finite.

Now we take the bowing friction into account, as shown by the solid lines in figure 5.3. (The dashed lines in figure 5.3 correspond to the solid ones in figure 5.2). The following discrepancies occur: as soon as the rounded corner attempts to move past the bow, sticking friction works to hold the string onto the bow longer, and so to straighten out the curved part of the string (figure 5.3a). Sticking friction achieves this by means of two waves propagating away from the bow in opposite directions, and superimposed on the displacement shown by dashed lines. The superimposed waves require the additional force

$$\delta R = 2Z\delta v. \tag{5.1a}$$

If the curvature is constant, the change in velocity, and consequently the sticking friction, increase linearly with time until δR reaches the maximum value ΔR. At this time, sliding friction begins (figure 5.3b).

If we simplify this construction by using a straight line instead of the characteristic line of friction schematically represented in figure 1.2, then δR, too, falls linearly during the interval of sliding friction, as does velocity v, though more rapidly than when there is no friction. If, in anticipation of notation to be introduced later, we designate the velocity of the wave arriving at the bow from the bridge as v_4, and the velocity of the wave

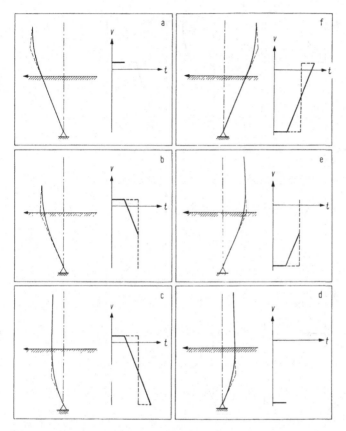

Figure 5.2
Displacements of the string at various times, and the function of velocity that has
occurred at the point of bowing up to those times. Solid lines, rounded corners in
the Helmholtz motion; dashed lines, sharp corners in the Helmholtz motion.

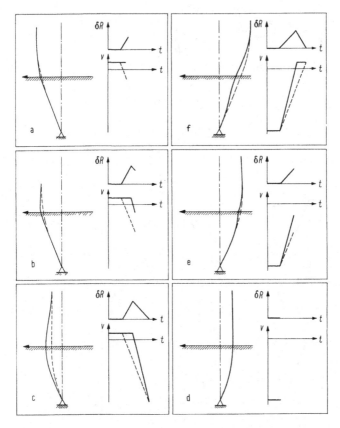

Figure 5.3
Displacements of the string at various times, and the function of velocity up to those times at the point of bowing, with the corresponding function of frictional force. Solid lines, with bowing pressure; dashed lines, without bowing pressure.

altered by friction after release as v_5, then on the one hand (5.1a) is transformed into

$$\delta R = 2Z(v_5 - v_4); \qquad (5.1b)$$

on the other hand, if we linearize the friction characteristic, δR may be represented as

$$\delta R = \Delta R \left(1 - \frac{v_{rel}}{\Delta v}\right) = \Delta R \left(1 - \frac{v_B - v_5}{\Delta v}\right). \qquad (5.2)$$

From these two formulas, it follows that the relationship between v_5 and v_4 is

$$v_5 - v_4 = \beta(v_5 - (v_B - \Delta v)) = \beta(v_5 - v_{g\,min}), \qquad (5.3)$$

in which

$$\beta = \frac{\Delta R}{2Z\Delta v} \qquad (5.4)$$

defines a dimensionless parameter that increases with the bowing pressure. We will consequently call this the *bow-pressure parameter.*

There exists, therefore, a linear relationship between the velocities v_4 of the arriving wave and v_5 of wave generated by bowing friction. The difference between these two velocities goes to zero when v_5 reaches the minimum value of v_4 (figure 5.3c).

The same relationship is repeated in mirror-image form when the curvature of the wave reflected from the bridge has the same radius as that of the wave propagating from above (figure 5.3d) ($v = v_2$). In this case, sliding friction brings about a more rapid increase in v ($= v_3$) (figure 5.3e) as the string's velocity approaches that of the bow. Sticking friction is achieved sooner; after it is achieved, δR falls once more until the string becomes straight once again—though somewhat displaced—at the point of bowing (figure 5.3f). In this case,

$$v_3 - v_2 = \beta(v_2 - v_{g\,min}) \qquad (5.5)$$

holds. This formula is analogous to (5.3). In each case, from these simple, though quantitatively thoroughly defined examples, we can determine the influence of friction on the motion of the string; lengthening the intervals of sticking friction, shortening the intervals of sliding friction, and steepen-

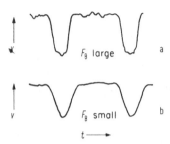

Figure 5.4
Recordings of velocity at the point of bowing: a, with large bowing pressure; b, with small bowing pressure (after Lazarus).

ing the slopes of the function of velocity as ΔR and the bowing pressure increase.

The importance of this influence can be determined only by working through an example with specific end impedances and specific assumptions about the damping of propagating waves, as in section 5.4. Our example will show, in addition, that this influence is significantly increased by an iteration effect: that is, by the repeated propagation of the rounded corners past the bow as the oscillation builds up.

In any case, the increase in the steepness of the slopes with increasing bowing pressure is in agreement with electrodynamic measurements of the function of velocity at the point of bowing. Lazarus (Cremer and Lazarus 1968) undertook such measurements; the results are illustrated in figure 5.4. Schelleng (1973) also published similar oscillograms, giving a larger number of examples. Our upper oscillogram corresponds to a larger bowing pressure, and our lower oscillogram to a smaller bowing pressure. These time functions make it clear that the overtone content is greater for a higher bowing pressure (curve a).

Of relevance to our problem of self-sustained oscillation by means of the alternation of sticking and sliding friction, rounded corners give us for the first time a function of frictional force over time that attains the maximum value of sticking friction at both capture and release. The force function consists of two narrow impulses at either end of the intervals of sliding and sticking friction (Lazarus 1970).

5.3 Method of the round trip of the rounded corner[1]

Increasing losses at higher frequencies tend to round off the corners more and more, and so to broaden the ramps in the velocity function. On the other hand, the ramps become steeper as they pass by the bow at capture and release. This suggest that conditions exist under which these tendencies compensate one another as the velocity ramps circulate back and forth along the string, that is, making "round trips," leading to a periodic wave within the string.

In order to analyze these conditions, it is helpful to take the terms representing oscillations of the string as a Helmholtz motion to be a first approximation, to which corrective terms are added. The latter have the character of additional waves, at least when the ramps are narrow and their changes are impulse-like. Consequently, it is appropriate for us to call these terms *corrective waves*.

Among these waves, the one which propagates along with the Helmholtz motion is most important to us; along with the Helmholtz motion, it generates the ramp whose propagation we wish to consider. Consequently, we will call this the *primary* corrective wave. The word "primary" implies, however, neither a temporal priority nor a cause-and-effect relationship.

From our observations in connection with figure 5.3, we know that the primary wave constitutes the difference between the step of Helmholtz motion and a ramp of finite width, and that this primary wave can be altered as it passes the bow only if, at the same time, another wave corresponding to the same change is generated, propagating in the opposite direction. We will call this other wave the *secondary wave*.

We certainly must expect that the secondary wave, as it is reflected from the bridge, will contribute to the time function of force at the bridge, and will have an effect on the tone color. In our "method the round trip of the rounded corner" or "method of circulating velocity ramps," however, we assume that the secondary wave is either weakened or out of phase at the moment of capture or release to such a degree that it has no significant effect on the conditions we will set up for the circulation of the ramps.

For this reason, we clearly must exclude long ramps, especially those in which departing and reflected waves begin to overlap and so lessen the

---

[1] For this very fertile formulation, the author is indebted to his English colleagues McIntyre and Woodhouse.

duration of the interval in which the velocity is $v_B - \Delta v$; this condition appears to be the case even in the lower curve of figure 5.4.

It will also help in setting up our conditions if we abandon our description of the velocities v in terms of values that change with position along the string as in figure 3.6; rather, we will describe the velocities in terms of the differences characteristic of the ramps:

$$u_k = v_k - v_{k\,\mathrm{min}}. \tag{5.6}$$

We will also still assume that the ramps begin suddenly, and we will choose the point at which a specific ramp begins as its zero point in time.

Let us consider a ramp $u_1(t)$ that propagates away from the bow after release of the string. This ramp begins with Δv, and sinks to zero; in other words, as seen at any particular moment, the ramp appears in the range $0 < x < x_B$, as a descending ramp, since it propagates in the $-x$ direction.

This ramp is reflected from the bridge, weakened and of opposite polarity. The reflected ramp can be defined by a convolution integral

$$u_2 = \int_0^t \frac{du_1}{dt'}(t')\psi_0(t - t')dt' + (u_2(0) \equiv 0). \tag{5.7}$$

In this integral, $\psi_0(t)$ represents the reflected wave which would generate a unit Helmholtz step at the bridge ($x = 0$). Consequently we will call ψ_0 the *reflection response* of the bridge. There is no formal objection to our including changes in the wave as it propagates along the string in this reflection response; the problem is made more difficult only in that the distance which the wave travels is then $2x_B$.

The wave u_3, which leaves the bow after capture of the string, undergoes changes at the nut or the finger. We may also account for changes as this wave propagates; such changes are greater in this case, since the path is longer. We account for all of these changes in the wave u_3 with the convolution integral

$$u_4 = \int_0^t \frac{du_3}{dt'}(t')\psi_l(t - t')dt' + (u_4(0) \equiv \Delta v). \tag{5.8}$$

These are the relationships which were missing from our observations derived from figure 5.3. We cannot expect, given the boundary conditions at the nut and the bridge, that the ramps will simply be reflected as straight lines of reduced slope; they will be more distorted than this.

On the other hand, the changes shown in figure 5.3 that result from sliding friction remain unaltered; in other words, we can take it as an approximation that

$$u_3 = \frac{u_2}{1 - \beta}, \quad 0 < u_3 < \Delta v \qquad (5.9)$$

and

$$u_5 = \frac{u_4}{1 - \beta}, \quad 0 < u_5 < \Delta v. \qquad (5.10)$$

It may be easy to grasp these relationships when they are represented graphically; yet they are nonlinear, being proportional only in the range below Δv, and leveling off to a constant value at Δv. This nonlinearity is of paramount importance in determining the behavior of the system; it describes the transition from sliding to sticking and vice versa. It also makes the relationships (5.9) and (5.10) difficult to describe analytically. It is therefore no surprise that there are no explicit formulas in which parameters of the ramps (for example, their average slopes) appear as functions of the bow-pressure parameter β.

However, a computer makes it possible to examine these problems, as well as the difficult behavior of ψ_0 and ψ_l. With the computer, it is not necessary to linearize the friction characteristic. Since Δv is given once we know the point of bowing and the bowing speed, we can find the relationship of the relative velocities $\Delta v - u_3$ and $\Delta v - u_5$ to Δv; we can also find the relationship of bowing friction δR to the range ΔR which corresponds to Δv [see (5.4)]. In this way we obtain figure 5.5, which shows the relevant part of a characteristic of friction between bow and string measured by Lazarus (1972), in the form

$$\frac{\delta R}{\Delta R} = \varrho\left(\frac{u_{3,5}}{\Delta v}\right). \qquad (5.11)$$

In this formula, the function ϱ takes on the limiting values $\varrho(0) = 0$ and $\varrho(1) = 1$. We next combine this equation with the relationship shown in (5.1a), in which we substitute $(v_3 - v_2)$ and $(v_5 - v_4)$ for δv, and introduce δR as defined in (5.5). The relationship now takes on the form

$$\beta\frac{\delta R}{\Delta R} = \frac{u_{3,5} - u_{2,4}}{\Delta v}, \qquad (5.12)$$

and so we obtain

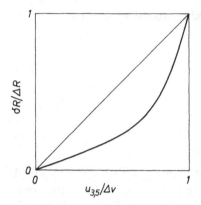

Figure 5.5
A normalized friction characteristic (after measurements by Lazarus).

$$\frac{u_{2,4}}{\Delta v} = \frac{u_{3,5}}{\Delta v} - \beta \varrho \left(\frac{u_{3,5}}{\Delta v}\right). \tag{5.13}$$

In this formula, the friction characteristic appears as the inverse of the function $u_{3,5}$ $(u_{2,4})$. When the bow-pressure parameter is small, its influence is slight. The curve in figure 5.6 for $\beta = 0.1$ deviates only slightly from the straight line

$$u_{3,5} = u_{2,4}/(1 - \beta), \tag{5.14}$$

which is drawn as a dashed line.

When β is larger, on the other hand, the deviations become so great that approximating the curve with a straight line is out of the question. Furthermore, $u_{3,5}$ may take on more than one value for a given value of $u_{2,4}$. In section 5.7, we will return to a discussion of the "instabilities" connected with this behavior.

If, in sequence, we

transform u_1 into u_2 using (5.7),
transform u_2 into u_3 using (5.9)
                or perhaps   (5.13),
transform u_3 into u_4 using (5.8), and finally
transform u_4 into u_5 using (5.10)
                or perhaps   (5.13),

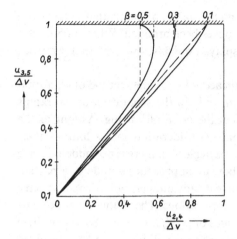

Figure 5.6
Behavior of the velocity difference u in the ramps ahead of (abscissa) and behind (ordinate) the bow for various bow-pressure parameters.

we will described a full round trip of the ramps. If, in doing this, we succeed in choosing u_1 such that u_5 exhibits the same form, then we will have found the periodic solution which we are seeking.

We can even take a time delay into account. We need require only that

$$u_5(t) = u_1(t - \Delta T),   \tag{5.15}$$

where ΔT represents an alteration in the duration of the period compared to the formula

$$T_0 = \frac{2l}{c},   \tag{5.16}$$

which corresponds to a string with rigid terminations. This duration of the period is, however, not exactly that to be expected even in the case of the plucked string if the ends are terminated other than rigidly.

Generally, the ramps cannot be assumed to be identical; rather, at the beginning of the second sequence

$$u_5 = u_1'   \tag{5.17a}$$

and at the beginning of the third

$$u_5' = u_1''   \tag{5.17b}$$

are appropriate, until finally $u_1^{(N)}$ and $u_5^{(N)}$ differ so slightly that a stationary condition has for all practical purposes been attained. When a computer is used these iterations can be displayed quickly, once the first sequence has been programmed.

In the solution by approximation according to the method of the round trip, the point of bowing is evident only in the separation of the ramps u_3 and u_5 in the velocity function at the point of bowing. As long as the secondary waves are not taken into consideration, or as long as their effect on the force at the bridge can be neglected, there is no sudden change in the tone color as the point of bowing approaches a node of a partial. This feature of the method may have disadvantages, but it may also have the advantage of not dragging in complications. The changes in tone color are hardly observable when the values of (x_B/l) are small; and when these values are large, the change in tone color is still insignificant enough so that instructions to the performer relative to them are not included in written music.

5.4 Introducing a simple frequency-dependent bridge impedance into the model

Since specific terminating impedances have a significant effect on the results of the method described in section 5.3, we must choose an example that allows us to show the influence of bowing pressure quantitatively. In connection with our present elementary observations, we are interested in the simplest possible example consistent with the decrease in decay times with increasing frequency noted in section 5.2.

We may simplify our calculations by neglecting losses during propagation, by considering only an open string, and by modeling the nut as a perfectly reflecting point of support.

The resulting frequency dependence then depends only on the impedance of the bridge. It might seem reasonable to consider such an impedance as real, as in the Raman model, though frequency-dependent. A real but frequency-dependent impedance is, however, not realizable in a system which includes resistive losses, masses, and compliances; and the system of the violin bridge and body can be analyzed as such a system in the frequency range of interest. In this frequency range, the radian frequency ω always occurs in a product with $j = \sqrt{-1}$; in other words, the impedance is always a function of $j\omega$, so that the real part (the resistance) can be frequency-dependent only if an imaginary part (the reactance) is also present.

The simplest model that results in the desired frequency response for the absorption coefficient consists of a spring in parallel with the frequency-independent resistance of the Raman model.

In this model, the force generated by the spring is proportional to displacement, and so to velocity integrated over time. Consequently the total input force at the bridge is

$$F(0, t) = wv(0, t) + s \int v(0, t)dt, \qquad (5.18a)$$

where w represents the resistance and s represents the stiffness of the spring. In terms of vectors in the complex plane, this formula is

$$\underline{F}_0 = \left(w + \frac{s}{j\omega} \right) \underline{v}_0 = \underline{Z}_0 \underline{v}_0. \qquad (5.18b)$$

If a wave with the velocity vector $\underline{v}_1$ arrives at the bridge and a wave with the velocity vector $\underline{v}_2$ is reflected, then $\underline{v}_0$ is expressed as the sum

$$\underline{v}_0 = \underline{v}_1 + \underline{v}_2 \qquad (5.19a)$$

and $\underline{F}_0$ as the sum

$$\underline{F}_0 = \underline{F}_1 + \underline{F}_2 = Z(\underline{v}_1 - \underline{v}_2). \qquad (5.19b)$$

From the quotients of (5.19b) divided by (5.19a), and taking (5.18b) into account, we obtain

$$\frac{\underline{v}_1 - \underline{v}_2}{\underline{v}_1 + \underline{v}_2} = \frac{\underline{Z}_0}{Z}. \qquad (5.20)$$

From this formula, we obtain the *reflection factor*, which we relate in this case to the velocity:

$$r = \frac{\underline{v}_2}{\underline{v}_1} = \frac{Z - \underline{Z}_0}{Z + \underline{Z}_0} = \frac{m'c - w - s/j\omega}{m'c + w + s/j\omega}. \qquad (5.21)$$

The square of the magnitude of the reflection factor is the *reflection coefficient*: this is the ratio of reflected to incident power. The *absorption coefficient* that we seek is the difference between unity and the reflection coefficient:

$$\alpha = 1 - |\underline{r}|^2 = \frac{4m'cw}{(m'c + w)^2 + s^2/\omega^2}. \tag{5.22a}$$

We can also express this as

$$\alpha = \frac{2\varepsilon - \varepsilon^2}{1 + \left(\dfrac{1}{\omega\tau_0}\right)^2}, \tag{5.22b}$$

by introducing a time constant for the bridge,

$$\tau_0 = \frac{w + Z}{s}, \tag{5.23}$$

and a parameter characterizing the asymptotic limiting value of the absorption coefficient for high frequencies,

$$\varepsilon = \frac{2Z}{w + Z}. \tag{5.24}$$

We will use this parameter again when we formulate the response function; we may call it the *damping parameter*.

The resulting frequency dependence of the energy decay of the string,

$$\frac{\tau}{2} = \frac{2l}{\alpha c} = \frac{2l}{c(2\varepsilon - \varepsilon^2)} \left[1 + \left(\frac{1}{\omega\tau_0}\right)^2\right] \tag{5.25}$$

agrees so well with the frequency response measured by Reinicke that we can even assign values for ε and τ_0, namely,

$$\varepsilon = 0.05, \quad \tau_0 \approx 10^{-4} \text{ s}. \tag{5.26}$$

(The corresponding curve is the dashed line shown in figure 2.5.)

In the examples in section 5.5, we will choose the higher but still physically representative values $\varepsilon = 0.125$ and 0.25, in the interest of making the drawings clearer.

The value of τ_0 derived from Reinicke's observations should not be confused with the decay time τ of the plucked string. It is significant that this value of τ_0 is smaller than the period of oscillation of 23×10^{-4} s used by Lazarus in his experiments (Cremer and Lazarus 1968; see our figure 5.4). The storage element of the impedance Z_0, namely the spring, must be regarded as "unloaded" after one period. Besides, τ_0 is even smaller than 1/10 period; even if we take $x_B/l = 1/10$, we do not have to concern ourselves with superpositions of the ramps u_1 and u_2 at the point of bowing.

If the impedance of the bridge is simple, then so is the reflection response of the bridge. First, by adding (5.19a) and (5.19b), taking (5.18b) into account, we obtain the relationship between $\underline{v}_1$ and $\underline{v}_0$:

$$2Z\underline{v}_1 = \left[(Z + w) + \frac{s}{j\omega}\right]\underline{v}_0, \tag{5.27a}$$

or

$$\varepsilon\underline{v}_1 = \left[1 + \frac{1}{j\omega\tau_0}\right]\underline{v}_0. \tag{5.27b}$$

Returned to the time domain, and differentiated, this leads to

$$\varepsilon\frac{dv_1}{dt} = \frac{dv_0}{dt} + \frac{v_0}{\tau_0}. \tag{5.28}$$

If $v_1(t)$ represents a unit step, then $v_0(t)$ begins with a step smaller by a factor of ε; however, in the adjacent range in which dv_1/dt vanishes, a valid solution to the homogeneous differential equation

$$\frac{dv_5}{dt} + \frac{v_0}{\tau_0} = 0 \tag{5.29}$$

leads to an exponential decay:

$$v_0(t) = \varepsilon e^{-t/\tau_0} \tag{5.30}$$

In the case of other functions $v_1(t)$, it is necessary to substitute differentials $\dot{v}_1(t')dt'$ for this and to obtain the motion of the bridge from a convolution integral

$$v_0(t) = \int_0^t (\dot{v}_1(t'))\varepsilon e^{-(t-t')/\tau_0}dt'. \tag{5.31}$$

This more general form allows us to account for the waves approaching the bridge in Helmholtz motion which do not consist only of periodic steps $-\Delta v$; in between the steps, these waves must rise linearly again and again, since there can be no static displacement of the string. The Helmholtz motion leads to a periodic motion of the bridge,

$$v_0(t) = \Delta v\left(\frac{\varepsilon\tau_0}{T} - \varepsilon e^{-t/\tau_0}\right), \tag{5.32}$$

which has no arithmetical average within a single period. For our chosen value of τ_0 and our duration of the period of the fundamental oscillation $T = 23 \times 10^{-4}$ s, the first term inside the parentheses is so small that it can usually be neglected despite its theoretical importance (Cremer 1974, p. 130).

If (5.30) is substituted into (5.29a), the result obtained for the reflected response of a unit step is

$$v_2(t) = -v_1(t) + v_0(t)$$

$$= -(1 - \varepsilon e^{-t/t_0}) = \psi_0(t). \tag{5.33}$$

This begins with a unit step in the negative direction whose magnitude is diminished by the amount ε; this diminished response decays exponentially.

5.5 Influence of bowing pressure in the chosen example

The simplicity of our chosen example is evident in our ability to calculate the average slope of a ramp systematically from one round trip. The ramp we examine is the one that, as u_1, propagates away from the position of bowing toward the bridge. Its slope is determined by values of τ_0, ε, and β. We assume u_1 to slope downward at the rate $-\Delta v / t_1$; t_1 is the variable to be found. Therefore,

$$u_1 = \Delta v \left(1 - \frac{t}{t_1}\right). \tag{5.34}$$

We now turn to figure 5.7, which is arranged in the same cyclic format as figure 5.2. The string is shown vertically in the middle, with the bridge at the bottom and the nut at the top. In the drawing at the lower left, the condition described in (5.34) is illustrated. Time is normalized to the decay time τ_0, the only time constant occurring in the accompanying formulas:

$$\vartheta = \frac{t}{\tau_0} \tag{5.35}$$

As a result of this normalization, the variable to be found, t_1, becomes a ramp parameter ϑ_1; like β and ε, this parameter is dimensionless.

Substituting u_1 from the convolution integral (5.7), in conjunction with

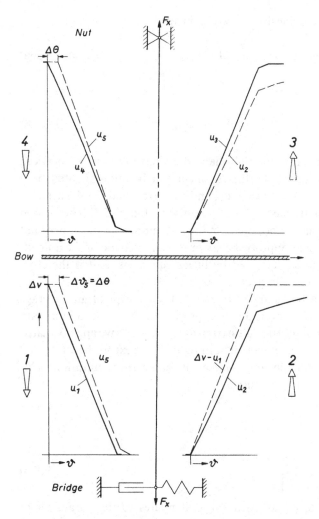

Figure 5.7
A cycle of the ramp-shaped waves u for a corresponding inclination of u_1, using a simple frequency-dependent model of the bridge; $\varepsilon = 0.25$; $\beta = 0.136$; $\vartheta_1 = 1$.

(5.33), the function u_2 consists of two segments

$$u_2 = \Delta v \left(\frac{\vartheta}{\vartheta_1} - \frac{\varepsilon}{\vartheta_1} (1 - e^{-\vartheta}) \right), \quad 0 < \vartheta < \vartheta_1 \tag{5.36a}$$

and

$$u_2 = \Delta v \left(1 - \frac{\varepsilon}{\vartheta_1} (1 - e^{-\vartheta_1}) \right) e^{-(\vartheta - \vartheta_1)}, \quad \vartheta_1 < \vartheta. \tag{5.36b}$$

This function is shown in the right-hand drawings in figure 5.7 between the bridge and the bow. The first segment can be approximated by a straight-line ramp, less steep than u_1, and so corresponds to the arriving rounded-off corner defined in connection with figure 5.3. The second segment, however, has no boundary in time; it approaches its upper limit asymptotically. Such asymptotic behavior is characteristic of the unloading of stored energy as described by linear equations, and is therefore characteristic of all input impedances of bridges.

The absence of a boundary to u_2 makes the concept of an "average slope" inapplicable; however, due to capture, the absence of a boundary is no longer evident in the transition from u_2 to u_3, which we approximated by (5.9). The ramp at the upper right in figure 5.7, next to the string and between the bow and the nut, once again has a finite duration. This duration, however, is longer than ϑ_1:

$$\vartheta_3 = \vartheta_1 + \Delta\vartheta_3, \tag{5.37a}$$

where

$$\Delta\vartheta_3 = \ln\left[\frac{\varepsilon}{\beta\vartheta_1} (1 - e^{-\vartheta_1}) \right]. \tag{5.37b}$$

Figure 5.8 shows $\Delta\vartheta_3$ (dashed lines) as it relates to the parameter of bowing pressure β, for ramp parameter ϑ_1 equal to 0.2, 1, 2, 3, and 4.

The transition from u_3 to u_4 is shown at the top left next to the string, between the nut and the bow. Since we assume an ideal reflection at the nut, u_4 consists simply of a reversal of the rising ramp u_3 into a falling one.

As indicated in the approximation (5.10), this falling ramp may be steepened by sliding friction after the delayed release at the transition from u_4 to u_5. (The function u_5 is shown as a dashed line next to u_4 and u_1.) But in this case there occurs only a finite lessening of the duration of the

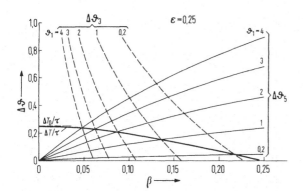

Figure 5.8
Parametric curves for evaluating the inclination of u_1 and the alteration in the period of a cyclic round trip of the ramp.

ramp, $\vartheta_4 = \vartheta_3$, by the amount $\Delta\vartheta_5$. This lessening can be calculated from the relationship

$$\frac{\Delta\vartheta_5}{\vartheta_1} - \frac{\varepsilon}{\vartheta_1}(1 - e^{-\Delta\vartheta_5}) = \beta(1 - \beta), \tag{5.38}$$

which cannot, however, be solved directly for $\Delta\vartheta_5$. We can, however, calculate $\Delta\vartheta_5$ from this relationship. It is included in figure 5.8 as solid lines, once more in terms of the ramp parameter ϑ_1 equal to 0.2, 1, 2, 3, and 4.

The ramps u_1 and u_5 are shown next to each other between the bow and the bridge at the left bottom of figure 5.7, in order to indicate how they differ. The derivatives du_1/dt and du_5/dt differ considerably; the requirement of (5.15) that the sequence be closed is not fulfilled. It is, however, possible to choose ϑ_1 so that the average slopes agree. This choice corresponds to the substitution of a falling straight line for u_5:

$$\frac{\Delta v}{t_5} = \frac{\overline{du_5}}{dt} = \frac{du_1}{dt} = \frac{\Delta v}{t_3}. \tag{5.39}$$

The requirement for a periodic sequence may then be approximated by

$$t_5 = t_4 - \Delta t_5 = t_3 - \Delta t_5 = t_1 + \Delta t_3 - \Delta t_5 = t_1,$$

and so

$$\Delta\vartheta_5 = \Delta\vartheta_3. \tag{5.40}$$

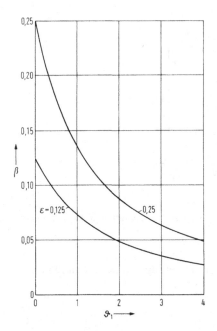

Figure 5.9
The bow-pressure parameter β as it depends on the inclination parameter of ramp 1 for various damping parameters ε.

Then the curves of $\Delta\vartheta_3(\beta, \vartheta_1)$ cross those of $\Delta\vartheta_5(\beta, \vartheta_1)$ at points that determine the relationship between the ramp parameter ϑ_1 and the bow-pressure parameter β. The bow-pressure parameter is shown in figure 5.9 for the damping parameter $\varepsilon = 0.25$, as in figure 5.8, and also for one-half this value, $\varepsilon = 0.125$. In both of these cases, an increase in bowing pressure corresponds to a decrease in the duration of the ramps. The bowing pressure must be increased as losses increase, if the ramp parameter is to remain constant. This dependence suggests that the relationships $\beta(\vartheta_1)$ are different in the case of impedances more representative of an actual violin. The trend of the curves, however, should be the same.

Above all, the ramp resulting from $u_1 - u_2$ will appear in the force at the bridge instead of a jump of amount $2\Delta v$, since the jump is transformed into a falling straight line with an asymptotic approach to the ascending straight line.

We may compare the change in ϑ_1 in figure 5.9 between 0 and 4 with the delays in release by the bow shown by the crossing points in figure 5.8.

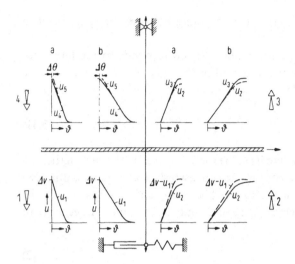

Figure 5.10
An iterated cycle of the ramp-shaped wave u, taking into account the friction
characteristic: a, for a high bow-pressure parameter $\beta = 0.136$; b, for a low bow-pressure
parameter $\beta = 0.088$.

These delays reach a maximum value of only 0.22. We recall that, in
connection with figure 5.3, we saw how the propagation of two rounded
corners past the bow led to a decrease in the duration of the interval
of sliding friction as bowing pressure increased; but now we state that
the iteration of round trips, discussed in section 5.3, must be considered
as well. As the string is bowed, the effects illustrated in figure 5.3 repeat
themselves over many periods, until a stationary condition is attained.

Figure 5.10 shows the results of an iteration carried out on a computer.
The drawings are arranged the same way as in figure 5.7, and the friction
characteristic is taken into account; its value in the left drawing of each
pair is $\beta = 0.136$, and in the right drawing $\beta = 0.888$. The curves corresponding to $\beta = 0.136$ are surprisingly similar to those in figure 5.7, in
which the ramp is approximated as a straight line.

The iteration effect makes it possible that the ultimate behavior of the
delay of release is to decrease rather than to increase with increasing
bowing pressure. This does not contradict our other statements, because
the ramp becomes much steeper due to the iteration.

It is clear that the maxima of the vertical distances $u_3 - u_2$ and $u_5 - u_4$

differ in the same way as the corresponding bow-pressure parameter in figure 5.10.

Also, we see that the release delay Δt_5 corresponds to the term representing a lengthening of the period in (5.15), which we added in case it might become needed:

$$\Delta \vartheta_5 = \Delta\Theta = \frac{\Delta T}{\tau_0}. \tag{5.41}$$

If we extrapolate $\Delta\vartheta_5(\beta)$ to its value for $\beta = 0$, we obtain the lengthening of the period for free oscillation—in other words, we obtain the duration of the period for the plucked string. The spring stiffness has exactly the same effect as a slight increase Δl in the length of the string. The equivalent stiffness is then

$$s = F_x/\Delta l. \tag{5.42}$$

Consequently, we obtain

$$\Delta T_0 = T_0 \frac{\Delta l}{l} = \frac{2\Delta l}{c} = \frac{2F_x}{cs} = \frac{2Z}{s} = \varepsilon\tau_0. \tag{5.43}$$

This maximum value is shown in the left margin of figure 5.8. Since ΔT falls as the bowing pressure increases, this represents a rise in the fundamental frequency. Even with the high values $\varepsilon = 0.25$, $\tau_0 = 10^{-4}$ s, and $T = 23 \times 10^{-4}$ s, the maximum increase in frequency through ΔT_0 is only 0.23%, considerably less than a quarter tone.

It is characteristic of string instruments that tuning is affected little by excitation, such as the bowing pressure here. This result is in accord with experience, and represents a considerable advantage of string instruments over wind instruments.

5.6 Influence of oscillations of the string perpendicular to the bow[2]

There are other corrective waves besides those already described. A rotation of the bridge may generate waves which propagate along the string in the x, z-plane perpendicular to the bow. Figure 5.11 (from a paper by Gough 1981), shows how the notch P for the g string moves

---

[2] This section, new with the English-language edition, is based on a suggestion by G. Weinreich (1982).

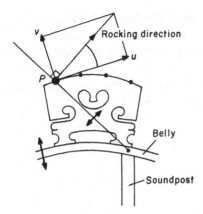

Figure 5.11
A schematic sketch of the rocking motion of the bridge, with motion of the left foot
prevented by the soundpost ($v = v_z, u = v_y$) (after Gough).

not only in the direction of bowing but also perpendicular to this direction,
assuming that the bridge rotates around its right foot. Helmholtz (1862)
had already made this assumption, apparently because he regarded the
bridge as being held rigidly by the soundpost at the position of the right
foot. In fact, this assumption is at best an approximation valid only in
certain frequency ranges. (We will examine this issue thoroughly in
chapter 9). Nonetheless, even if both feet of the bridge move, the overall
motion of a bridge, regarded as a rigid plate, is unambiguously defined;
the motion is a rotation around a center which changes depending on
the frequency. Indeed, Gough discussed such behavior of the bridge.
(There is a further complication when the motions of the two feet of the
bridge are not in phase; in this case, the center shifts position periodically
with the frequency of excitation, even if this frequency is fixed.)

In the following observations as well, we are interested only in the
principles which govern physical behavior. Consequently, we assume the
geometric relationships illustrated in figure 5.11 to be independent of
frequency, and we describe the kinematics of the bridge as

$$v_z(0)/v_y(0) = \tan \vartheta = 1/2, \tag{5.44}$$

where ϑ is the angle between the bow and the direction of motion of the
notch in the bridge. This formula holds as an approximation for all
four strings.

We cannot define a relationship between the forces in the y and z directions by means of dynamics. The rotational motion can, in fact, be generated by either an $F_z(0)$ or an $F_y(0)$ alone. For this purpose, the effects of $F_z(0)\sin\vartheta$ and $F_y(0)\cos\vartheta$ are equivalent; they generate the same rotational moment. It follows from this, however, that the loading of the bridge by a wave propagating away from it in the x, z-plane,

$$F_z(0) = -Zv_z(0), \tag{5.45}$$

would require a force in the y-direction

$$F_y(0) = Z\tan^2\vartheta\, v_y(0) = \frac{Z}{4}v_y(0) \tag{5.46}$$

on the string. In (5.18b), we designated the bridge impedance by $\underline{Z}_0$. We will retain this impedance for the y-direction in order to maintain consistency with our previous analysis; however, we must add another $Z/4$. The terminating impedance of the string is, then, raised to

$$\underline{Z}_0^+ = \underline{Z}_0 + \frac{Z}{4}, \tag{5.47}$$

where the superior $+$ sign indicates that the y and z directions, perpendicular to one another, are both taken into consideration.

This new complication does not prevent us from continuing to use our previous calculations, with the parameters ε and τ_0; however, they must be interpreted somewhat differently:

$$\tau_0 = (w + 5Z/4)/s, \tag{5.48}$$

$$\varepsilon = 2Z/(w + 5Z/4). \tag{5.49}$$

We can, however, continue to use the numerical values in (5.26), since these are based on Reinicke's measurements (Reinicke 1965; see figure 2.5). A power transfer from the x, y-plane to the x, z-plane certainly occurred in the experiments in which these measurements were taken.

The process of reflection at the bridge in the x, y-plane—that is, as in our simplified model of the transition from u_1 to u_2 in figure 5.7—is also retained.

The transition from u_2 to u_3 is, however, altered physically: a force in the z-direction—the direction of bowing pressure—is generated by the transverse wave in the x, z-plane; and it is the bowing pressure that determines the maximum attainable sticking friction in the y-direction.

Another factor increases the possibility that this value, decisive in the transition from sliding to sticking friction and back, will be influenced: namely, the wave in the z-direction generated by the motion of the bridge is exactly synchronized with the wave reflected in the y-direction, and so with the Helmholtz motion and the primary corrective wave. The wave in the z-direction is therefore fundamentally different from the secondary corrective waves, which can arrive at the bow at the moment of capture or release only after many reflections, weakened and altered in shape. For this reason, we have already decided to ignore the secondary corrective waves in our analysis.

The v_z wave accompanies the step of the Helmholtz motion up to reflection at the nut, and furthermore continues during many periods; each new impulse generated at the bridge is consequently superimposed on the previous one. The "phase matching" that occurs in this case is characteristic of reasonance phenomena involving sinusoidal oscillations (Cremer and Müller 1982, vol. 2).

However, quantitative experimental results regarding the v_z motions of bowed strings will be necessary in order to determine whether such resonant reinforcement actually occurs. Experiments to decide this issue have not yet been carried out.

There is, however, an experiment which anyone can carry out on any string instrument. In section 3.6 we showed that oscillations of a string plucked in the direction of the bow decay rapidly when the bow is resting on the string next to the point at which it is plucked. Damping is also very strong if plucking is perpendicular to the bow resting on the string; the damped tone pulse in fact seems even more brief. (Nonetheless, its pitch is still subjectively not very different from that of the string plucked without the bow's resting on it.) The rapid decay may result from the parallel bow hairs being moved in differing phase and so rubbing on one another. In any case, this is dry friction, a nonlinear process. The degree of damping may be influenced by varying amounts of rosin on the bow hairs.

For our present purposes, we will attempt to analyze these losses as a viscous frictional resistance, such as the one we used in our simplified model of the bridge. We designate the new resistance as Z_B, the resistance of the bow. This resistance may also include losses resulting from transverse waves propagated in both directions in the bow hairs. Since the bow hairs are under much less tension than the strings, the mismatch between characteristic impedances should be great enough that these losses will be much less significant than frictional losses.

The additional bowing pressure

$$\Delta F_{zB} = Z_B v_{zB} \qquad (5.50)$$

results from a given z-component of velocity v_{zB} at the bow and must be subtracted from the static bowing pressure F_{zB0}, since the latter, seen as exerted by the string on the bow, is directed upward; the v_y and v_z processes at the bridge, on the other hand, point in negative directions, as they are generated by a reflection process involving a change of sign.

The force ΔF_{zB} exerted by the string on the bow is equal to the difference between the transverse forces, F_{z1} in the segment of the string from which the wave arrives and F_{z2} in the remainder of the string beyond the bow. The force F_{z1} consists of an arriving component F_{z+} and a reflected component F_{z-}, while F_{z2} corresponds only to the wave that continues to propagate beyond the bow (see the analogous situations discussed in section 5.4 and below in section 6.4):

$$\Delta F_{zB} = F_{z1} - F_{z2} = F_{z+} - F_{z-} - F_{z2}. \qquad (5.51)$$

We may express the forces as the corresponding velocities, as we have already done in section 5.4. Noting that $v_{zB} = v_{z2}$, we obtain

$$v_{zB}Z_B = Z(v_{z+} - v_{z-} - v_{zB}) \qquad (5.52a)$$

or

$$v_{z+} - v_{z-} = [(Z_B + Z)/Z]v_{zB}. \qquad (5.52b)$$

On the other hand,

$$v_{z1} = v_{zB} \qquad (5.53)$$

is also composed of an arriving and a reflected component:

$$v_{z+} + v_{z-} = v_{zB}. \qquad (5.54)$$

Adding the last two equations leads immediately to the relationship between v_{zB} and the velocity output at the bridge v_{z+}:

$$v_{zB} = (2Z/(Z_B + 2Z))v_{z+}, \qquad (5.55)$$

and so to the change in bowing pressure that is to be found:

$$\Delta F_{zB} = (2Z_B Z/(Z_B + 2Z))v_{z+}. \qquad (5.56)$$

We now choose the—as yet undertermined—value of Z_B to represent the observed high damping of the string when plucked perpendicularly to the bow. The impedance Z_B appears in the middle term of the power-balance equation

$$Zv_{z+}^2 = Zv_{z-}^2 + Z_B v_{zB}^2 + Zv_{zB}^2, \tag{5.57}$$

where the three terms represent the three destinations of the arriving power: the string before the bow, by reflection; the bow hair, through the point of bowing; and the string beyond the bow, by transmission past the point of bowing. We next write down the ratio of bow losses and incident power:

$$\frac{Z_B v_{zB}^2}{Zv_{z+}^2} = \frac{4Z_B Z}{(Z_B + 2Z)^2} = \frac{2}{Z_B/2Z + 2 + 2Z/Z_B}; \tag{5.58}$$

we see that when

$$Z_B = 2Z \tag{5.59}$$

this ratio reaches its highest value, $1/2$, and that $1/4$ of the arriving power is reflected and $1/4$ is transmitted. In this case,

$$F_{zB} = Zv_{z+} = F_{z+}, \tag{5.60}$$

a result easy to keep in mind.

On the other hand, we note that the static bowing pressure F_{zB0} can be derived from the y-component of velocity at the bridge. In figure 5.7, the maximum difference in sliding friction appears twice as a maximum value of the differences of $u(\vartheta)$, namely, as

$$\Delta R = 2Z(u_3 - u_2)_{\max} \tag{5.61a}$$

and

$$\Delta R = 2Z(u_5 - u_4)_{\max}. \tag{5.61b}$$

Now, $(u_3 - u_2)_{\max}$ is only slightly larger than $u_3 - u_2$, which occurs slightly earlier as the corner at $\vartheta = \vartheta_1$ passes the bow. If the characteristic of sliding friction is linearized, this value results directly from the difference between $(\Delta v - u_1)$ and u_2. This difference represents the velocity of the notch of the bridge in the y-direction:

$$\Delta R \approx 2Z(\Delta v - v_{y\max})(1/(1 - \beta) - 1) = \Delta R(1 - v_{y\max}/\Delta v)/(1 - \beta). \tag{5.62}$$

From this, however, it follows that the bow-pressure parameter β is defined by

$$\beta = v_{y\,max}/\Delta v, \quad \text{or} \quad \Delta R = 2Zv_{y\,max}, \tag{5.63}$$

another very simple relationship. Since $v_{y\,max}$ is, according to (5.44), twice as great as $v_{z\,max}$, it follows in connection with (5.60) that

$$\Delta R = 4|\Delta F_{zB\,max}|. \tag{5.64}$$

We must also take into consideration the relationship expressed in (4.39), between the maximum difference of sliding friction ΔR in the y-direction and the static bowing pressure F_{zB0} in the z-direction. For the expected maximum relative correction to the bowing pressure, we obtain

$$|\Delta F_{zB\,max}/F_{zB0}| = (\mu_{h\,max} - \mu_{g\,min})/4. \tag{5.65}$$

In this equation, the value of $\mu_{h\,max}$ is always approximately 1, while $\mu_{g\,min}$ depends on the relevant part of the characteristic line of sliding friction and therefore on Δv. In ranges in which the characteristic of sliding friction can be approximated as a straight line, $\mu_{g\,min}$ cannot fall below 0.2. Consequently, the maximum relative correction given by (5.65) is 20%. This correction would deserve to be taken into consideration if it could be supported by experiment. However, it can be seen that the corresponding decrease in β leads to a relatively lesser slope of u_3 relative to u_2 and hence delays capture.

The opposite situation occurs in the transition from u_4 to u_5 as expressed in (5.10). The change in sign of the corresponding impulse of v_z as the wave is reflected at the nut adds the corresponding force ΔF_{zB} to the static bowing pressure; β is increased and u_5 becomes steeper relative to u_4; release is advanced somewhat. However, the effect is not as great, since the power of the wave arriving at the bow from the nut is only 1/4 that of that which is generated at the bridge.

We will limit ourselves at this point to these general theoretical observations.

The relationship between increasing bowing pressure and steepening of the ramp of velocity, described in the last few sections, should hold with no significant change if waves in the x, z-plane are also taken into consideration.

5.7 The flattening effect

The slight dependence of the fundamental frequency on the bowing pressure ceases altogether when the bow-pressure parameter reaches a high value; that is, when the relationship between the velocities of the ramps $u_{3,5}$ occurring at capture and release and the velocities of the arriving ramps $u_{2,4}$ leads to curves with reversed sections. These sections do not occur in reality, since u_2 increases steadily from 0 to Δv while u_4 falls from Δv to 0. We recall that the line at the ordinate 1 in figure 5.6 corresponds to the regime of sticking friction. (For this reason, this line is shaded on one side.) In this regime, $u_{3,5} = \Delta v$ can be correlated to any $u_{2,4}$. Consequently, it becomes understandable that u_2, increasing, has only to make a step upward as it becomes vertical in order to attain capture; and correspondingly, u_4, decreasing, will attain release with a downward step, without taking the roundabout route along the curve u_5 (u_4). Both steps are shown in figure 5.6 as dashed lines.

The possibility of such steps was pointed out long ago by Friedländer (1953). He discussed the relationships in figure 5.6 with perhaps even greater physical insight, as points at which the characteristic of sliding friction $\delta R(u_3)$ crosses the straight line $\delta R = 2Z(u_3 - u_2)$ [see (5.1b)]. It is appropriate in such graphic solutions to choose the oridinates so that the more complicated parameter, here the characteristic of sliding friction, remains constant. For this reason, Friedländer divided the characteristic of friction by the bowing pressure. In this way, he obtained an intersecting straight line whose slope is inversely proportional to the bowing pressure. As long as the bowing pressure is low, the slope of this line is great and there is only one crossing point. If the bowing pressure becomes too great, there are multiple solutions; the upper limit in figure 5.6 becomes the vertical part of the characteristic of friction. The step to this part at capture (see figure 5.12) no longer requires that the maximum value of sticking friction be attained; release, however, begins at the maximum value of sliding friction just as it did before. The step following release, to the stable segment of the characteristic of friction, is by far the greater.

More recently, Schelleng (1973) also has referred to these "instabilities."

McIntyre, Schumacher, and Woodhouse (1977), however, were first to recognize that these steps must lead to a decrease in the fundamental frequency, which can in fact be observed at high values of bowing pressure.

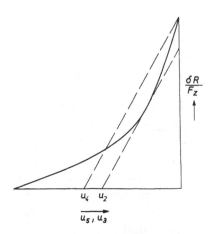

Figure 5.12
A graphic solution for the relationship of the velocities u on either side of the bow with high bowing pressure (after Friedländer).

Since the pitch one semitone lower than d, for example, is called d flat, this process is called the *flattening effect*. In order to explain this effect in the simplest possible way, these authors make use of the method of the round trip. In figure 5.13, the functions of u_4 and u_2 from figure 5.10, with the higher value $\beta = 0.136$ of the bow-pressure parameter, are shown as dashed lines. The overall function in figure 5.13 corresponds to the combination of u_4 and u_2 at the point of bowing. If we now assume that the bow-pressure parameter suddenly increases to $\beta = 0.3$ after release, u_3 is obtained as shown in the corresponding curve of figure 5.6. The steepness, even at the beginning of this curve, is greater than in figure 5.10; the slope of the curve increases until it ends with a small jump upward to capture, as its vertical tangent is reached. These changes are not very striking when compared with u_3 in figure 5.10. Above all, however, it should be noted that the accompanying shortening of the ramp ϑ_3 makes no difference in the fundamental pitch.

On the other hand, if we use the curved characteristic line in figure 5.6, which corresponds to $\beta = 0.3$, then u_5 begins with a large step. The end of this step is distinctly marked by a corner. Above all, however, there results a much larger delay in release than in figure 5.10, and we know from our discussion in section 5.5 that this leads to a lengthening of the period, hence a flattening.

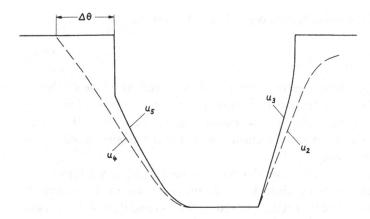

Figure 5.13
Velocity over time at the point of bowing with steps at capture and release (after McIntyre, Schumacher, and Woodhouse).

The final value of $\Delta\Theta$ is determined only after iteration; the cited authors considered this. However, the iteration must repeatedly make use of the ambiguous characteristic in figure 5.6; doubtless, then, the basic features of the final behavior of the ramp u_5 are similar to those shown in figure 5.13.

The cited authors consequently recommend that an attempt be made to realize this effect in a high position on the g string; McIntyre, trained as a concert violinist, is most likely responsible for this suggestion, which may have been based on the especially great losses near the primary resonance of the violin body and on the smaller values of τ corresponding to the frequent reflections in the shortened string. Bowing over the finger-board is also recommended, as a low bowing pressure is sufficient there to bring about Helmholtz motion.

To be sure, it is desirable to avoid the flattening effect—that is, to remain below the limit of bowing pressure at which it can occur—in the interest of producing a pleasing musical tone (McIntyre, Schumacher, and Woodhoose 1977).

5.8 Upper limit of bowing pressure

We are also interested in obtaining the simplest possible theoretical representation of the bowing process. We therefore set as an upper limit the bowing pressure above which the jumps of the flattening effect can occur. This bowing pressure is expressed in figure 5.12 by

$$\frac{2Z}{F_z} = \left(\frac{\delta R}{F_z}\right)'_{\text{max}}.$$ (5.66a)

This relation shows clearly that the result is influenced by the maximum slope of the normalized characteristic of friction, the bowing pressure, and only one other variable, the characteristic impedance of the string. Δv is not relevant, and consequently neither is the bowing speed v_{B} nor the point of bowing x_{B}.

The list of relevant variables remains unchanged if we regard the audible limit as being given by a certain difference in the maximum inclination of the characteristic of sliding friction and that of the straight lines which intersect it:

$$\frac{2Z/F_z}{(\delta R'/F_z)_{\text{max}}} = \gamma < 1.$$ (5.66b)

Bowing speed and point of bowing, however, play a part in other definitions of the upper limit of bowing pressure, which, as Schumacher (1979) correctly emphasizes, limits the range of validity of Raman's representation and even of Helmholtz's.

In both of these representations it is assumed that the kinematic alternation between sticking and sliding, tied to a step Δv, is not affected by the maximum sliding friction ΔR. But the alternation would be affected if

$$\Delta R > 2Z\Delta v$$ (5.67a)

or

$$\beta > 1.$$ (5.67b)

The bowing pressure, the maximum coefficient of sticking friction, and the characteristic impedance of the string are also relevant here; additionally, $\Delta v = v_{\text{B}} l/x_{\text{B}}$.

When the author examined the possible limit given by (5.67a) at an earlier date, this limit proved to be so far above the value given in Völker's first observations (Völker 1961) that he decided not to publish this relation.

Schelleng (1973)[3] was first to publish this limit; he regarded it as the

---

[3] Note Schelleng's footnote 10, which includes a factor of 2 that does not appear in his text.

limit at which the step Δv is reflected from the bow. This concept is identical in meaning to the "not releasing" of the string discussed in the present work.

Schelleng regarded it as most significant that $F_{z\,max}$ is proportional to Δv, and consequently falls as $(x_B/l)^{-1}$ for a given value of bowing speed. Schelleng contrasted this tendency to the decrease of $F_{z\,min}$ as $(x_B/l)^{-2}$ which results from Raman's model (see section 4.6); Raman's model cannot be applied to all impedances, so we might better draw on Lazarus's measurements (see figure 4.5). Still, Schelleng obtained a range between $F_{z\,max}$ and $F_{z\,min}$ in which it is possible to produce an acceptable tone. This range shrinks as the bow is brought closer to the bridge, a result in agreement with Schelleng's own experience as a performer. However, Schelleng himself admitted that these limits of musical acceptability require further experimental study.

In a more recent investigation, McIntyre, Schumacher, and Woodhouse (1981) examine a further upper limit of bowing pressure connected with the finite width of the bow and which manifests itself as an increase in noise.

References

Cremer, L., 1974. *Acustica* **30**, 119.

Cremer, L., and Lazarus., H., 1968. ICA Tokyo, 1968, N-2-3.

Cremer, L., and H. A. Müller, 1982. *Principles and Applications of Room Acoustics.* Tr. by T. J. Schultz. London: Applied Science.

Friedländer, F. G., 1953. *Proc. Cambridge Phil. Soc.* **49**, 516.

Gough, C. E., 1981. *Acustica* **49**, 124.

Lazarus, H., 1970. *Technischer Bericht des Heinrich-Hertz Institutes*, Berlin-Charlottenburg, no. 117.

Lazarus, H., 1972. Dissertation, Technical University of Berlin.

McIntyre, M. E., R. T. Schumacher, and J. Woodhouse, 1977. *Catgut Acoust. Soc. Newsletter* no. 28, 27.

McIntyre, M. E., R. T. Schumacher, and J. Woodhouse, 1981. *Acustica* **49**, 13 *and* **50**, 294.

Müller, H. A., 1962. *Untersuchungen zur Physik der Geige* [*Investigations of the Physics of the Violin*] unpublished ms. of work carried out at the violin-builders' school in Mittenwald.

Reinicke, W., 1965. Thesis work, Institute for Technical Acoustics. Technical University of Berlin. Reported on in Cremer and Lazarus (1968).

Schelleng, J., 1973. *J. Acoust. Soc. Amer.* **53**, 26.

Schumacher, R. T., 1979. *Acustica* **43**, 109.

Trendelenburg, W., 1925. *Die Natürlichen Grundlagen der Kunst des Streichinstrumenten-spiels.* [*The Natural Basis of the Art of String-Instrument Playing*]. Berlin.

Völker, E. J., 1961. Thesis, Institute for Technical Acoustics, Technical University of Berlin.

Weinreich, G., 1982. *Catgut Acoust. Soc. Newsletter* no. 37.

6 Accounting for the Torsion of the String

6.1 Generation of torsional waves by bow-friction forces

Before we incorporate the secondary waves into our solutions, it is necessary for us to deal with another phenomenon that we have not as yet examined. Helmholtz assumed that the velocity of the string in the direction of bowing, at the point of bowing, is the same as the bow velocity:

$$v_+(x_B) = v_B. \tag{6.1}$$

In recordings of displacement, it is difficult to observe departures from this assumption. But if, as in the oscillograms shown in figure 5.4, the string's velocity at the point of bowing is recorded instead, it can be seen, surprisingly, that $v(x_B)$ is by no means constant, especially in the upper oscillogram, made under conditions of high bowing pressure. Sticking does not appear to be interrupted by short intervals of sliding; if this were the case, the discrepancies would consist only of dips in the curve below a constant maximum value.

The true explanation of this observation is as follows (Cremer and Lazarus 1968, Schelleng 1973): What we record electrodynamically is the velocity in the middle of the string: that is, at its center of gravity. We will designate this velocity as v_S or, as before, simply as v. This is the variable that we have already used in the differential equations of string motion, and also in the boundary conditions and in the laws of reflection that apply to them. However, this variable must be distinguished from velocity at the surface of the string. Here the sliding friction force is determined by the relative velocity. Furthermore, the absence of relative velocity amounts to sticking friction. We will denote the velocity at the surface by v_R. Velocity at the center and at the outside of the string differ by an amount which can be expressed in terms of an angular velocity w:

$$v_R = v_S + wa, \tag{6.2}$$

where a is the radius of the string.

The velocity v_S oscillates, as does v_R with its alternation between sticking and sliding. Consequently, w varies over time. This variation, however, is inseparably bound up with the excitation of the torsional waves.

As early as 1896, Cornu proved experimentally that a bowed string oscillates in torsion as well as transversely. He fixed a mirror 1 mm × 2 mm

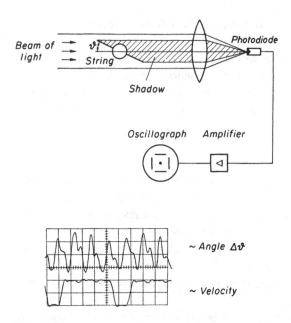

Figure 6.1
Top, arrangement for recording torsional waves. Bottom, an oscillogram of angle
corresponding to torsion, with another showing the corresponding transverse velocity
(after Tippe).

to a cello string, with the plane of the mirror parallel to the string and
perpendicular to the direction of bowing. A beam of light reflected from
the mirror oscillated in the direction of the string in response to transverse
waves, and perpendicularly to the string in response to torsional waves.
Together these produced a Lissajous figure which could be photographed,
but was hard to interpret.

Today, there exist much simpler electronic means of recording torsional
waves separately from other types of waves. Figure 6.1 shows the method
used by Tippe (1967). To the string of a monochord he attached a light-
weight small stiff blade or disk, tilted at an angle of less than 45° to the
string's transverse motion. Torsional waves caused this to modulate a
beam of light falling on a photocell. Tippe's observations showed particularly marked torsional oscillations when the length of the string led
to a natural resonance in torsion. It is clear that the variations of transverse
velocity in the regime of sticking friction exhibit the same periodicity as
the torsional waves. To what degree such resonant reinforcement occurs

on an actual instrument—or whether the torsional waves travel past the bridge and the nut nearly without reflection and are absorbed beyond them—has not yet been determined.

Apart from their losses, which may be very different, torsional waves obey the same laws as transverse waves. The appropriate kinematic variables for torsional waves are

angular velocity w instead of the velocity v;

angle of rotation χ instead of the displacement η.

As a dynamic variable, then,

angular moment M instead of the transverse force F

is appropriate. Transverse force is proportional to the rate of change of displacement; but the angular moment is proportional to the rate of change of the angle of rotation:

$$M = T\frac{\partial \chi}{\partial x}. \tag{6.3}$$

T is called the *torsional stiffness*. It appears in our analogy in place of the tension on the string F_x. For a homogeneous string of radius a, the torsional stiffness is

$$T = G\pi a^4/2; \tag{6.4}$$

whose G is the shear modulus. In the case of transverse waves, the differential increase in F subjects the element of mass $m'dx$ to the acceleration $\partial^2\eta/\partial t^2$; here, for the element with the mass moment of inertia $\Theta'dx$, we obtain as a basic dynamic equation

$$\frac{\partial M}{\partial x}dx = \Theta'dx\frac{\partial^2 \chi}{\partial t^2}; \tag{6.5}$$

in other words, the torsional oscillations obey the wave equation

$$T\frac{\partial^2 \chi}{\partial x^2} = \Theta'\frac{\partial^2 \chi}{\partial t^2}. \tag{6.6}$$

The corresponding propagation speed

$$c_T = \sqrt{\frac{T}{\Theta'}} \tag{6.7}$$

is markedly greater than that for transverse waves,

$$c_S = \sqrt{\frac{F_x}{m'}}. \tag{6.8}$$

This is apparent in figure 6.1 in that the there evidently recorded lowest natural frequency of torsion,

$$f_{T_1} = \frac{c_T}{2l}, \tag{6.9}$$

is three times the fundamental frequency of the transverse wave,

$$f_{S_1} = \frac{c_S}{2l}. \tag{6.10}$$

Evidently, then, in this example

$$\frac{c_T}{c_S} = 3. \tag{6.11}$$

If the string is homogeneous, the expression for the moment of inertia per unit length, which occurs in place of the mass per unit length, simplifies to

$$\Theta' = \varrho\pi a^4/2, \tag{6.12}$$

where ϱ is the density of the material of which the string is made. The speed c_T is then expressed as

$$c_T = \sqrt{\frac{G}{\varrho}}. \tag{6.13}$$

This is the propagation speed of all torsional waves within the material under consideration. For steel, it is 317,000 cm s^{-1}.

Since the ratio c_S/c_T plays an essential role in what is to follow, we will define the parameter

$$\zeta = c_S/c_T, \tag{6.14}$$

For a steel e string with a length of approximately 32 cm, and tuned to $f_1 = 660$ Hz,

$$c_S = 2lf_1 = 42{,}200 \text{ cm s}^{-1}, \tag{6.15}$$

and $\zeta = 0.13$.

For a gut e string, we have to set $\zeta = 0.5$, according to Schelleng's (1973) observations. Since c_S has the same value, this means that the much lower shear modulus leads to a propagation speed of torsional waves lower than in a steel string, despite of the lower density. For a gut cello A string, Schelleng obtained $\zeta = 0.36$; for a metal-wound, and consequently inhomogenous, steel A string, he obtained $\zeta = 0.2$.

6.2 Analysis of the problem with impulse excitation

It may be possible to assume that the torsional waves, excited at release and capture by the same impulse-like frictional forces δR, have damped out before they arrive after reflection from the bridge, despite their greater propagation speed. Even when overlaps occur with reflected waves, however, synthesis using impulse processes remains possible and appropriate (see section 8.6). Consequently, with torsional waves, just as with transverse waves, we can calculate that the impulse moments corresponding to δR,

$$\delta M = \delta R a, \tag{6.16}$$

send two impulse-shaped waves of equal magnitude propagating away from the bow in either direction. Their angular velocity, analogous to

$$\delta R = 2m'c_S \delta v_S, \tag{6.17}$$

satisfies

$$\delta M = \delta R a = 2\Theta' c_T \delta w. \tag{6.18}$$

From these two relationships, and from the kinematic relationship (6.2), it follows that the velocity generated at the bow is

$$\delta v_R = \delta v_S + a\delta w = \delta R \left[\frac{1}{2m'c_S} + \frac{a^2}{2\Theta'c_T} \right]. \tag{6.19}$$

The terms in brackets are reciprocal impedances, or *admittances*. Since these occur only in connection with forces of excitation of impulse form exerted at a point, they can be called *point-impulse admittances*. The first of the two characterizes the transverse waves excited by δR; the second, the torsional waves. The resulting effective point-impulse admittance at the bow is their sum; so we see that torsion increases the compliance of the string. The force δR necessary to generate a transverse wave of velocity δv_S remains the same. At the bow, however, there appears the greater impulse velocity

$$\delta v_R = \delta v_S \left[1 + \frac{a^2 m'}{\Theta'} \frac{c_S}{c_T} \right]. \tag{6.20a}$$

Since δR is unchanged, the additional impulse velocity requires that additional power be expended in order to generate the torsional waves.

For a homogeneous string, the additional term corresponds to 2ζ; consequently, (6.20a) can also be written

$$\delta v_R = \delta v_S [1 + 2\zeta]. \tag{6.20b}$$

According to Schelleng's (1973) measurements $\Big($ see his table IV, in which $\zeta = \dfrac{V}{V'}, \dfrac{a^2 m'}{\Theta'} \dfrac{c_S}{c_T} = \dfrac{Z}{Z'} \Big)$, the second term in (6.20a) is so close to 2ζ, even in the case of metal-wound strings, that (6.20b) is valid for them as well. In order, however, to account for all possible inhomogeneities, we would do better to expand the definition of ζ, given earlier in (6.14), to

$$\zeta = \left(\frac{a^2 m'}{2\Theta'} \right) \frac{c_S}{c_T}, \tag{6.21}$$

which agrees with the earlier definition when applied to homogeneous strings. It is basic to this relationship that ζ does not decrease along with a; a^2 is also included in Θ'. It can, therefore, *not* be assumed that the effect of torsion can be neglected if the string is sufficiently thin.

We now wish to examine how the method of the round trip in section 5.3 is changed when we consider the differences between v_S and v_R; we also wish to examine which of the two velocities (and which of the corresponding quantities u_S and u_R) we should use in reformulating equations (5.7) through (5.13).

We must first establish that the reflection processes at the ends of the string, defined in (5.7) and (5.8), are related to u_S. We note that (5.12) is equivalent to the dynamic relationship (6.17), if we substitute

$$\delta v_S = u_{S3,5} - u_{S2,4}$$

into (5.12) in an appropriate way:

$$\beta \delta R = \Delta R \frac{u_{S3,5} - u_{S2,4}}{\Delta v}. \tag{6.22}$$

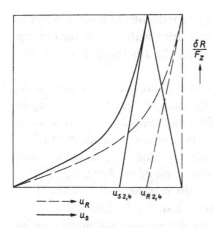

Figure 6.2
Dependence of normalized friction force on the velocity. Dashed lines, at the surface of
the string; solid lines, in the middle of the string.

On the other hand, the function ϱ in (5.11) is related to the values of u_R
at the bow:

$$\delta R = \Delta R \varrho \left(\frac{u_{R\,3,\,5}}{\Delta v} \right)$$

$$= \Delta R \varrho \left(\frac{u_{S\,3,\,5}/(1 + 2\zeta)}{\Delta v} \right). \tag{6.23}$$

Solution of (6.22) and (6.23) is best accomplished graphically, in the
manner shown in figure 5.12. We give two such solutions in figure 6.2.
The dashed lines represent the functions incorporating u_R. The characteristic of friction, shown here as a dashed line, remains the same as
before; but with u_R instead of u_S, the slope of the lines corresponding
to (6.22),

$$\delta R = \frac{\Delta R}{\beta(1 + 2\zeta)}(u_{R\,3,\,5} - u_{R\,2,\,4}) \tag{6.24}$$

is not as great as before.

The line used in figure 6.2 is the limit (for $\delta R = \Delta R$) above which the
flattening effect can occur. Since the slope is determined by the product
$\beta(1 + 2\zeta)$, this limit is reached when the bow-pressure parameter is
smaller, and so when bowing pressure is lower. In figure 6.2, the value

of ζ is 0.13, representing a steel e string; the reduction in bowing pressure is by a factor of 0.79. In the case of a gut e string with $\zeta = 0.5$, the bowing pressure necessary to initiate the flattening effect is reduced to one-half of its previous value.

There is another possibility, illustrated in figure 6.2 by the solid lines; we avoid u_R entirely. We instead use the straight lines corresponding to (6.22), but we use a friction characteristic that depends on u_S. It follows, therefore, that we must diminish all of the abscissas dependent on u_R by the factor $1/(1 + 2\zeta)$. This is also true for the value $u_{S2,4}$ on the line that is once again chosen to represent the same limit of the flattening effect.

As quantitatively significant as this shift may be, it brings about no basic change in the relationship between the sliding friction and u_S.

The most evident difference between the dashed and solid characteristics of friction is that the regime of sticking friction begins earlier, at $u_S = \Delta v/(1 + 2\zeta)$. The vertical straight line, which arrives at the regime of sticking friction sooner, is transformed into a straight line with a rising slope $\Delta v - u_S$:

$$\delta R = \Delta R \frac{1 + 2\zeta}{2\zeta} \left(1 - \frac{u_S}{\Delta v}\right). \tag{6.25}$$

It follows that $\delta R(u_S)$ becomes a characteristic line whose slope is always finite. As a function of u_S or v_S, δR now becomes an "impressed" force, given by the current value of v_S.

As we will see again in section 8.9, this mathematical treatment of the bowed string in the regimes of sticking and sliding friction, unified over an entire period, is of great importance. It is precisely because of the processes described here that the average velocity of a point on the string as recorded electrodynamically is not the same as the bowing speed.

6.3 Longitudinal compliance of the bow hairs

Even the assumption we have been making, that velocity is known at the surface of the string in the regime of sticking friction, is by no means self-evident. The only velocity that we can count on as known is that generated by the performer's right hand at the "frog" of the bow. For all practical purposes, this same velocity can also be ascribed to the point of the bow. It is by no means certain, however, that the bow hairs produce this same velocity at the surface of the string.

Just as impulse forces between the bow and the string can produce impulses of torsion in the string, they can also generate longitudinal waves in the bow hairs.

Longitudinal waves in one-dimensional structures such as beams, strings, and in this case hairs, are subject to the same relationships as the transverse waves in the string which we described in chapter 2. We will use x as the coordinate of position along the bow hairs, though the bow travels in the y-direction in our usual set of coordinates relative to the violin. In our mathematical formulas, the displacement ξ in the x-direction will appear in place of the transverse motion of the string. The element of mass $m'dx$ undergoes an acceleration $\partial^2\xi/\partial t^2$. This acceleration is generated by the local change in a longitudinal force proportional to the specific extension $\partial\xi/\partial x$:

$$F_x = ES\frac{\partial\xi}{\partial x}, \tag{6.26}$$

where E is the modulus of elasticity and S is the cross-sectional area.

The dynamic relationship

$$\frac{\partial F_x}{\partial x}dx = ES\frac{\partial^2\xi}{\partial x^2}dx = m'\frac{\partial^2\xi}{\partial t^2}dx \tag{6.27}$$

leads here, too, to the wave equation

$$ES\frac{\partial^2\xi}{\partial x^2} = m'\frac{\partial^2\xi}{\partial t^2}. \tag{6.28}$$

The speed of propagation of these longitudinal waves is, consequently, given by

$$c_L = \sqrt{\frac{ES}{m'}}, \tag{6.29a}$$

and, in the case of a homogeneous material, by

$$c_L = \sqrt{\frac{E}{\varrho}} \tag{6.29b}$$

where ϱ, as above, is the density of the material.

For the same material, c_L is always greater than c_T, the speed of propagation of torsional waves. In what follows, we use the subscript S for variables relative to strings and the subscript H for those relative to bow hairs, in order to distinguish the two sets of variables.

From the analogy expressed in (6.26) and (6.28), we also obtain the characteristic impedance applicable to the propagation of waves:

$$Z_H = \sqrt{S_H E_H m'_H} = m'_H c_{LH}. \tag{6.30}$$

The bow as well—with the exception of the parts immediately adjacent to the frog and the point—is excited by the impulse force δR, so that waves propagate in both directions away from the string. The difference between the absolute speed v_H of the hairs at the point of bowing and the forced speed v_B at the frog is given by

$$v_H - v_B = \delta v_H = \frac{-\delta R}{2m'_H c_{LH}}, \tag{6.31}$$

where $v_B > v_H$ at all times; hence the negative sign.

It is appropriate to call $2m'_H c_{LH}$ the *point impulse impedance of the bow hairs*.

On the friction characteristic, relative velocity determines the frictional force; but the relative velocity is now no longer the difference between the velocity of the bow and the velocity at the surface of the string ($v_S + aw$), but rather the difference between the velocity of the bow hairs and the velocity at the surface of the string:

$$v_{rel} = v_H - v_S - aw = v_B + \delta v_H - v_S - aw. \tag{6.32}$$

Like δv_S in (6.17), the quantity $-\delta v_H$ is proportional to δR. Consequently this correction can be expressed in terms of v_S, as aw was earlier; in other words, in terms of the quantity that governs all other laws relating to the round trip of the rounded corner. Analogously to (6.20a), (6.32) can be rewritten as

$$v_{rel} = v_B - v_S \left(1 + \frac{a^2 m'_S c_S}{\Theta'_S c_{TS}} + \frac{m'_S c_S}{m'_H c_{LH}} \right). \tag{6.33}$$

The way of illustrating the characteristic line of friction that was introduced in figure 6.2, with the abscissa ($v_B - v_S$) or u_S, reflects no change in underlying theory. Merely, as Schumacher (1975) has already pointed out, in addition to the 2ζ which accounts for torsion, there occurs a further corrective term equal to the ratio between the characteristic impedances for transverse waves in the string and longitudinal waves in the bow hairs.

This latter impedance must be determined experimentally. Figure 6.3

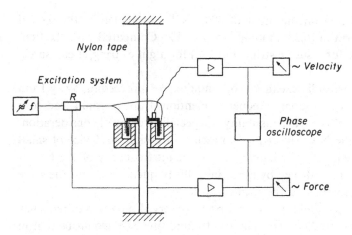

Figure 6.3
Arrangement for measuring the longitudinal impedance of the bow (after Ottmer).

illustrates the measuring apparatus used by Ottmer (1967). He suspended a nylon ribbon, such as is used for mechanical bowing, rigidly at both ends and mounted a voice coil (of the type used in loudspeakers) in its middle. Alternating current through the voice coil induced sinusoidal exciting force, whose frequency could be varied between 125 and 5,000 Hz. The amplitude and phase of the resulting velocity was measured using a piezoelectric transducer. Taking the internal impedance of the oscillator into account, it was possible to derive curves of the admittance of the ribbon when it was excited in the middle. Unfortunately, the measurements exhibited a large random variation.

Schumacher (1975) too, used variable-frequency electrodynamic excitation. In his experiments, however, the coil was coupled to the bow through a needle attached to the bow hairs, and a photoelectric transducer was used to observe the displacement of the needle. Schumacher's apparatus corresponded better to normal circumstances, since it used a complete, standard bow; with it he was able to change the point of excitation and also to excite transverse oscillations in the bow hairs.

We need only know that the two author's results agree on a lowest longitudinal resonance of the bow between 1,500 and 2,000 Hz for a length of approximately 62 cm. From this, we can conclude that the speed of longitudinal propagation is between 190,000 and 250,000 cm s^{-1}. Schumacher concludes from this that the characteristic impedance Z_{LH} is approximately 30,000 g s^{-1}. He gives the characteristic impedance of a g string as being on the order of 300 g s^{-1}; consequently, the second corrective term in (6.33) amounts to only 1%. Compared with the first, which is 0.26 for a steel e string and 1 (!) for a gut e string, it can surely be neglected.

However, periodic excitation by impulses δR proceeding away from the string at the resonant frequency mentioned above could lead to a much greater effect. This possibility is especially worthy of consideration, because, to the bow, the string represents an excitation device of small internal impedance. The longitudinal resonant frequency of the bow is consequently not affected by the string; this frequency remains the same wherever along the bow the string may lie.

Such an effect could, however, first be observed only in the frequency range of 1,500 to 2,000 Hz. The most interesting case would be that in which the fundamental frequency of the string excites the bow, and would be observed only in high positions on the e string. The present author does not know of any difficulties in bowing in this range which might result from resonance with the longitudinal natural oscillatory mode of the bow hairs. Such difficulties are possible in theory but evidently are prevented by the difference in characteristic impedances or by damping.

We may therefore feel justified in omitting the characteristics of the bow from the theory of periodic self-excited oscillations of the bowed string; but this conclusion in no way implies that players find all bows alike. The characteristics of bows are expressed to a much greater degree in low-frequency transverse resonances. Not only the tension on the bow hairs, but also the wooden parts of the bow, affect these resonances. These oscillations can lead to a periodic alteration in bowing pressure; this can even go beyond the limits defined above. Every player, and particularly every beginner, knows this "trembling" of the bow. Additionally, the bow plays a great role in all changes from one note to another and in the transient processes characteristic of all bowing styles. The cited work by Schumacher includes some preliminary observations on the question of differences among bows.

6.4 Transformation of transverse waves into torsional waves by reflection from the bow in the regime of sticking friction

The generation of torsional waves at the moments of capture and release is significant only in that the principal statements in the previous chapter need not be modified, at least at the low bowing pressures assumed in that chapter. On the other hand, torsional waves do play a decisive role in connection with the secondary waves, as these are reflected from the bow.

As they are reflected, the transverse waves are partially transformed into torsional waves. This phenomenon was incorporated in Schelleng's (1973) electrical model and, to an even greater degree, in the precise solutions of the problem of the bowed string (Schumacher 1979, McIntyre and Woodhouse 1979) that will be discussed in sections 8.3, 8.4, and 8.9. Nonetheless, a separate discussion of this process is necessary in order to understand the fate of the secondary waves (Cremer 1979).

The secondary waves are to be superposed on the Helmholtz motion and the primary corrective wave; superpositions of the latter two already take into account the motion of the bow. Consequently, the only additional boundary condition for the secondary waves reflected from the bow in the regime of sticking friction is

$$v(0) - aw(0) = 0. \tag{6.34}$$

(The change of sign with respect to (6.2) is appropriate since the arriving transverse wave, not the bowing force, determines the direction of excitation.)

In addition to this kinematic condition, there occurs the dynamic condition

$$a[F_1(0) - F_2(0)] = M(0), \tag{6.35}$$

where F_1 represents the force of the transverse wave on the side of the bow from which it arrives, and F_2, the force of the remaining wave on the opposite side. The latter stands in the simple relationship

$$F_2 = m'c_S v_2 \tag{6.36}$$

to the corresponding velocity, given by v_2. On the side from which the wave arrives, F_1 and

$$v_1 = v_2 = v(0) \tag{6.37}$$

are each composed of an arriving and a departing component. Here, also,

$$F_1 = F_{1+} + F_{1-} = m'c_S(v_{1+} - v_{1-}), \tag{6.38}$$

$$v_1 = v_{1+} + v_{1-}. \tag{6.39}$$

Since the torsional moment leads to waves of equal strength in both directions, the relationship between $M(0)$ and $w(0)$ is given by

$$M(0) = 2\Theta' c_{\mathrm{T}} w(0). \tag{6.40}$$

We can therefore replace the dynamic variables with kinematic ones. Consequently, we can write, instead of (6.38), (6.34), and (6.35),

$$v_{1+} + v_{1-} = v_{2+}, \tag{6.41}$$

$$v_{2+} - aw = 0, \tag{6.42}$$

$$v_{1+} - v_{1-} - v_{2+} = \frac{2\Theta' c_{\mathrm{T}}}{a^2 m' c_{\mathrm{S}}}(aw) = \frac{1}{\zeta}(aw). \tag{6.43}$$

We may divide all the variables that are to be found by the term v_{1+}, a given, and so eliminate v_{1+}. In other words, we introduce a *reflection factor*

$$r = \frac{v_{1-}}{v_{1+}}, \tag{6.44}$$

and a *transmission factor*

$$t = \frac{v_{2+}}{v_{1+}}, \tag{6.45}$$

and, finally, a ratio of velocities that we will call the *transformation factor*

$$t_{\mathrm{T}} = \frac{aw}{v_{1+}}. \tag{6.46}$$

The conditions (6.41) through (6.43) then become

$$r - t = -1, \tag{6.47}$$

$$t - t_{\mathrm{T}} = 0, \tag{6.48}$$

$$-r - t - \frac{1}{\zeta} t_{\mathrm{T}} = -1. \tag{6.49}$$

From these it follows that

$$r = \frac{-1}{1 + 2\zeta}, \tag{6.50}$$

$$t = t_{\mathrm{T}} = \frac{2\zeta}{1 + 2\zeta}. \tag{6.51}$$

Even in the case of a steel e string for which ζ has been calculated as 0.13 [see the discussion of (6.14) in section 6.1], $|r|$ amounts to 0.794. The corresponding absorption coefficient α is $1 - |r|^2 = 0.37$. Even this is much higher than any absorption coefficient that can be ascribed to losses at the bridge, nut, or finger based on the decay of plucked strings. This is even more markedly so when we substitute the value for a gut e string, $\zeta = 0.5$. In this case $|r| = 0.5$ and $\alpha = 0.75$!

However, the power in the transverse waves and in the torsional waves does not simply disappear; both types of waves return after reflection from the bridge and nut, and the torsional waves are partially retransformed into transverse waves as they are reflected from the bow. The "law of reciprocity" applies to this transformation: the percentage of power in an arriving transverse wave which is transformed into a torsional wave is the same as that in an arriving torsional wave which is transformed into transverse waves. We can call this percentage the *transformation coefficient*

$$\tau_T = 1 - r^2 - t^2 = \frac{4\zeta}{(1 + 2\zeta)^2}. \tag{6.52}$$

Therefore, τ_T is greater by a factor of $1/\zeta$ than the *transmission coefficient*

$$\tau = t^2 = \frac{4\zeta^2}{(1 + 2\zeta)^2}; \tag{6.53}$$

In other words, the largest part of the power that is not reflected is transformed into torsional waves. In the case of the steel e string this is 33% and in the case of the gut e string 50% of the arriving energy. Even if the torsional waves were to be completely reflected from the ends of the string, we only have to be concerned, after the next reflection from the bow, with a transverse wave whose power is reduced by a factor of τ_T^2.

6.5 Measurement of the reflection and transmission factors of the bow in the regime of sticking friction

Due to the importance of transverse and torsional waves, the present author suggested that H. Lazarus and P. Eisenberg investigate them (Eisenberg 1967, Lazarus and Eisenberg 1969).

These researchers placed a string in a magnetic field concentrated at the point of observation. They did not use the string itself as an electrical

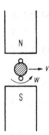

Figure 6.4
Electrodynamic arrangement for measuring transverse velocity and angular velocity (after Lazarus and Eisenberg).

conductor, but rather glued conductive wires to its upper and lower side (see figure 6.4). The sum of the voltages induced in these corresponded to transverse velocity v; the difference, to the angular velocity w.

Lazarus and Eisenberg also separated transverse and torsional waves by using short pulsed tones as the excitation. The envelope of such pulses is a Gaussian curve; such functions are called "Gaussian tones." The authors were consequently able to distinguish the impulses generated at the bow according to their separation in time, based on their differing speeds of propagation. The strings used in the experiment were 30 m long, and were composed either of braided steel wire 3 mm in diameter or of braided plastic wire 3 mm or 1 mm in diameter. Though the apparatus was much different from that of a normal string instrument, this made no fundamental difference in the results. At the top in figure 6.5 are shown the positions of the velocity transducer, the bow, and the points at which the motion of the string is sensed.

With this apparatus, Lazarus and Eisenberg were able to record both types of wave simultaneously, by sensing the outputs of the two conductive wires. An example is given at the bottom of figure 6.5. The upper oscillogram corresponds to sensing of the string motion at a point between the support (that takes the place of the bow) and the input transducer. Here, the transverse wave from the transducer arrives first; then, the torsional wave proceeding from the bow; after this, first the torsional wave reflected from the input transducer and then the transverse wave reflected from the bow. The lower oscillogram, corresponding to sensing behind the bow, shows similar results. Here, too, the torsional wave generated at the bow arrives first; then the torsional wave reflected

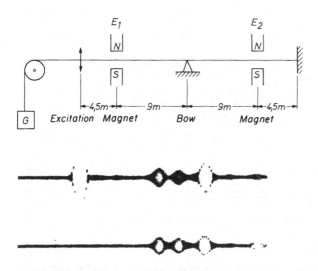

Figure 6.5
Top, arrangement for investigating reflection and transmission of transverse waves and
their transformation into torsional waves; bottom, examples of oscillograms (after
Lazarus and Eisenberg).

from the end of the string; and only after these the transverse wave
transmitted past the bow. From the separation of the impulses in the
oscillogram and the distances indicated in figure 6.5, it can be determined
that the ratio between the two speeds of propagation was only 1 : 3. In this
case, the string was the 3 mm plastic one, under a tension of 554 N.

Examples of measured reflection and transmission factors are given in
figure 6.6 as a function of the carrier frequency of the impulse; the string
and the tension were those just mentioned. The measured magnitudes of
r and t show only a slight frequency dependence at low frequencies. Also,
in agreement with (6.50) and (6.51),

$$|r| + |t| \approx 1. \tag{6.54}$$

The values of these factors shown in figure 6.6 correspond to the credible
ratio $\zeta = 1/4$. This somewhat smaller value, compared with the ratio of
measured speeds of propagation $c_S/c_T = 1/3$, is easily explained by an
inhomogeneity of the string such that its density is somewhat greater at
the outside than at the inside.

It is notable, however, that $|r|$ becomes smaller and $|t|$ becomes greater

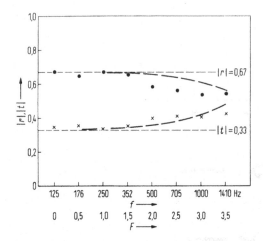

Figure 6.6
Measured reflection and transmission factors as they depend on frequency (after Lazarus and Eisenberg). Dashed line, values calculated as in section 7.2.

as frequency increases. In the next chapter, we will prove that this effect, as expected, has to do with bending stiffness.

For now, we will be satisfied with the statement that, at low frequencies, the values given by (6.50) represent maximum reflection factors, and that the values given by (6.51) represent minimum transmission factors.

6.6 Reflection, transmission, and transformation in the regime of sliding friction

The intervals of sliding friction are generally short in comparison with those of sticking friction; still, it is important for us to examine the processes which are to be expected when a transverses wave arrives at the bow during an interval of sliding friction.

The boundary conditions (6.41) and (6.43) remain completely unchanged. On the other hand, the velocity $(v_{2+} - aw)$ is now relative; it decreases along with the force $(F_1 - F_2)$. This relationship can be linearized, since this relative velocity corresponds only to a small variation along the friction characteristic. Besides, we are here interested only in a glance into fundamental processes, and in rough calculations; and we concern ourselves only with bowing pressures far below the limit at which the flattening effect begins. Consequently, we replace the falling characteristic with a linear relationship:

$$\frac{v_2 - aw}{\Delta v} = -\frac{(F_1 - F_2)}{\Delta R} = -\frac{(F_1 - F_2)}{2\beta Z \Delta v},$$ (6.55)

where β is the bow-pressure parameter defined in (5.4). If we introduce the dimensionless factors given in equations (6.44) through (6.46), then (6.55) leads to

$$(t - t_T) = \frac{1}{2\beta}(-1 + r + t).$$ (6.56)

Along with (6.47) and (6.49), which are valid without any changes, this leads to

$$r = \frac{\beta}{1 - \beta(1 + 2\zeta)},$$ (6.57)

$$t_T = \frac{-2\zeta\beta}{1 - \beta(1 + 2\zeta)},$$ (6.58)

$$t = \frac{1 - 2\zeta\beta}{1 - \beta(1 + 2\zeta)} > 1.$$ (6.59)

As we have already discussed in section 1.3, a falling characteristic always corresponds to an oscillatory instability; therefore, it is not surprising that $t > 1$. In other words, the wave leaving the bow is of greater amplitude than the wave that arrived there. This effect could, however, lead to significant amplification only if it were to repeat again and again. But whenever a secondary wave happens to reach the bow during an interval of sliding friction, then either a chief wave from the direction of the bridge has preceded it or a chief wave from the direction of the nut follows it during the interval of sliding friction $(2x_b/c)$. In both cases, the bow returns to the regime of sticking friction very soon after the arrival of the secondary wave. Only after many weakening reflections does the secondary wave again arrive at the bow during the time window of sliding friction. This phenomenon can lead to subharmonics (McIntyre, Schumacher, and Woodhouse 1981).

The denominators in (6.57) through (6.59) can go to zero, but this is of no practical importance, because the formulas for the corresponding values of β are for other reasons no longer valid in this case.

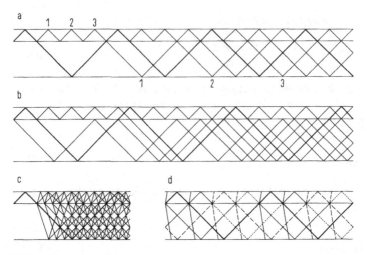

Figure 6.7
Space-time diagram of the corrective waves: a, for $x_B/l = 1/4$; b, for $x_B/l = 1/4.5$; c, as in (a) but incorporating torsional waves; d, as in (c) but assuming that the torsional waves are completely absorbed at the ends.

If we assign $\beta = 0.1$ and $\zeta = 0.13$, as a practical example, then the resulting values are

$$r = 0.11, \quad t_T = -0.03, \quad \text{and} \quad t = 1.17. \tag{6.60}$$

These values differ only slightly from $r = t_T = 0$ and $t = 1$, which correspond to a bow transparent to the arriving transverse waves.

6.7 Fate of the secondary waves

Now we have studied the generation of torsional waves, which is inevitable, and their transformation into transverse waves, which is in its turn possible; and so we are ready to deal to some extent with the question about the fate of the secondary waves that was posed at the beginning of this chapter.

We first turn to the use of graphic timetables such as those of figure 3.15. Once more, we let time proceed from left to right, and represent position by the vertical axis; the bridge is at the top, the nut at the bottom (see figure 6.7a–d).

If only the primary corrective wave were present, propagating together with the Helmholtz velocity step, then we would see only the single thick zigzag line in our space-time diagrams. All other lines represent secondary waves.

We first (in diagrams a and b) consider only the transverse waves generated at capture and release; we regard their reflection from the bow as complete, corresponding to an earlier stage of the theory of the violin string.

Diagram a shows what can be called a "building-up process." First, we see the wave generated at capture, which is reflected back and forth between the bow and the bridge. At the input of the bridge, which is of primary interest to us—this wave generates the impulses 1, 2, and 3. These become progressively weaker, one after the other. The simple fraction 1/4 was chosen for x_B/l in diagram a; consequently, after three reflections from the bridge and two reflections from the bow, the corrective wave arrives at the bow at the moment of release, and could conceivably be strong enough to have an influence on this process. If we had chosen $x_B/l = 1/10$, this wave would have already undergone nine reflections at the bridge and eight at the bow.

The fate of the wave which propagates toward the nut or the finger at the moment of release is similar. (On this side of the bow, the damping during propagation should not be neglected.) These waves, too, fill the interval of sticking friction with intermediate reflection; in this case, the waves arriving at the bow are weaker the sooner they arrive; as Schelleng (1973) pointed out, the more reflections they must have undergone and the greater the distance they must therefore have traveled. Again, the secondary wave, having undergone three reflections from the end of the string at the nut or finger and two reflections from the bow, coincides with the primary wave at the bow; this time, at the moment of release.

If, however, as in diagram b, the ratio $x_B/l = 1/4.5$, then the primary and secondary waves do not coincide until later, after the secondary waves has become much weaker. Also, the spacing of the secondary waves in time is only half as great.

In diagram a, the secondary wave propagating between the bow and the nut (or finger) was shielded by the bow from propagating further toward the bridge. In diagram b, this wave is seen arriving at the bow during the interval of sliding friction, so that it makes its way through to the bridge.

The network of secondary waves becomes considerably more dense if we include in the diagram the generation of torsional waves and their transformation into transverse waves, and vice versa; both of these transformations occur at the bow, which is not at the center of the string (diagram c). In fact, we obtain a lattice with finite spacing only because we have chosen a speed of propagation for the torsional waves precisely three times as great as that for the transverse waves. (This lattice would be the same if off-center support of the strings at the ends led to transformations there; however, it has not been proven that transformations do occur there.) If we had chosen another ratio, particularly one with an irrational value, we could have obtained a lattice of any desired density. In such a case, it would become almost meaningless to trace an individual secondary wave; a statistical treatment would be more appropriate (Cremer 1974, p. 130).

A deterministic analysis of the torsional waves is practical only when $x_B = l/n$ and n is not too great, or when the damping is so heavy that the secondary waves reflected from the ends of the string and the bow can be neglected after one or at most two periods. Heavy damping is, in fact, also a prerequisite in applying the method of the full round trip. This possibility is illustrated in diagram d; the torsional waves are assumed to be so heavily damped that their transformation back into transverse waves can be neglected. This assumption, however, seems questionable in the light of Tippe's observations of resonances on the monochord (Tippe 1967). Unfortunately, no measurements on an actual stringed instrument are available at this time, though it would seem probable that the torsional waves are more heavily damped than on the monochord.

Given the assumption of heavy damping, it is, however, possible to trace the secondary waves which arise when the method of the full round trip is applied; especially, it is possible to incorporate the effect of the secondary waves on the input force at the bridge. This force modifies the curve for $F(0)$, a sawtooth with sloped transitions representing the average of u_1 and u_2, shown in figure 6.8. In the interest of simplicity, the ramps, and their differences, are based on the mathematical approximation in figure 5.7. The asymmetry assumed in figure 5.7 leads to a fundamental difference between the impulse δR_h at capture and the impulse δR_g at release, making it easier to distinguish them.

The value $\zeta = 0.5$, representative of a gut e string, is applied to the generation of torsional waves in figure 6.8. For this reason, the waves generated at release (bottom row in figure 6.8) arrive at the bow with

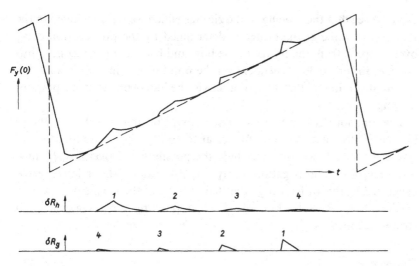

Figure 6.8
Function of force at the bridge with a finite ramp resulting from the primary corrective
wave, and with added impulses corresponding to the secondary corrective waves.

the same strength as those generated at capture (second row from bottom).

The calculation has also been very much simplified by assuming that
the reflection of the secondary waves from the bridge is ideal, that is,
a simple doubling of force; the reflection at the bridge certainly plays a
much smaller part in weakening the secondary waves than does the
reflection at the bow. This simplifying assumption, however, leaves out
changes due to the frequency dependence of losses at the bridge, change
which lead to an increasing rounding off of the impulses. There would,
however, have been no point in carrying out a convolution with the
reflection response at the bridge so as to include these losses if the same
were not done in the case of the reflection at the bow, where the experiments of Lazarus and Eisenberg indicate that there is also a decrease in
the reflection factor with frequency.

This particular attenuation of the peaks of the impulses of the secondary
waves is, besides, an additional basis for the assumption that these waves
could affect capture and release only when bowing pressure is very high.

The addition of the secondary impulses δR_h and δR_g in figure 6.8 is
based on their values as determined according to the method of the round
trip; it represents a significant step in improving our model in that we

can now see that the spacing of the zigzags which we have included in the function of force at the bridge is determined by the propagation delay over the path from the bridge to the bow and back again, or $2x_B/c$. Thus changes in tone color dependent on the point of bowing are included in our model, a model that would otherwise be independent of the point of bowing.

The function $F_y(0)$ in figure 6.8, with ramps of finite width, also shows how these decrease the step of force at the bridge from $2\Delta v$ to, at most, $2\Delta v(1 - t_1/T)$. If we are to exclude the possibility of overlaps, t_1 must not exceed the propagation delay $2x_B/c = Tx_B/l$. The relative error generated by this difference in the step of force at the bridge is, however, small at best, even at the limit of t_1/T corresponding to bowing of an open string near the bridge.

References

Cornu, M. A., 1896. *J. de phys.* **3**, 5.1.

Cremer, L., 1974. *Acustica* **30**, 119.

Cremer, L., 1979. *Acustica* **42**, 133.

Cremer, L., and H. Lazarus., 1968. ICA Tokyo, 1968, N-2-3.

Eisenberg, P., 1967. Unpublished student work, Technical University of Berlin.

Lazarus, H., and P. Eisenberg, 1969. Lecture, Physicists' Convention, Heidelberg.

McIntyre, M. E., R. T. Schumacher, and J. Woodhouse, 1981. *Acustica* **49**, 13 *and* **50**, 294.

McIntyre, M. E., and J. Woodhouse, 1979. *Acustica* **43**, 93.

Ottmer, D., 1967. Thesis, Technical University of Berlin.

Schelleng, J., 1973. *J. Acoust. Soc. Amer.* **53**, 26.

Schumacher, R. T., 1975. *Catgut Acoust, Soc. Newsletter* no. 24 (November).

Tippe, W., 1967. Unpublished student work, Technical University of Berlin.

7 Accounting for the Bending Stiffness of the String

7.1 A general look at the properties of a string with bending stiffness[1]

We have already pointed out repeatedly that the corners used in chapters 2, 3, and 4 as part of the description of $\eta(x, t)$ are compatible with the wave equation of the string

$$F_x \frac{\partial^2 \eta}{\partial x^2} = m' \frac{\partial^2 \eta}{\partial t^2},$$ (7.1)

but that these corners are not physically realizable, due to the bending stiffness of the string. This stiffness may be small, but it is always present to some degree.

Also, we have already shown in section 2.5 that damping effects, increasing with frequency, lead to the rounding of all corners, and that the consequences of damping are in most cases overwhelmingly more significant than those of bending stiffness.

However, the discrepancy between the calculated and the measured reflection factor from the bow in figure 6.6 leads us to take up the problem of bending stiffness again.

Suspicion that this discrepancy is related to bending stiffness arises from its increase with frequency. The shorter the wavelength, the greater the curvature of the string, even if transverse velocity is kept the same. The curvature is proportional to the second spatial derivative of small displacements. Furthermore, the bending moment

$$M = B \frac{\partial^2 \eta}{\partial x^2}$$ (7.2)

increases with curvature in a string of bending stiffness B. For a homogenous string of radius a, the bending stiffness is given by

$$B = E \frac{\pi a^4}{4},$$ (7.3)

where E is the modulus of elasticity.

If, as is the case when a wave is present, this bending moment changes along the length of a string (or beam) which is at first under no tension, the momentary balance of forces on an element of the string $m' dx$ leads to a transverse force which varies from one point to another:

---

[1] This and the following section are based upon Cremer (1979a and 1979b).

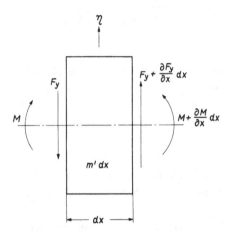

Figure 7.1
Choice of directions for forces and moments of an element of a string with bending
stiffness.

$$F_y dx = -\frac{\partial M}{\partial x} dx. \tag{7.4}$$

The negative sign follows from the relationships shown in figure 7.1.
We have used the same signs for the moment and its increase as in (7.2):
a positive curvature (concave toward the top) and a sign for the force
corresponding to that in figure 2.1.

The increase in transverse force is responsible also for the acceleration
$\partial^2\eta/\partial t^2$ of the element $m'dx$:

$$\frac{\partial F_y}{\partial x} dx = -\frac{\partial^2 M}{\partial x^2} dx = -B\frac{\partial^4\eta}{\partial x^4} dx = m'dx\frac{\partial^2\eta}{\partial t^2}. \tag{7.5}$$

It follows from this that the equation for bending waves is

$$-B\frac{\partial^4\eta}{\partial x^4} = m'\frac{\partial^2\eta}{\partial t^2}. \tag{7.6}$$

This equation differs basically from the wave equations for transverse
waves in a string without bending stiffness, for torsional waves, and for
longitudinal waves, in that it contains the fourth rather than the second
spatial derivative.

In the stretched string, forces given by longitudinal tension F_x and
inclination $\partial\eta/\partial x$ occur, as well as transverse bending forces:

$$F_y = F_x \frac{\partial \eta}{\partial x} - B \frac{\partial^3 \eta}{\partial x^3}, \tag{7.7}$$

and so the wave equation for a string with bending stiffness is

$$F_x \frac{\partial^2 \eta}{\partial x^2} - B \frac{\partial^4 \eta}{\partial x^4} = m' \frac{\partial^2 \eta}{\partial t^2}. \tag{7.8}$$

This equation reduces to (7.1) as frequency is decreased. It is, then, certainly appropriate to assume that this equation has solutions in the form of sinusoidal waves of radian frequency ω and wave number k. We set

$$\eta = \text{Re} \{ \hat{\eta} e^{j(\omega t \pm kx)} \}; \tag{7.9}$$

for the relationship $\omega(k)$, called the *dispersion equation*, we obtain

$$\omega^2 = \frac{F_x}{m'} k^2 + \frac{B}{m'} k^4. \tag{7.10}$$

In optics, the word "dispersion" means the separation of white light into its spectral colors. In acoustics, too, this term is applied to systems in which the phase speed is no longer independent of frequency or, consequently, of the wave number k. The phase speed as defined by (7.10) is independent of frequency if the second term on the right side can be neglected.

The frequency dependence can be of importance even when, as in the case of the string with bending stiffness, the second term does not make the largest contribution, but rather amounts only to a correction term.

In such cases, it is possible to substitute into the equation the value of k^2 which would be obtained if the correction term were not present; here, this is the value corresponding to the simple wave equation:

$$k^2 = \frac{m'}{F_x} \omega^2 = \frac{\omega^2}{c_S^2}. \tag{7.11}$$

The corrected relationship is, then,

$$\omega^2 = c_S^2 k^2 \left(1 + \frac{B\omega^2}{F_x c_S^2} \right), \tag{7.12}$$

and the phase speed is

$$c = c_S \left(1 + \frac{1}{2} \frac{B}{F_x} \frac{\omega^2}{c_S^2} \right). \tag{7.13}$$

We note that the variables ω and c_S, as they appear in the correction term, are always givens in this problem, since a given fundamental frequency should always be obtained with a string of a given length. It follows that whether bending stiffness must be considered depends on the quotient (B/F_x), just as it did in (7.8). B is proportional to a^4; but on the other hand, for a given c_S, F_x is proportional to the mass, and so to a^2. The correction term, then, increases as a^2. The influence of bending stiffness, in contrast to that of torsion, can be negligible if the string is very thin. For this reason, the monochord strings used by Krigar–Menzel and Raps (1891) could show surprisingly sharp corners. Certainly, the reason that wound strings are used for lower fundamental frequencies is to keep the bending stiffness small. The change of direction of the strings in the x, z-plane as they pass over the bridge gives an indication of how sharply they can bend.

In a string rigidly supported at its ends, the increase in phase speed with frequency results in the natural resonances no longer being harmonically related to one another (Rayleigh 1877, vol. 1, sec. 137). This inharmonicity is often considered a basic attribute of strings with bending stiffness. When a string is stretched over an actual instrument, however, the frequency dependence of the bridge input impedance would in any case lead to the natural frequencies not being harmonically related.

Just as basic and often even more evident is another influence of bending stiffness, which can be detected at low frequencies—even in the limiting case of static deformation, corresponding to $\omega = 0$. It is a well-known feature of the equation for waves in bending, (7.5), that one solution, with a sinusoidal time function, consists of "boundary fields" that decrease exponentially with increasing distance from the location of a disturbance (see, e.g., Cremer and Heckl 1973, p. 123):

$$\eta = \operatorname{Re}\{\hat{\eta}_j e^{(j\omega t \pm k'x)}\}. \tag{7.14}$$

(Here the subscript j on η indicates the change from k to jk'). This is true also in the case of the wave equation of a string with bending stiffness. If we solve (7.10) for k^2:

$$k^2 = -\frac{F_x}{2B} \pm \sqrt{\left(\frac{F_x}{2B}\right)^2 + \frac{m'\omega^2}{B}}, \tag{7.15}$$

we see that only the $(+)$ sign before the root corresponds to a real k; that is, to a wave motion in the sense of the expression in (7.9). The $(-)$

sign always corresponds to an imaginary solution $jk = k'$. If we once again consider the second term under the root as a corrective term, we obtain

$$k' = \sqrt{\frac{F_x}{B}} \left(1 + \frac{B}{2F_x} \frac{\omega^2}{c_S^2}\right). \tag{7.16a}$$

The corrective term is the same as in (7.13); consequently, it must be taken into consideration (or, on the other hand, it may be neglected) in the same frequency ranges as in (7.13). If $\omega = 0$, then

$$k' = \sqrt{\frac{F_x}{B}}. \tag{7.16b}$$

It is easiest to see that this equation represents a static-limit condition if we set the dynamic term on the right side in (7.8) equal to zero. The solutions of the resulting equation

$$F_x \frac{d^2\eta}{dx^2} - B\frac{d^4\eta}{dx^4} = 0 \tag{7.17}$$

have two components. One consists of particular solutions for which the second term vanishes; that is, it consists of straight lines

$$\eta = \eta_0 \pm \gamma x, \tag{7.18}$$

such as are characteristic of a string without bending stiffness, with forces applied to it at discrete points. The other component consists of solutions to the total differential equation

$$\Delta\eta = \Delta\eta_0 e^{\pm\sqrt{F_x/B}x}; \tag{7.19}$$

superimposing this component removes the corners at the points where the forces are applied.

Figure 7.2 shows a simple example. Halfway between two "bridges," a string is loaded with the force F_y. A string without bending stiffness would take on a triangular displacement described by the angle γ, which could be expressed with sufficient accuracy by the force ratio

$$\gamma = \frac{F_y}{2F_x}. \tag{7.20}$$

(In the drawing it has been necessary, as is always the case in problems

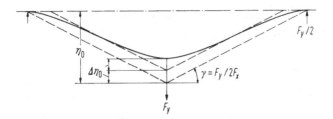

Figure 7.2
Static displacement of a string with bending stiffness when loaded in the middle.

involving strings, to increase the displacement compared to the length of the string.) Now we take bending stiffness into account. We see that the same inclination is present in the segments of the string where bending stiffness does not play a role in the solution. The solutions to be super-imposed as in (7.19) have their greatest value $\Delta\eta_0$ at the points where the forces are applied. Since these solutions must prevent the occurrence of corners, $\Delta\eta_0$ follows from

$$\frac{d\Delta\eta}{dx} = \Delta\eta_0 \sqrt{\frac{F_x}{B}} = \gamma = \frac{F_y}{2F_x}; \tag{7.21}$$

consequently,

$$\Delta\eta_0 = \frac{F_y}{2F_x} \sqrt{\frac{B}{F_x}}. \tag{7.22}$$

It can be seen how the maximum displacement of the string without bending stiffness,

$$\eta_0 = \frac{F_y}{2F_x} \cdot \frac{1}{2} \tag{7.23}$$

is decreased by $2\Delta\eta_0$, to

$$\eta = \frac{F_y}{2F_x} \left(\frac{1}{2} - 2\sqrt{\frac{B}{F_x}} \right). \tag{7.24}$$

This correction is of no interest whatever in the case of static displacements when the span and the forces are great and the bending stiffness is low. For our dynamic problems, however, this correction has proven to be of fundamental importance.

7.2 Reflection from the bow at the outside of a string of finite thickness, taking bending stiffness into account

We now return to the problem of reflection discussed in section 6.4, and expand the discussion to account for bending stiffness. Because of the small amplitude of all deformations, we can assume that torsion and bending are superimposed without having an effect on one another, even when we use a fixed coordinate system for both of these deformations.

Consequently, we can adopt our previous boundary conditions (6.34), (6.35), and (6.37) in combination with (6.40):

$$v_1 = v_2, \tag{7.25}$$

$$v_2 = aw, \tag{7.26}$$

$$F_1 - F_2 = \frac{M}{a} = \frac{2\Theta' c_\mathrm{T}}{a^2} w. \tag{7.27}$$

There are, however, two new boundary conditions. The absence of corners, due to bending stiffness, requires that

$$\frac{\partial v_1}{\partial x} = \frac{\partial v_2}{\partial x}, \tag{7.28}$$

and if we regard the bow as having only an infinitesimal width, it can transfer no moment, so that we can regard the curvatures of the string on either side of the bow as being equal:

$$\frac{\partial^2 v_1}{\partial x^2} = \frac{\partial^2 v_2}{\partial x^2}. \tag{7.29}$$

In the light of an actual bow's finite width, this assumption is, to be sure, questionable; but there is no way to replace it with another equally simple one. It may be valid to assume that any possible transfer of moments by the bow is small.

These two additional boundary conditions, however, require that some additional variables be defined: namely, the amplitudes of the velocity of the boundary fields mentioned earlier. Between the bow and the bridge ($x < 0$), these increase with x:

$$\hat{\underline{v}}_{1j} e^{k'x}. \tag{7.30}$$

Behind the bow ($x > 0$), these decrease as x increases:

$$\hat{\underline{v}}_{2j}e^{-k'x}. \tag{7.31}$$

We relate these amplitudes of velocities to that of the arriving wave $\hat{\underline{v}}_{1+}$ as well, by introducing the additional dimensionless factors

$$\underline{r}_j = \frac{\hat{\underline{v}}_{1j}}{\hat{\underline{v}}_{1+}} \tag{7.32}$$

and

$$\underline{t}_j = \frac{\hat{\underline{v}}_{2j}}{\hat{\underline{v}}_{1+}}, \tag{7.33}$$

where r and t indicate the ratios on the reflection and transmission sides, respectively, and the subscript j indicates the transition from k to $-jk'$. Consequently, we can describe the resulting velocity field before the bow as

$$\hat{\underline{v}}_{1+}(e^{-jkx} + \underline{r}e^{+jkx} + \underline{r}_je^{k'x}), \tag{7.34}$$

and the one behind the bow as

$$\hat{\underline{v}}_{1+}(\underline{t}e^{-jkx} + \underline{t}_je^{-k'x}). \tag{7.35}$$

The boundary conditions (7.24) and (7.29) now take on the form

$$1 + \underline{r} + \underline{r}_j = \underline{t} + \underline{t}_j, \tag{7.36}$$

$$-1 - \underline{r} + \left(\frac{k'}{k}\right)^2 \underline{r}_j = -\underline{t} + \left(\frac{k'}{k}\right)^2 \underline{t}_j. \tag{7.37}$$

From their difference, it follows that

$$\underline{r}_j = \underline{t}_j; \tag{7.38}$$

the boundary fields prove to be symmetrical. Consequently, (7.36) simplifies to

$$1 + \underline{r} = t. \tag{7.39}$$

If this is added to the condition (7.28),

$$-1 + \underline{r} + \frac{k'}{jk}\underline{r}_j = -\underline{t} - \frac{k'}{jk}\underline{t}_j, \tag{7.40}$$

it then follows, taking (7.38) into consideration, that

$$\underline{r}_j = \underline{t}_j = -\frac{jk}{k'}\underline{r}.\tag{7.41}$$

Since there is certainly a reflected wave, boundary fields are also present.

Taking (7.38), (7.39), and (7.41) into consideration, (7.26) takes on the form

$$\underline{t} + \underline{t}_j = \underline{t}_T;\tag{7.42}$$

but we can also express $\underline{t}_T$ in terms of $\underline{r}$:

$$\underline{t}_T = 1 + \underline{r}\left(1 - \frac{jk}{k'}\right).\tag{7.43}$$

In examining the remaining and most difficult boundary condition (7.27), we must account for the fact that the forces F_1 and F_2 also have boundary fields. We derive these most simply if we turn to the relationship

$$\underline{F} = -j\omega m' \int \underline{v}\,dx,\tag{7.44}$$

which is valid even for strings with bending stiffness. This relationship results from the proportionality of the acceleration of the mass $m'dx$ to the difference between the transverse forces. (The opposite sign compared with figure 7.1 is to make the phase of the F and v terms agree; so these terms represent a power transfer in the x-direction.)

For the force field normalized to $\underline{\hat{v}}_{1+}$ between the bow and the bridge, we obtain

$$\underline{\hat{F}}_1 = \omega m' \underline{\hat{v}}_{1+}\left[\frac{1}{k}e^{-jkx} - \frac{r}{k}e^{+jkx} - \frac{j}{k'}\underline{r}_j \cdot e^{k'x}\right],\tag{7.45}$$

and for the one behind the bow, we obtain

$$\underline{\hat{F}}_2 = \omega m' \underline{\hat{v}}_{1+}\left[\frac{t}{k}e^{-jkx} + j\frac{t_j}{k'}e^{-k'x}\right].\tag{7.46}$$

If we substitute these expressions into (7.27), divide by $\omega m' \underline{\hat{v}}_{1+}$ and multiply by k, we obtain

$$\left(1 - \underline{r} - \frac{jk}{k'}\underline{r}_j - \underline{t} - \frac{jk}{k'}\underline{t}_j\right) = \frac{2\Theta'c_T}{a^2 m'c_S}k\underline{t}_T = \frac{c_S k}{\omega \zeta}\underline{t}_T;\tag{7.47}$$

or, if we express t, r_j, t_j and t_T in terms of r, we obtain

$$r = \frac{-1}{1 + \dfrac{2\omega\zeta}{c_s k}\left[1 + \left(\dfrac{k}{k'}\right)^2\right] - j\dfrac{jk}{k'}}. \tag{7.48}$$

Up to this point, we have set no conditions on the frequencies of acceptable solutions. We have permitted all solutions of the dispersion equation, even a wave purely in bending—through this would be a valid solution only at not too high frequencies.

In accord with (7.13), we now restrict the phase speed of the stretched string with bending stiffness by substituting

$$\frac{\omega}{k} = c_s(1 + \tfrac{1}{2}(\varkappa(\omega)^2). \tag{7.49}$$

In the correction term in this formula, we have expressed the quantities which do not depend on the radian frequency by the coefficient

$$\varkappa = \sqrt{\frac{B}{F_x}}\frac{1}{c_s}. \tag{7.50}$$

Consequently,

$$\frac{k}{k'} = \frac{\omega}{c_s}\frac{1}{[1 + \tfrac{1}{2}(\varkappa\omega)^2]}\sqrt{\frac{B}{F_x}}\frac{1}{[1 + \tfrac{1}{2}(\varkappa\omega)^2]}.$$

$$= \varkappa\omega[1 - (\varkappa\omega)^2]. \tag{7.51}$$

Here, however, the correction already is proportional to the third power of the quantity $(\varkappa\omega)$, which we have up to now assumed to be small. We can more or less neglect the correction term if we restrict ourselves to

$$\varkappa\omega < 0.5. \tag{7.52}$$

On the other hand, we cannot ignore the quadratic terms; when we later examine the magnitudes of the resulting expressions, the linear term ($j\omega\varkappa$) also appears as a quadratic term. But the dispersion, too, is expressed in the quadratic terms. If the boundary fields play a role, they also do so in the quadratic terms, though to a smaller extent. Consequently, we arrive at the simplified form of (7.48),

$$r = \frac{-1}{1 + 2\zeta[1 + \tfrac{3}{2}(\omega\varkappa)^2] - j\omega\varkappa} \tag{7.53a}$$

and, similarly, we obtain, as in (7.39),

$$t = \frac{2\zeta[1 + \frac{3}{2}(\omega\varkappa)^2] - j\omega\varkappa}{1 + 2\zeta[1 + \frac{3}{2}(\omega\varkappa)^2] - j\omega\varkappa}. \tag{7.54a}$$

As $(\omega\varkappa) \to 0$ these expressions reduce to (6.50) and (6.51).

If we restrict ourselves to quadratic terms in $(\varkappa\omega)$, we obtain

$$r = \frac{1}{\sqrt{(1 + 2\zeta)^2 + (1 + 6\zeta + 12\zeta^2)(\omega\varkappa)^2}} \tag{7.53b}$$

and

$$t = \frac{\sqrt{(4\zeta^2) + (1 + 12\zeta^2)(\omega\varkappa)^2}}{\sqrt{(1 + 2\zeta)^2 + (1 + 6\zeta + 12\zeta^2)(\omega\varkappa)^2}} \tag{7.54b}$$

for the magnitudes shown in figure 6.6.

We saw in section 6.5 that 0.25 was the appropriate value of ζ in the low-frequency limit. If we now substitute the value $\varkappa = 6 \times 10^{-5}$, which also agrees with experimental results, we obtain the curves shown as dashed lines in figure 6.6. The inequality in (7.52) is, then, exceeded only slightly at the highest frequency. The failure of measurement to agree with calculation is certainly more due to fluctuations in the former. The plastic strings used by Lazarus and Eisenberg (see chapter 6) had a very high degree of damping of propagating waves. This had to be measured first and then later compensated for mathematically in order to determine the reflection and transmission factors. Nonetheless, the agreement seems good enough to confirm the hypothesis that the measured effects have to do with bending stiffness.

7.3 Point impedance of a string with bending stiffness

There is an even more important question regarding our having neglected bending stiffness up to this point: namely, when we account for the bending stiffness, what change occurs in the relationship between the frictional force δR exerted at a point, and the corresponding increase in velocity δv? Previously, we expressed this relationship through the real point-impulse impedance

$$2m'c_S = \frac{\delta R}{\delta v}. \tag{7.55}$$

We know that the effect of the corrections due to bending stiffness is frequency dependent; so we first examine the point impedance which corresponds to a sinusoidal force function. This impedance first generates a wave propagating to both sides; the amplitude of its transverse velocity is $\hat{\underline{v}}_+$ ($x > 0$). However, there must be another symmetrical boundary condition: the inclination must go to zero at $x = 0$. (This also eliminates the corner at the bow.) We must, therefore, require that

$$\frac{d}{dx}(\hat{\underline{v}}_+ e^{-jkx} + \hat{\underline{v}}_j e^{-k'x}) = 0. \tag{7.56}$$

This leads to

$$\hat{\underline{v}}_j = -\frac{jk}{k'}\hat{\underline{v}}_+. \tag{7.57}$$

In order to calculate the resulting force in the region $x > 0$, we once more make use of (7.44), and we obtain

$$\hat{\underline{F}}(+0) = -j\omega m'\left(\frac{\hat{\underline{v}}_+}{-jk} + \frac{\hat{\underline{v}}_j}{-k'}\right)$$

$$= \frac{\omega m'}{k}\hat{\underline{v}}_+\left(1 + \left(\frac{k}{k'}\right)^2\right). \tag{7.58}$$

If we divide this by

$$\hat{\underline{v}}(+0) = \hat{\underline{v}}_+\left(1 - \frac{jk}{k'}\right), \tag{7.59}$$

and if we also note that the driving force must move both segments of the string the same way, we obtain the point impedance that we seek:

$$\frac{2\underline{F}(+0)}{\underline{v}(+0)} = 2\omega m'\left(\frac{1}{k} + \frac{j}{k'}\right). \tag{7.60a}$$

This relationship holds throughout the entire range in which the dispersion equation (7.10) applies, right out to the limiting case of pure bending waves—with the exception of the very highest frequencies, for which the wavelengths are comparable to the thickness of the string (see e.g., Cremer and Heckl 1973).

Even, however, if we restrict ourselves to the approximations of interest in connection with the violin string, equations (7.13) and (7.6b), the point impedance is given by

$$\frac{2\underline{F}(+0)}{\underline{v}(+0)} = 2m'c_S\left[1 + \frac{(\omega\varkappa)^2}{2} + j(\omega\varkappa)\right]. \tag{7.60b}$$

This is still frequency dependent enough that we cannot define a point-impulse impedance for impulses or steps of any chosen width.

In this formula, the correction term in the real component is based on the dispersion (the increase in phase speed with frequency. As a result, an impulse generated at $x = 0$ degrades as it propagates; the higher-frequency components travel faster. This degradation of impulses is one more factor in reducing the systematic influence of the secondary waves on capture and release. Consequently, bending stiffness lends more credibility to the assumptions behind the method of the round trip.

We should, however, at this point examine the actual values of $\varkappa$ that apply to the strings used on string instruments. To this end, we rewrite $\varkappa$ for homogeneous strings as

$$\varkappa = \sqrt{\frac{B}{F_x}}\frac{1}{c_S} = \sqrt{\frac{B}{m'}}\frac{1}{c_S^2} = \sqrt{\frac{E}{\varrho}}\frac{a}{2c_S^2} = \frac{c_{LS}a}{2c_S^2}. \tag{7.61}$$

The value of c_S is established by the fundamental frequency to be expected for a string of a given length. The only other variables which play a part in setting the value of $\varkappa$ are, then, the material of the string, as it determines the speed of propagation of longitudinal waves c_{LS}, and the axial radius of inertia of the string, $a/2$. For a steel e string, the rate of propagation is approximately 500,000 cm s^{-1}; the radius is at most 0.15 mm, and the value of c_S, as calculated in section 6.1, is 42,200 cm s^{-1}. The result is $\varkappa \approx 2 \times 10^{-5}$ s. Even if it were appropriate to include frequencies at the upper limit of hearing, approximately 16,000 Hz, in problems having to do with the tone color of violins, $\omega\varkappa$ would still be only 0.2 and the correction to the phase velocity only 2%. Such strings, in other words, require no attention to their dispersion.

On the other hand, the experiments of Lazarus and Eisenberg (see chapter 6) have shown that bending stiffness can begin to have an effect in a 3-mm plastic string at a frequency of only 500 Hz. Schelleng's[2] measurements of $\varkappa$ furthermore prove that this type of string is not very different from the type used in cellos. According to these measurements, $\varkappa$ is approximately 32 times as great for a metal-wound c string as for a

---

[2] Schelleng (1973), table III. The quantity B in Schelleng's paper corresponds to $\varkappa^2(2\pi f_1)^2/2$ here.

homogeneous steel violin e string. The limit mentioned above falls to 500 Hz in the case of the cello c string.

The question of whether it is necessary to consider the frequency-dependent terms of the point impedance also depends in this case on whether the spectrum of the impulse of force includes frequencies above this limit to an appreciable extent. Instead of an abrupt step of force, whose spectral function from 0 to F_∞ is given by

$$\hat{\underline{F}} = \frac{F_\infty}{j\omega},\qquad(7.62)$$

we may use a "soft" transition in force, the formula for which might be

$$F = F_\infty(1 - e^{-t/\tau_F}).\qquad(7.63)$$

We then obtain the spectral function

$$\hat{\underline{F}} = \frac{F_\infty}{j\omega(1 + j\omega\tau_F)},\qquad(7.64)$$

which falls off much more rapidly. If, in this formula, we set $\tau_F = 20\varkappa$, then at the radian frequency $\omega = 0.2/\varkappa$, the spectrum of (7.64) is already weakened to 0.24 of the spectrum of (7.62). In other words, a change whose magnitude is in any case only 2% in the real part of the point impedance would be evaluated as one of only 0.5% in the function of velocity. Consequently, we can ignore dispersion—at least over the short propagation distances in which we are interested. As in (7.53), we can express the transverse displacement of the propagating wave as

$$\eta(x,t) = v_\infty\left[\left(t - \frac{x}{c_S}\right) - \tau_F(1 - e^{-(t-x/c_S)/\tau_F})\right].\qquad(7.65)$$

This wave is shown as a solid line in figure 7.3 for the moment in time given by $t_1 = x_1/c_S$. From the drawing, it can be seen that the onset of the propagating wave is not a corner, but rather a curve whose radius is $c_S^2\tau_F/v_\infty$. Since the ordinates must be increased in figure 7.3 by a factor of $v_\infty t_1$, this radius appears as $c_S^2\tau_F/v_\infty^2 t_1$.

On the other hand, a corner continues to exist at the point where the force is applied, $x = 0$, at least at this stage of the analysis. It would seem obvious to remove this corner, as in figure 7.2, by subtracting an exponential boundary field

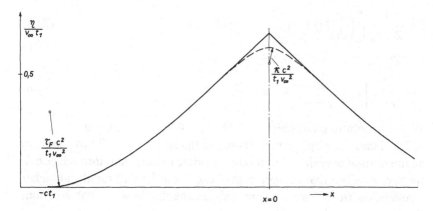

Figure 7.3
Dynamic deformation of a string with a soft onset of force and rounding of the corner
due to superimposed boundary fields related to bending stiffness.

$$\Delta\eta = -\Delta\eta_0 e^{-k'x} \tag{7.66}$$

whose slope at $x = 0$ is equal and opposite to that of the straight line defined by v_∞/c_S. Then, instead of the corner, there would be a curve with the radius $c_S^2 \varkappa/v_\infty$. This radius, again, appears in figure 7.3 as $c_S^2 \varkappa/v_\infty^2 t_1$. In order for this curve to be clear in the drawing, its radius was chosen as 1/5 that of the wavefront. With this ratio, however, the amount of rounding off at $x = 0$ is so great that an additional bending moment is hardly necessary. It is valid to add a static displacement to the two propagating waves, since the quantity k' is more or less as frequency independent as is the phase speed.

The above synthesis has been developed only intuitively; but it can also be derived from (7.60a), the formula for the point impedance, if we ignore the correction for dispersion, which is justified if the force is applied smoothly:

$$\underline{F} = \underline{v}(2m'c_S)(1 + j\omega\varkappa). \tag{7.67a}$$

Transformed back into the time domain, this becomes

$$F = 2m'c_S\left(v + \varkappa\frac{dv}{dt}\right). \tag{7.67b}$$

For any "smooth" function of force, the solution to this equation is given by the convolution integral

$$v = \frac{1}{2m'c_S} \int_0^t \left(\frac{d}{dt'} F(t') \right) (1 - e^{j(t-t')/\varkappa}) dt'. \tag{7.68}$$

For the function in (7.63), this leads to

$$v = \frac{F_\infty}{2m'c_S} \left[1 - \frac{\tau_F}{\tau_F - \varkappa} e^{-t/\tau_F} + \frac{\varkappa}{\tau_F - \varkappa} e^{-t/\varkappa} \right]. \tag{7.69}$$

With this solution, not only $v = 0$ at $t = 0$, but also $dv/dt = 0$.

To obtain the displacement shown in figure 7.2, we will now perform an integration over time; but it is appropriate to note that, through (7.65), we have assumed that the spectrum of excitation falls off rapidly at higher frequencies. In the present case, this means that $\varkappa \ll \tau_F$. The solution, then, is composed only of

$$\eta(t) = \frac{F_\infty}{2m'c_S} t - \tau_F(1 - e^{-t/\tau_F}) - \varkappa(1 - e^{-t/\tau_F}). \tag{7.70}$$

If we now add a spatial dependence, we must note that only the first two terms of this equation are related to the real part of the point impedance; the power expended through this impedance is determined by the waves that propagate away on both sides. If we replace t by $(t - x/c_S)$, we obtain the expression (7.65).

The term multiplied by $\varkappa$ belongs to the imaginary part of the point impedance; this may be represented as the reactance of an equivalent inertial mass:

$$2c_S m' \varkappa = 2m'/k'. \tag{7.71}$$

This equivalent mass represents not only the effective oscillating mass but also the equally great potential energy of deformation in bending.

We need now only incorporate the function $e^{-k'x}$ into this last term. Our overall result is the same as the one shown in figure 7.3 for $t - t_1$:

$$\eta = \frac{F_\infty}{2m'c_S} \left[\left(t_1 - \frac{x}{c_S} \right) - \tau_F \left[1 - \exp \left\{ - \left(t_1 - \frac{\varkappa}{c_S} \right) \tau_F \right\} \right] \right.$$
$$\left. - \varkappa(1 - e^{-t_1/\tau_F}) e^{-k'x} \right]. \tag{7.72}$$

The value $\Delta\eta_0$ to be subtracted in order to remove the corner corresponds exactly to that which was needed in our earlier calculations:

$$\Delta\eta_0 = v_\infty \varkappa = \frac{v_\infty}{c_s k'}. \tag{7.73}$$

The convolution carried out here could be extended to functions of force of any degree of smoothness, thus accounting for the processes which arise at capture and release due to bending stiffness—though only in terms of a linearized friction characteristic.

In this process the rising segment is always followed by a falling one. Consequently, the corner in the solid line at $x = 0$ in figure 7.3 is eliminated. The rising and falling slopes of the impulses are determined by the time constants of the response at the bridge input. These time constants presumably follow the laws of similarity: that is, their ratio for celli as compared to violins is the same as the ratio between the periods of the fundamental oscillations. Consequently, it can be assumed that the limiting frequencies of the impulse spectra of these instruments are as the reciprocal of this ratio.

It is also to be noted that we are interested only in the waves which proceed from these impulses, not in the boundary fields; in other words, we are interested in the real part of the point impedance, which hardly changes, not in the imaginary part. For this reason, we can generally assume that the effect of bending stiffness at capture and release need not be considered. This question, however, is always decided by the curvatures appearing in the result. In the calculations leading to our result, it has been formally possible to use ideal steps and even ideal impulses temporarily as mathematical tools. However, the damping, increasing with frequency, leads in almost every case to curvatures such that no particular attention need be given to the effect of bending stiffness, at least in treatments employing the other simplifications that we must use.

References

Cremer, L., 1979a. *Acustica* **42**, 141.

Cremer, L., 1979b. *Catgut Acoust. Soc. Newsletter* no. 31, 12 and no. 32, 27.

Cremer, L., and M. Heckl, 1973. *Structure-Borne Sound*. Tr. by E. Ungar. New York: Springer.

Krigar-Menzel, O., and A. Raps. 1891. *Berl. Ber.* **32**, 613.

Rayleigh, Lord, 1877. *Theory of Sound*. London: Macmillan. Reprinted 1929.

Schelleng, J., 1973. *J. Acoust. Soc. Amer.* **53**, 26.

8 Toward Complete Solutions

8.1 Modeling the string as a lumped-constant transmission line

A review of the preceding four chapters holds out little hope of a simple, complete solution to the problem of the bowed string, even if we ignore its bending stiffness; indeed, even if we could ignore torsion.

On the other hand, as shown in chapter 1, a strict solution is possible if the string is replaced by a system of one kinematic degree of freedom— a simple oscillator. We simplified this system even further by taking the sliding friction to be a constant less than the maximum sticking friction. It would be possible—and, in fact, not especially difficult—to replace this step function of friction by a continuous characteristic. In such cases, it is always necessary to trade off the difficulty of calculation against the improvement in the result.

By distributing the mass of the string into 2, 3, or n evenly-spaced fractional masses (Cremer 1975), it is possible to bridge the gap between a system with one degree of freedom and a one-dimensional continuum. This process is shown at the left in figure 8.1 for the cases in which $n = 4$, 8, and 12.

This method was first developed by Lazarus (1972). We mention him first because, to the best of our knowledge, this was the first strict treatment of frictionally excited oscillators more complex than the simple oscillator.

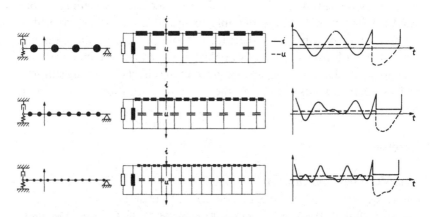

Figure 8.1
Description of the string as a lumped-parameter transmission line with 4, 8, and 12 elements. Left, mechanical system. Center, electrical circuit model corresponding to the F-i analogy. Right, functions of friction i (solid lines) and velocity u (dashed lines) at the point of bowing (after Lazarus).

Lazarus had studied communications engineering; this was doubtless of great use to him in grasping the problem. For the masses and the connecting elements of the string, seen as springs, he substituted an analogous electrical network; the impedance of the bridge, which we discussed in section 5.4, was modeled in a similar manner. Aside from the termination formed by the modeled impedance of the bridge, this system constitutes what is called a *lumped-constant transmission line* (Wagner 1919)—in other words, a chain of similar "two port" elements.

It is only necessary in practice to replace a mechanical system by an analogous electrical one when the latter proves to be an easily realizable and alterable aid to the study of the mechanical system. Today, the system is in most cases simulated on a computer; we will discuss computer simulations below.

For all persons who have learned to think in terms of electrical circuits, translation into electrical terms can make the problem easier to understand; also, the electrical analogy appears again and again in the literature relating to the violin (Schelleng 1963, 1973). For these reasons, we will now examine such analogous electrical circuits.

The basis of the analogy is this: on the one hand, in mechanics, a difference in force leads to acceleration of a mass

$$F_1 - F_2 = m\frac{dv}{dt} \tag{8.1a}$$

where there is no friction, but to velocity

$$F_1 - F_2 = rv \tag{8.2a}$$

where there is frictional resistance. On the other hand, a difference in velocity determines the increase of force with time across a spring:

$$v_1 - v_2 = \frac{1}{s}\frac{dF}{dt}. \tag{8.3a}$$

Also, a frictional resistance between two points moving at velocities v_1 and v_2 generates an equal force at both of these points:

$$v_1 - v_2 = \frac{1}{r}F. \tag{8.4a}$$

Analogously, the following equations apply to the differences between voltages at particular places in electrical networks, compared with the reference voltage u, and the currents i:

$$u_1 - u_2 = L\frac{di}{dt},$$ (8.1b)

where L is the inductance;

$$u_1 - u_2 = Ri,$$ (8.2b)

where R is the ohmic resistance;

$$i_1 - i_2 = \frac{1}{C}\frac{du}{dt},$$ (8.3b)

where C is the capacitance; and

$$i_1 - i_2 = \frac{1}{R}u = Gu,$$ (8.4b)

where G is the conductance.

In our juxtaposition of these equations, analogous quantities are, then,

force F and voltage u
velocity v and current i
mass m and inductance L
stiffness s and the inverse of capacitance, $1/C$
frictional resistance r and ohmic resistance R

The analogy is the same as that in Ohm's mind when, drawing on mechanics, he compared the flow of liquids in pipes to that of electrons in conductors. Forces and voltages usually take on a causal role, so their correlation in this F-u analogy seems to make very good sense.

However, it is just as possible to correlate (8.1a) → (8.3b), (8.3a) → (8.1b), and (8.2a) → (8.4b). Then the analogous quantities are

force F and current i
velocity v and voltage u
mass m and the inverse of capacitance, $1/C$
stiffness s and inductance L
frictional resistance r and conductance G

This F-i analogy has an advantage in terms of circuit topology. In structures built up out of masses, springs, and frictional resistances, it is the forces that branch apart like currents; on the other hand, the sum of velocity differences in the mesh of a network goes to zero with reference to the input, as do the voltage differences. Lazarus used the F-i analogy, and this is what is shown in the middle in figure 8.1. At the left in the figure, the impedance of the bridge appears as an inductance and a conductance in parallel, analogous to the mechanical system. In the resistance model, i.e., in the F-u analogy, a series circuit consisting of a capacitor and a resistor would have been necessary instead. At the nut, velocity goes to zero; with the F-i analogy shown, this corresponds to zero voltage difference—a short circuit. In the resistance model, a pair of unconnected terminals would appear instead.

In all three examples shown, the point of bowing is at $x_b/l = n_b/n = 1/4$, where n_b is the number of similar elements of the circuit between the point of bowing and the bridge. In contrast to the case of the simple oscillator of chapter 1, the point of bowing is here not at a mass but rather in the middle of an element of the string linking two masses. In order to clarify the boundary between similar elements of the transmission line, all of these elements of the string are represented by two coils in series. The frictional force, represented by i, is not exerted on a mass; therefore, velocity, represented by u, can go through an instantaneous step at release.

This happens for the same reason as the flattening effect described in section 5.6. Sliding friction begins with a step downward from the maximum value of sticking friction, and is assumed by Lazarus to be constant, as we assumed in section 1.1. In this case, the instability at release occurs even at the lowest bowing pressures, while capture is without a step even at the highest bowing pressure.

This step characteristic and the constancy of sliding friction are evident from the constancy of i during the interval of sliding friction, and the steps on either side.

It can also be seen that oscillations at the higher natural resonances of the transmission line are strong during the interval of sticking friction. These oscillations correspond here to the reflected secondary waves; it must, however, be noted that the natural frequencies of the transmission line are not harmonically related, even if the impedance of the bridge is infinite. The highest of these resonances, which appears clearly in the functions shown in figure 8.1 is approximately a fifth lower than nf_1, if f_1 represents the fundamental frequency of the string, taken in this model to be 100 Hz. Though the peaks at capture and release become more and more prominent as n is increased, it cannot be claimed that the function adequately represents the details of the forces generated by the string during the interval of sticking friction. Also, the number of calculations required, even at $n = 8$, is very great.

Lazarus attempted to keep the difficulty of calculation within tolerable limits by working in the frequency domain when discussing the processes during the intervals of sticking and sliding friction, idealized through the use of constant force. The electrical engineer thinks not only of circuits, but also of the frequency dependences that can be ascribed to them. The question does arise as to how a problem such as that of self-excited oscillation by means of friction can be studied at all in the frequency domain; capture and release, after all, depend on instantaneous values. Lazarus, however, makes use of the finite integral transform: in other words, the transform taken in only one interval of either sticking or sliding friction (specifically, Lazarus uses the general Laplace transform, which extends the concept of frequency to complex numbers). The limiting values of the interval over which integration takes place appear in this transform; therefore, boundaries can be set for this interval. Also, corresponding to the boundary conditions at capture and release we are discussing, these limiting values must agree. These boundary values and conditions, then, determine the duration t_h of the interval of sticking friction; this, the duration of the interval of sliding friction t_g, and their sum, the period T, are among the variables whose values are sought. Since the duration of the interval of sticking friction must be obtained by means of several transcendental equations, it is appropriate to look for it first near the values for Helmholtz motion. When the function of i over time found by means of the transforms is plotted, we sometimes see, however, that the maximum value of sticking friction is exceeded much earlier.

For another reason as well, calculations are difficult when the string is replaced by a lumped transmission line: transverse forces and velocities (currents and voltages) at all junctions along the transmission line must be found, even though only the quantities at the point of bowing and the function of force at the bridge are of interest. The paths between bridge and point of bowing, and between point of bowing and nut or finger, are of interest only as distortionless delays. These can be modeled simply in the frequency domain. Consequently, it may be possible to use the finite (Laplace) transform, and perhaps to use it to advantage, when the transition is made from a lumped to a continuous transmission line. Still, it seems unlikely that this method would be more elegant than the one to be described in the following sections.

8.2 Use of a computer as a restricted analog system

Among all of the techniques which lead to a complete or nearly complete solution to the problem of the bowed string, the one suggested by McIntyre and Woodhouse (1979) is mathematically the least difficult and—for related reasons—the most lucid in terms of the physics of the string.

Consequently, we will take a thorough look at this technique, especially since the publications of McIntyre and Woodhouse present only its principles. Also, the details are important if it is desired to actually use the technique. In his dissertation, written under McIntyre's supervision, Woodhouse (1977) points out that the technique is an extension of the approximation using the method of the pure round trip, which was explained in section 5.3. Though McIntyre's and Woodhouse's technique is capable of accounting for all secondary waves, we will base our presentation mostly on that method. We will, for example, once again regard the dynamic process of bowing as periodic, but superimposed on a static displacement generated by a minimum frictional force R_{min}.

When we look at the space-time diagram of all of the corrective waves in figure 6.7, it appears to be an exceptionally difficult task to account for the fate of all of the secondary waves. This would be so even if we, like the cited authors, at first ignored the torsion of the string in order to represent the process more clearly. Accounting for the secondary waves would, however, be difficult only if it were necessary to trace each of them separately; the present author confesses that the assumption that this would be necessary led him to attempt to use the method of the pure round trip without taking secondary waves into account.

The difficulty of calculation is reduced to a surprising degree when no attempt is made to trace each secondary wave separately. The simplification is apparent in McIntyre's and Woodhouse's use of a computer as an analogous system for the bowed string, in which they considered only those field quantities essential to the playing of the violin. These are the quantities at the bridge, $x = 0$, the nut or finger, $x = 1$, and the bow, $x = x_b$. The segments of the string in between these points are taken into consideration only insofar as they determine time delays. Losses and

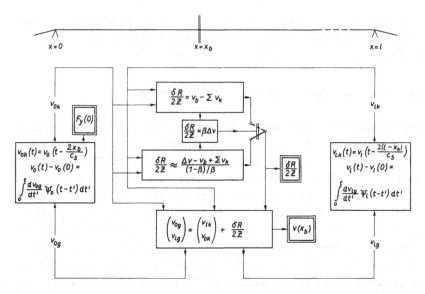

Figure 8.2
Signal-flow diagram for computer simulation of the periodic bowing process without accounting for torsion (based on work of McIntyre and Woodhouse).

possible corrections in phase velocities are taken care of in the reflection responses already introduced in (5.7) and (5.8). We must refer these to velocities themselves, not to the velocity differences defined in (5.6); and it is also appropriate for us to treat in the same way the time delays $2x_b/c_s$ and $2(l - x_b)/c_s$ between the velocities going from the bow to the ends, v_g, and those returning to the bow, v_k. In this way, for the two convolution integrals that correspond to $x = 0$ and $x = 1$, we obtain

$$v_0(t) - v_0(0) = \int_0^t \frac{dv_{0g}}{dt'}\psi_0(t - t')dt', \quad \text{with} \quad v_{0k}(t) = v_0\left(t - \frac{2x_b}{c_s}\right), \quad (8.5)$$

$$v_l(t) - v_l(0) = \int_0^t \frac{dv_{lg}}{dt'}\psi_l(t - t')dt', \quad \text{with} \quad v_{lk}(t) = v_l\left(t - \frac{2(l - x_b)}{c_s}\right). \quad (8.6)$$

In figure 8.2, these relationships are enclosed in boxes at the right and left, corresponding to the spatial relationships on the violin. All other relationships are related to the bow, and are in the middle. Figure 8.2 represents what is called a *signal-flow diagram*. It would have been more appropriate to this type of diagram if the boxes at both sides had appeared together at one side, as will be the case in figure 8.4 below; the relationships in both of these boxes take something from the bow, change its form, delay it, and return it to the bow.

The analysis is not at all changed by our being interested in recording the force generated during reflection at the bridge only; our interest in this force is indicated by the arrow proceeding from the "bridge box" to the box designated by $F_y(0)$. $F_y(0)$ is in this case determined by the difference between the wave $v_{0g}(t - x_b/c)$ arriving at the bridge and the wave $v_{0k}(t + x_b/c)$ reflected from it; t, as before, is the local time at the point of bowing:

$$F_y(0, t) = Z\left[v_{0g}\left(t - \frac{x_b}{c}\right) - v_{0k}\left(t + \frac{x_b}{c}\right)\right]. \qquad (8.7)^1$$

A good initial impulse would be obtained by beginning the function v_{0g} with a linear ramp from 0 to $\Delta v/2$ followed by a Helmholtz step or a ramp of constant inclination, leading to $-\Delta v/2$ and ending with another linear ramp from $-\Delta v/2$ to 0.

The first impulse to return from the bridge to the bow, v_{0k}, initially finds the interval of sliding friction to be in progress there; this condition is shown in figure 8.2 by the position of the "switch" indicated by a solid line. There is only one switch after the boxes, by means of which sticking (above) and sliding (below) are distinguished.

As we have already made clear in connection with figure 5.3f, however, a switchover to the regime of sticking friction occurs as soon as the additional frictional force has reached its maximum value, the upper limit of sticking friction:

$$\delta R = R_{h\,max} - R_{min} = \Delta R. \qquad (8.8)$$

We represent ΔR here by

$$\Delta R = 2Z\beta\Delta v, \qquad (8.9)$$

using the bow-pressure parameter β defined in (5.4).

---

[1] In the German edition, to this difference there was added the linear increase of force between the steps; the values of v were related only to the Helmholtz steps and the corrective waves generated by them, following from the observations on ramps in chapter 5. For the computer simulation discussed here, it is more appropriate to regard the Helmholtz motion as two opposing sawtooth curves (see section 3.3), thus incorporating the linear ramp in $F_y(0, t)$. This representation is also the one used by McIntyre and Woodhouse (1979).

In doing this, we exclude the overhanging curves of figure 5.6, as we already had done with (5.3); consequently, we also exclude the flattening effect described in (5.4).

In every case, the end of the following interval of sticking friction is set by the condition in (8.8). For this reason, between the two boxes representing the regimes of sticking and sliding friction, there is a box containing the equation

$$\frac{\delta R}{2Z} = \beta \Delta v, \tag{8.10}$$

which combines the conditions in (8.8) and (8.9). From this middle box, an arrow leads to the switch at the output of the boxes corresponding to sticking and sliding friction.

As long as only a v_k is arriving—whether it is a v_{0k} from the direction of the bridge of a v_{lk} from the direction of the nut or finger—δR is determined, in the cases of both sticking and sliding friction, exactly as in chapter 5. We can obtain these restricted conditions of signal flow in figure 8.2 if we close off the signal paths representing the secondary waves; it is easy to program this.

In fact, however, the calculations shown in figure 8.2 can account not only for waves returning to the bow for the first time, but for all of their reflections from the bow, the bridge, and the nut. It is necessary only to extend the conditions described above to take into account that waves are constantly arriving from both the bridge and the nut, and that all such waves in turn influence the waves leaving for the bridge and the nut.

In the language of communications engineering, we have here a two-port device, with input quantities v_{0k} and v_{lk} and output quantities v_{0g} and v_{lg}; this is true of both the regime of sliding friction and that of sticking friction.

In both of these regimes, it is true that the velocity of the string (at the center of its cross section) v_S is identical at the nut and the bridge sides of the bow:

$$v_S = v_{0k} + v_{0g} = v_{lk} + v_{lg}. \tag{8.11}$$

This quantity v_S is the second one which it is useful to observe on the screen of a computer while calculations are in progress, since this quantity is not only the basis of all classical representations of the bowing of the string, but is also measurable. Measurement can be made by means of electrodynamic sensing, as already mentioned several times, or by means of the method developed by Hancock (1977), in which two laser beams interfere with one another after one has been reflected from the string.

Furthermore, in both the regime of sliding friction and that of sticking friction, both waves leaving the bow—v_{0g} toward the bridge and v_{lg} toward the nut—differ from the waves v_{lk} and v_{0k} returning from these points in the same way: namely, by addition of the δv-waves radiated to both sides by the force δR:

$$v_{0g} = v_{lk} + \frac{\delta R}{2Z},\tag{8.12}$$

$$v_{lg} = v_{0k} + \frac{\delta R}{2Z}.\tag{8.13}$$

[Also note that the difference between these two equations is given by (8.11).] By adding v_{0k} to (8.12), right and left, and v_{lk} to (8.13), right and left, it is also possible to obtain the even more general relationship

$$v_S = \sum v_k + \frac{\delta R}{2Z}.\tag{8.14}$$

The difference between the regimes of sticking and sliding friction lies, then, only in the way that the additional frictional force is determined from the given quantities v_{0k} and v_{lk}, i.e., in the quantity $\delta R/2Z$, which can also be regarded as a frictional force normalized to $2Z$.

We begin with the regime of sticking friction, where the relationships are always linear.

In order to retain the condition

$$v_S = v_b,\tag{8.15}$$

even though the arriving velocities lead to only one resultant value

$$\sum v_k = v_{0k} + v_{lk} < v_b,\tag{8.16}$$

δR must be responsible for the difference:

$$\frac{\delta R}{2Z} = v_b - \sum v_k.\tag{8.17}$$

This condition is shown in the uppermost box representing the bow in figure 8.2.

It is more difficult to determine δR in the regime of sliding friction, if the "sag" in the friction characteristic is to be accounted for:

$$\delta R = \Delta R \varrho \left(1 - \frac{v_{\text{rel}}}{\Delta v}\right); \tag{8.18}$$

we already accounted for it in this way in the method of the pure round trip. Difficulty arises because a quantity representing a departing wave— that is, quantity to be found—always occurs along with a quantity representing an arriving wave—a given quantity—in the relative velocity between the bow and the string:

$$v_{\text{rel}} = v_{\text{b}} - v_{\text{S}} = v_{\text{b}} - v_{0k} - v_{0g} = v_{\text{b}} - v_{lk} - v_{lg}; \tag{8.19}$$

Consequently, we must find δR by combining (8.18) with one of the equations (8.12) or (8.13); or with (8.14), which includes both of these. This approach led, in the method of the pure round trip, either to the curves in figure 5.6 or to the graphic solutions in figure 5.12. In the latter case, only one friction characteristic, normalized to the bowing pressure but independent of it, need be stored in the computer. Also, it is easy for a computer to determine points of intersection with straight lines of different slopes. Consequently it is easier now to find the intersection of

$$\frac{\delta R}{2Z\beta\Delta v} = \varrho \left(1 - \frac{v_{\text{b}} - v_{\text{s}}}{\Delta v}\right) \tag{8.20}$$

and

$$\frac{\delta R}{2Z\beta\Delta v} = \frac{1}{\beta\Delta v}(v_{\text{s}} - \sum v_{\text{k}}). \tag{8.21}$$

The abscissa for both curves remains relative velocity, although, in this case, this is composed of a given quantity and one to be found. The straight line corresponding to (8.14) begins at the abscissa ($\delta R = 0$), at the point $\sum v_{\text{k}}$.

The points 0 and l, representing the ends of the string, can be exchanged in this sum, just as in the relationships (8.20) and (8.21).

For bowing pressures sufficiently below the boundary of the flattening effect, it is a good approximation to linearize the relationship in (8.20) as we did before; we replace the function ϱ by a straight line between the end points (0, 0) and (1, 1). (A basic nonlinearity remains in the problem: the corner between the regimes of sticking and sliding friction.) The quantity δR, in this case, can be found from the following relationships, which represent v_S in (8.20) and (8.21):

$$v_S = -\Delta v + v_b + \frac{\delta R}{2\beta Z} = \sum v_k + \frac{\delta R}{2Z} \tag{8.22a}$$

and thus

$$\frac{\delta R}{2Z} = \frac{\beta}{1 - \beta}(\Delta v - v_b + \sum v_k). \tag{8.22b}$$

This, instead of the general graphic relationship, is what we have shown in the lowest box representing the bowing forces. In every case, the time function of $\delta R/2Z$ is interesting enough to be shown on the screen, even though it cannot be measured directly.

Figure 8.3 shows four examples of recordings of functions generated by the computer.[2] In each, the function v_S over time is shown at the top; below this, the frictional force; and at the bottom, the input force at the bridge. The examples at the left take account of the secondary waves; those at the right correspond to the pure round trip method. (In connection with the latter set of examples, it should be noted that only the impulse at capture is to be seen in δR, because only $v_{l_g} - v_{0k}$ was used in determining δR.)

The differences between the upper sets of functions, corresponding to low bowing pressure, are minor. The secondary waves can hardly be seen in the velocity function, with its relatively wide ramps, or in the force function at the bridge. Only in the δR do we see reflected impulses between the initial impulses of the secondary waves.

In the lower examples, representing high bowing pressure, the secondary waves have an effect which cannot be ignored. Their contribution is to be seen clearly at the left and is absent at the right. This does not, however, mean that they significantly affect capture or release. The v_S-functions differ very little.

Certainly, the secondary waves have an effect on the tone color, as they affect the force at the bridge. It is, however, possible, as in figure 6.8, to derive the secondary waves from the impulses of δR at capture and

[2] The recordings shown were prepared by Woodhouse for a lecture given by the present author on October 3, 1978, at the Department of Applied Mathematics and Theoretical Physics at Cambridge University. I wish to thank both Dr. McIntyre and Dr. Woodhouse for many interesting discussions, and Dr. McIntyre for his correspondence.

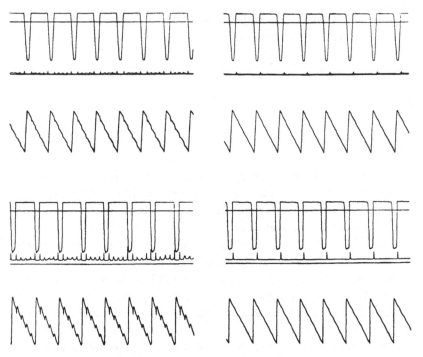

Figure 8.3
Examples of computer-generated time dependences: velocity of the string and frictional
force at the bow, and force at the bridge. Top, for low bowing pressures. Bottom, for high
bowing pressures. Left, including the secondary waves Right, excluding the secondary
waves (after Woodhouse).

release, and from the fate of their reflections; the secondary waves can
then be added to the solution. The exact time function is less important
than the resulting spectrum; consequently, it is possible to start with the
spectrum corresponding to the Helmholtz motion and the primary
corrective wave, subsequently adding the additional spectrum representing
a decaying series of secondary impulses separated from one another by
intervals that depend on distances from the point of bowing (in particular,
the interval $2x_b/c$). Physically, this separation would not be uninteresting.
If the program for a complete solution is available, however, this procedure would be unnecessary.

The statements in section 5.6 show that it is also possible to include
in the program the waves in the x, z-plane generated by the bridge, and

their influence on the bowing pressure. This has, however, not yet been tried; it would be appropriate only after measurements had been made of waves in the x, z-plane of an actual bowed string. Furthermore, the influence of waves in the x, z-plane proved, in the example in section 5.6, to be of no basic importance. Consequently, we refrain from this extension of the analysis.

8.3 Accounting for torsion in the restricted analog system

On the other hand, it is a relatively simple task to extend the simulation to include the secondary torsional waves. This extension is even more important than accounting for the sag in the friction characteristic, for the sag becomes less important the lower the bowing pressure; no similar statement can be made about torsion.

The difficulty of calculation is approximately doubled. We see this if we note that torsional waves, too, now depart and arrive; their reflection at the bridge and the nut (or finger) is described by relationships that include a shorter delay time, but that are otherwise analogous to those in (8.5) and (8.6) for transverse waves:

$$w_0(t) = \int_0^t \frac{dw_{0g}}{dt'}\chi_0(t-t')dt', \quad \text{with} \quad w_{0k}(t) = w_0\left(t - \frac{2x_b}{c_T}\right), \quad (8.23)$$

$$w_l(t) = \int_0^t \frac{dw_{lg}}{dt'}\chi_l(t-t')dt', \quad \text{with} \quad w_{lk}(t) = w_l\left(t - \frac{2(l-x_b)}{c_T}\right). \quad (8.24)$$

(Here the possibility of a transformation of transverse into torsional waves and vice versa has not been taken into account; we still lack the requisite experiments. But the necessary relations, should they prove to be necessary, can be included at any time.) Instead of these relationships in the time domain—and the corresponding ones for transverse waves—we have shown relationships in the frequency domain in the boxes in figure 8.4. This time, all of the boxes representing the ends of the strings are shown at the right side. The equations are

$$\underline{v}_{0k}/\underline{v}_{0g} = \underline{r}_{0S}e^{-2jk_Sx_b}, \quad (8.25)$$

$$\underline{v}_{lk}/\underline{v}_{lg} = \underline{r}_{lS}e^{-2jk_S(1-x_b)}, \quad (8.26)$$

$$\underline{w}_{0k}/\underline{w}_{0g} = \underline{r}_{0T}e^{-2jk_Tx_b}, \quad (8.27)$$

$$\underline{w}_{lk}/\underline{w}_{lg} = \underline{r}_{lT}e^{-2jk_T(1-x_b)}. \quad (8.28)$$

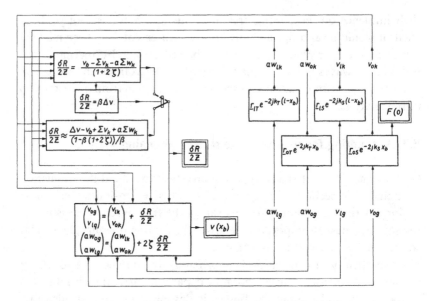

Figure 8.4
Signal-flow diagram for computer simulation of the bowed string accounting for torsion (based on work of McIntyre and Woodhouse).

We choose this representation not only because it is shorter, but also because it is the best-known physical representation, and because any large computer is equipped to transform a relationship from the frequency domain into the time domain. Above all, however, quantities in the frequency domain can be measured more precisely than the corresponding ones in the time domain.

We now have four input quantities at the bow:

$$v_{0k}, \quad v_{lk}, \quad w_{0k}, \quad w_{lk};$$

it is appropriate for us to normalize them all to the same dimension, by multiplying the last two by the radius a:

$$v_{0k}, \quad v_{lk}, \quad aw_{0k}, \quad aw_{lk}.$$

These four input quantities are related to the four output quantities

$$v_{0g}, \quad v_{lg}, \quad aw_{0g}, \quad aw_{lg}.$$

Consequently, it would seem likely that the difficulty of calculation would be quadrupled, since we have gone from a 2 × 2 matrix to a 4 × 4 matrix. However, this is not the case.

The output quantities v_{0g} and v_{lg} are obtained in the same way as they were previously, through (8.12) and (8.13).

The new output quantities (aw_{0g}) and (aw_{lg}) are obtained through the additional, analogous equations

$$(aw_{0g}) = (aw_{lk}) + \frac{\delta R a^2}{2\Theta' c_T} = (aw_{lk}) + 2\zeta \frac{\delta R}{2Z}, \tag{8.29}$$

$$(aw_{lg}) = (aw_{0k}) + 2\zeta \frac{\delta R}{2Z}. \tag{8.30}$$

Now the quantity aw represents the difference between the velocity v_R at the bow and the velocity v_S at the cross-sectional center of the string:

$$aw = v_R - v_S = a(w_{0k} + w_{0g}) = a(w_{lk} + w_{lg}). \tag{8.31}[3]$$

Thus just as with the transition from (8.12) and (8.13) to (8.14), we can go from (8.29) and (8.30) to

$$aw = a \sum w_k + 2\zeta \frac{\delta R}{2Z}. \tag{8.32}$$

If we add this equation to (8.14), we obtain

$$v_R = v_S + aw = \sum v_k + a \sum w_k + (1 + 2\zeta)\frac{\delta R}{2Z}. \tag{8.33}$$

The difficulty of calculating the force δR has also only approximately doubled.

In the regime of sticking friction, δR must account this time for the difference between v_b and $(\sum v_k + \sum aw_k)$ at the surface of the string. It should also be noted that δR is exerted against the lesser point-impulse impedance $2Z/(1 + 2\zeta)$. Both of these considerations lead to the same result as if we simply set $v_R = v_b$ in (8.33). We obtain

$$\frac{\delta R}{2Z} = \frac{v_b - \sum v_k - \sum aw_k}{(1 + 2\zeta)}. \tag{8.34}$$

In the regime of sliding friction, too, it is to be noted that the relative velocities now depend on v_R:

---

[3] In contrast to the treatment of reflection from the bow in chapter 6, the sign of aw is here positive in the direction of the force produced by the bow.

$$v_{rel} = v_b - v_R = v_b - v_S - aw. \tag{8.35}$$

Below the threshold of the flattening effect, we now obtain, instead of (8.20),

$$\frac{\delta R}{2Z\beta\Delta v} = \varrho\left(1 - \frac{v_b - v_R}{\Delta v}\right); \tag{8.36}$$

the intersection of this with the straight line that follows from (8.33),

$$\frac{\delta R}{2Z\beta\Delta v} = \frac{v_R - \sum v_k - \sum aw_k}{\beta\Delta v(1 + 2\zeta)}, \tag{8.37}$$

must now be found. This time, the two sums $\sum v_k$ and $\sum aw_k$ determine the point where the straight lines begin at the abscissa.

The limit within which it is appropriate to linearize has now decreased by the factor $1/(1 + 2\zeta)$, due to our accounting for torsion; nonetheless, linearization is still of interest to us. Once again, we need only replace (8.36) by the straight line

$$\frac{\delta R}{2Z\beta\Delta v} = 1 - \frac{v_b - v_R}{\Delta v}, \tag{8.38}$$

in combination with (8.37); corresponding to the derivation of (8.22b) from (8.20) and (8.21), this leads to

$$\frac{\delta R}{2Z} = \frac{\beta}{1 - \beta(1 + 2\zeta)}\{\Delta v - v_b + \sum v_k + \sum aw_k\}, \tag{8.39}$$

and to what is written in the corresponding box in figure 8.4 instead of the graphic description.

If, on the other hand, the expression for $(\delta R/2Z)$ from the equation (8.34) is substituted into (8.12) and (8.29), then the result in the regime of sticking friction is

$$v_{0g} = \frac{v_b - v_{0k} + 2\zeta v_{lk} - aw_{0k} - aw_{lk}}{1 + 2\zeta} \tag{8.40}$$

and

$$(aw_{0g}) = \frac{2\zeta v_b - 2\zeta v_{0k} - 2\zeta v_{lk} - 2\zeta aw_{0k} + aw_{lk}}{1 + 2\zeta}; \tag{8.41}$$

in the regime of sliding friction the result is

$$v_{0g} = \frac{\beta(\Delta v - v_b) + \beta v_{0k} + (1 - 2\zeta\beta)v_{lk} + \beta a w_{0k} + \beta a w_{lk}}{1 - \beta(1 + 2\zeta)} \tag{8.42a}$$

and

$$(aw_{0g}) = \frac{2\zeta\beta(\Delta v - v_b) + 2\zeta\beta v_{0k} + 2\zeta\beta v_{lk} + 2\zeta\beta a w_{0k} + (1 - \beta)a w_{lk}}{1 - \beta(1 + 2\zeta)}. \tag{8.42b}$$

If we compare the coefficients of v_{0k} and v_{lk} in (8.40) and (8.42a) here with those of equations (6.50), (6.51), (6.57), and (6.59), we see that these coefficients represent the reflection and transmission factors derived in chapter 6. In the same way, the coefficients of v_{0k} and v_{lk} in (8.41) and (8.42b) represent the transformation factors derived in (6.51) and (6.58).[4]

Since v_{0k} and v_{lk} can be pure secondary waves, these coefficients could have, in fact, simply been substituted from the earlier calculations.

Also, we can conclude that the coefficients of w_{0k} and w_{lk}, which we have not yet interpreted, represent reflection, transmission, and transformation factors for arriving torsional waves characterized by an amplitude aw, which are not discussed in chapter 6.

The coefficients of v_b and $(\Delta v - v_b)$ can also be derived as those of exciting functions according to the boundary conditions given earlier in sections 6.4 and 6.5.

The derivation used in this section, however, allows a better transition to the more general considerations in the next few sections.

8.4 Theory of pulse synthesis

The methods discussed in the previous two sections were particularly well adapted to periodic motions of the bowed string; but a more general method is now to be described. This method can be used whenever any linear system is excited at a fixed point by an external force, and when only one field variable, generated as a result of the excitation, is of interest at any one time. Usually, as in the present case, the variable to be found is a kinematic variable at the point of excitation. This condition is met, in particular, if the output variable, such as, in this case, velocity, has a feedback effect on the exciting force.

[4] The change of sign already mentioned should be noted here.

For now, we will want to consider the exciting force $F(t)$ as predetermined, beginning at $t = 0$. We can regard $F(t)$ as being made up of a sequence of narrow, rectangular pulses, like the lathes of a wooden fence, each having width dt' and height F. That is, $F(t)$ is made up of a sequence of impulses of strength $F(t')dt'$. We need only know the velocity function that results from a single impulse whose strength is 1; this function is called the *velocity impulse response.*

We know that in the case of the string all pulses are exerted against the point-impulse impedance $2Z$; therefore, it is appropriate to introduce the normalized force

$$f(t) = \frac{F(t)}{2Z}.$$ (8.43)

In doing this, we also, in turn, normalize the pulse responses $g(t - t')$ corresponding to a pulse whose duration is Δt and whose strength is $\int_0^{\Delta t} f(t')dt' = 1$. We use the symbol g, as do McIntyre and Woodhouse (1979) and Schumacher (1979), along with their name for such a function: a *Green's function.*[5]

Once this normalization has been carried out, the velocity function to be found is obtained by integration:

$$v(t) = \int_0^t g(t - t')f(t')dt'.$$ (8.44)

The simplest example of a Green's function having to do with the string is that of the Raman model, loaded at either end with a real impedance w. Each reflection, then, brings about a weakening by the reflection factor (related to velocity)

$$r = -\frac{w - Z}{w + Z}.$$ (8.45)

In this case too, the Green's function consists of ideal pulses, which we have represented at the top in figure 8.5 as rectangles whose width Δt we have chosen at will. Corresponding to our normalization, the area of the

[5] Green introduced this function only for potential fields excited at one point in space. The expression was later extended to wave fields excited at discrete points in space; but not usually, as here, to pulse excitations, which are also related to discrete points in time.

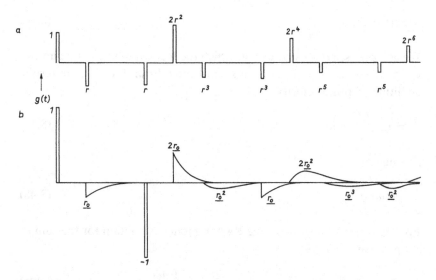

Figure 8.5
Examples of Green's functions. Top, for the Raman model. Bottom, for a spring and a resistance in parallel with respect to force, but only at the bridge.

first impulse is $g(0)\Delta t = 1$. The function $g(0)$ is, then, inversely proportional to Δt. Thus we see also that the chosen normalization of the pulse response leads to a quantity whose dimension is 1/time. We need only require that Δt be smaller than any of the time intervals necessary to represent the pulse response sufficiently accurately. The time resolution of the recording apparatus or the computer determines such limits.

These limits, i.e., the realizable pulse width, determine how far a pulse peak rises above the rest of the recorded function. In figure 8.5, top, the reflected pulses still have the chosen width Δt, but their height decreases as the powers of r.

After a time delay $2x_b/c$, a reflection, reversed in polarity and weakened by a factor of $|r|$, returns from the bridge; after a delay of $2(l - x_b)/c$, an identical reflection returns from the finger. After one full period, both reflections, each reflected one more time, come back together at the point of excitation. From this moment on, the process repeats itself; the impulses are weakened by a factor of r^2 in each period.

We can see from this little example that the contributions become smaller and smaller with time.

Generally,

$$\lim_{t\to\infty} g(t) = 0. \tag{8.46}$$

We can now see one more way in which the Raman model fails: namely, when we excite the string with a step of force from 0 to f_∞ there results an ultimate, finite velocity

$$v_\infty = f_\infty \cdot \int_0^\infty g(t)dt, \tag{8.47}$$

as long as

$$\int_0^\infty g(t)dt \neq 0. \tag{8.48}$$

For the Green's function of the Raman model, we obtain for the sum of the positive impulses

$$1 + 2r^2(1 + r^2 + r^4 + \cdots) = 1 + \frac{2r^2}{1 - r^2} = \frac{1 + r^2}{1 - r^2}; \tag{8.49a}$$

and for the sum of the negative impulses, we obtain

$$-2r(1 + r^2 + r^4 + \cdots) = \frac{-2r}{1 - r^2}. \tag{8.49b}$$

Together, these lead to the finite value

$$\frac{1 + r^2}{1 - r^2} - \frac{2r}{1 - r^2} = \frac{1 - r}{1 + r}. \tag{8.49c}$$

There is, then, a final velocity:

$$v_\infty = \frac{F_\infty}{2Z} \frac{1 - r}{1 + r}, \tag{8.49d}$$

i.e., according to (8.45), this find velocity is

$$v_\infty = \frac{F_\infty}{2w}. \tag{8.49e}$$

(Note that in (8.49a–d) we are interested in the magnitude of r, whereas the minus sign was necessary in (8.45) in order to yield the string modes.) This corresponds exactly, however, to the loading of the two ends of the string with resistances w. The Raman model has the additional disadvantage that, in contrast to the string, it has no well-defined rest position.

This situation is not altered if we distribute the damping along the propagation path; in doing this, we only formally replace r by $e^{-\mu l}$. In both cases, the losses would also have to be frequency-dependent, at least at $\omega = 0$; here, $r = 1$, $w = \infty$, or $\mu = 0$.

Since, in our model of the bridge in section 5.4, this transition takes place at the division of forces between a resistance and a spring, we give the corresponding Green's function as a second example, at the bottom of figure 8.5.

The impedance consisting of the spring and resistance has no effect on the pulse at $t = 0$. Since we regard the nut as an ideal reflector, as we did in section 5.4, the pulse returns once, unaltered, from the nut. On the other hand, the reflection factor at the bridge is complex:

$$\underline{r} = \frac{Z - w - s/j\omega}{Z + w + s/j\omega} = \frac{-(1 - j\omega(Z - w)/s)}{1 + j\omega\tau}; \tag{8.50}$$

we use the time constant

$$\tau = \frac{w + Z}{s} \tag{8.51}$$

of the bridge from (5.23); there, it is called τ_0.

This time, however, we wish to simplify the reflection process even further by setting

$$w = Z. \tag{8.52}$$

Then

$$\underline{r} = \frac{-1}{1 + j\omega\tau}, \tag{8.53a}$$

and the time function of the reflected pulse becomes

$$-\frac{1}{\tau}e^{-t/\tau}. \tag{8.53b}$$

This "softening" of the pulse reflected from the bridge repeats over and over. The effect of the repeated reflections increases as the nth power of $\underline{r}$:

$$\underline{r}^n = \frac{(-1)^n}{(1 + j\omega\tau)^n}, \tag{8.54a}$$

and the corresponding time function is

$$\frac{(-1)^n}{n!}\frac{t^{n-1}}{\tau^n}e^{-t/\tau}. \tag{8.54b}$$

These pulse functions are shown in figure 8.5, bottom, above and below the abscissa; they soon begin to overlap. From this example, which incorporates losses at the bridge that increase with frequency, it can be seen how the pulses become more and more rounded off. As they do, the difference between the conditions of reflection at the bridge and at the nut becomes less and less significant. This is important, because the initial pulses are kinematically even less realizable than the steps of velocity and the step responses accounted for by the Helmholtz motion. This limitation, however, does not prohibit the formal application of pulse synthesis. The question of the appropriateness of the results obtained without considering the resistance of the string to sharp deformations, particularly its bending stiffness, depends on the results. But these need no longer contain any pulses or steps, since $F(t)$ ultimately no longer exhibits the properties of an ideal pulse or step function.

Furthermore, it is notable that the peak values of the pulses decrease more rapidly than the order of the reflection.

Finally, it can be seen that (8.48) applies to this Green's function; this must be the case for any system with a well-defined rest position. First, the two ideal pulses oppose one another. All further pulses of positive polarity appear twice more at half strength and of negative polarity. There is no need to carry out the tedious integration of (8.48); the sums of the areas under all positive and all negative pulses are equal. (The graphic representation in figure 8.5 has a slight inaccuracy: the reflected pulses are drawn as if the duration of the pulse excitation were infinitesimal. This inaccuracy has no effect on the mathematical conclusions.)

8.5 Periodic pulse synthesis

Up to now, we have been interested above all in the periodic phenomena of the bowed string. In whatever way the function of force $F(t)$ comes to approximate a periodic function, the decay of the pulse response with time allows us to assume that the variable to be found also ultimately becomes periodic, and in a way independent of the specific transitional process. For this reason, we can think of the periodic response as having been generated by a periodic force.

The force function can also be represented by "fence laths" of height $F(t')$ and width dt'; that is, pulses that, in contrast to those discussed in the previous section, repeat periodically. These generate a periodic Green's function. This function can be obtained by dividing the Green's function into sections

$\Delta g_1, \quad 0 < t < T,$

$\Delta g_2, \quad T < t < 2T,$

$\cdots$ (8.55)

$\Delta g_n, \quad (n - 1)T < t < nT,$

and superposing these:

$$k(t) = \sum_1^\infty \Delta g_n[t - (n - 1)T].$$ (8.56)

We are interested even in the simple case of this superposition in which the period of excitation is equal to the duration of a full round trip of a pulse on the string:

$$T = \frac{2l}{c}.$$ (8.57)

In figure 8.6, top, the transition from $g(t)$ to $k(t)$ is carried out for the example given in figure 8.5. The Raman model (figure 8.6, top) leads to a positive pulse—its height given by the sum in (8.49)—and two negative pulses, each of which is half as high as the smaller sum in (8.49b).

Once more, we can see that, unless additional conditions are stated, no definite rest position results; if there were one, then the relationship

$$\int_0^T k(t)dt = 0$$ (8.58)

would have to hold. This follows from (8.48) and (8.56), but also results simply, physically, from the fact that $k(t)$ must represent a periodically oscillating velocity without a constant term.

The situation is different when the impedance of the bridge consists of a resistance and a spring (figure 8.6, bottom). In this case as well, two ideal pulses occur; the one at the point of excitation at $t = 0$, and its reflection from the nut, which arrives after $2(l - x_b)/c$.

The result to be expected in summing the remaining series of reflections would be some sort of a rounded pulse. Instead, the sums to be calculated

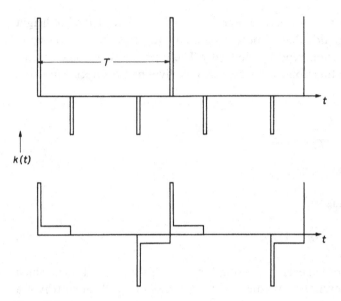

Figure 8.6
Examples of periodic Green's functions. Top, for the Raman model. Bottom, for a spring and resistance in parallel with respect to force, but only at the bridge.

for $t = 0$, $t = 2x_b/c$, and $t = 2(l - x_b)/c$,

$$\frac{e^{-t/\tau}}{\tau}\left(1 + \frac{t}{\tau} + \frac{1}{2!}\left(\frac{t}{\tau}\right)^2 + \frac{1}{3!}\left(\frac{t}{\tau}\right)^3 + \cdots\right) = \frac{1}{\tau} \tag{8.59}$$

result in step functions which increase by $2/\tau$ at $t = 0$ and decrease by $1/\tau$ at $t = 2x_b/c$ and at $t = 2(l - x_b)/c$. Taking into consideration the additional condition in (8.58), we obtain the function $k(t)$ shown in figure 8.6, bottom. If $\delta(t)$ is a unit impulse and $\sigma(t)$ a unit step, this function can be represented analytically as

$$k(t) = \delta(0) + \frac{1}{\tau}\left(\sigma(0) - \sigma\left(\frac{2x_b}{c}\right)\right)$$

$$- \delta\left(\frac{2(l - x_b)}{c}\right) - \frac{1}{\tau}\left(\sigma\left(\frac{2(l - x_b)}{c}\right) - \sigma\left(\frac{2l}{c}\right)\right). \tag{8.60}$$

The periodic pulse synthesis always takes the form

$$v(t) = \int_0^T k(t - t')f(t')dt'. \tag{8.61}$$

This convolution integral in the time domain corresponds to a vector equation in the frequency domain, valid for any partial tone of order n:

$$\underline{v}_n = \underline{A}_n \underline{F}_n = 2Z\underline{A}_n \cdot \underline{f}_n. \tag{8.62}$$

In this formula, $\underline{A}_n$ is the admittance of the string at the point of excitation, when excitation is periodic (not a single impulse), for the nth partial. The expression $2Z\underline{A}_n$, then, represents the vector of the nth partial tone of $k(t)$.

Since it can often be simpler to define $k(t)$ by this second procedure, we will go through it step by step, confirming the solution we derived in the time domain.

To determine the complex admittance $\underline{A}(\omega)$, we first observe separately the impedances of the two segments of the string on either side of the point of excitation; these impedances will ultimately be added together. We express velocity in terms of two components. One propagates from the point of excitation (here $x = 0$) to the bridge (here $x = -x_b$). We express this as $\underline{v}_+ e^{jkx}$. The other component returns from the bridge; we express this as $\underline{v}_+ \underline{r}_0 e^{-jkx} \cdot e^{-j2kx_b}$ (k with the factor j always designates the wave number). We treat force in exactly the same way, noting that $\underline{F}_+ = Z\underline{v}_+$, but that $\underline{F}_- = -Z\underline{v}_-$. In this way, we obtain the impedance at the point of excitation ($x = 0$) for this segment:

$$\frac{\underline{F}(-0)}{\underline{v}(-0)} = Z\frac{1 - \underline{r}_0 e^{-2jkx_b}}{1 + \underline{r}_0 e^{-2jkx_b}}. \tag{8.63a}$$

In exactly the same way, for the segment between the point of excitation and the nut:

$$\frac{\underline{F}(+0)}{\underline{v}(+0)} = Z\frac{1 - \underline{r}_l e^{-2jk(l-x_b)}}{1 + \underline{r}_l e^{-2jk(l-x_b)}}. \tag{8.63b}$$

Since the input force of both segments is to be added, the two components of the impedance must be added together. The inverse of this sum is the admittance $2\underline{A}Z$ to be found:

$$
\begin{aligned}
2\underline{A}Z &= \frac{2}{\dfrac{1 - \underline{r}_0 e^{-2jkx_b}}{1 + \underline{r}_0 e^{-2jkx_b}} + \dfrac{1 - \underline{r}_l e^{-2jk(l-x_b)}}{1 + \underline{r}_l e^{-2jk(l-x_b)}}} \\[2ex]
&= \frac{1 + \underline{r}_0 e^{-2jkx_b} + \underline{r}_l e^{-2jk(l-x_b)} + \underline{r}_0\underline{r}_l e^{-2jkl}}{1 - \underline{r}_0\underline{r}_l e^{-2jkl}}.
\end{aligned} \tag{8.64}
$$

If we consider a periodic excitation whose partial tones satisfy

$$k_n l = 2\pi l/\lambda_n = n\pi, \tag{8.65}$$

then (8.64) simplifies to

$$2\underline{A}_n Z = \frac{1 + \underline{r}_0 e^{-2jk_n x_b} + \underline{r}_l e^{-2jk_n(l-x_b)} + r_0 r_l}{1 - \underline{r}_0 \underline{r}_l}. \tag{8.66}$$

This expression is valid for any reflection factor; with the particular reflection factors

$$\underline{r}_{0n} = \frac{-1}{1 + j\omega_n \tau} \tag{8.67a}$$

and

$$\underline{r}_{ln} = -1, \tag{8.67b}$$

it takes on the specific form

$$2\underline{A}_n Z = \frac{1}{j\omega_n \tau}\left\{ (1 + j\omega_n \tau) - e^{-2jk_n x_b} - (1 + j\omega\tau)e^{-2jk_n(l-x_b)} + e^{-jk_n l} \right\}$$

$$= 1 + \frac{1 - e^{-2jk_n x_b}}{j\omega_n \tau} - e^{-2jk_n(l-x_b)} - \frac{e^{-2jk_n(l-x_b)} - e^{-2jk_n l}}{j\omega_n \tau}. \tag{8.68}$$

Each term of (8.68) describes a spectral component. In the numerator of this formula, we have made use of (8.64). Thus each term in (8.68) corresponds to a term in the function $k(t)$ given by (8.60).

8.6 Accounting for torsion in pulse synthesis

As we already have discussed in sections 6.2 and 8.3, the excitation force is exerted on the surface of the string, at a distance a from the center. In addition to the transverse velocity v_S at the center of the string, an angular velocity w is generated, so that the velocity v_R at the point of excitation consists of two components:

$$v_R = v_S + aw. \tag{8.69}$$

In pulse synthesis, both components are excited by the same force F; when the problem is analyzed in terms of sinusoidal tones, this corresponds to the addition of two admittances

$$\underline{A}_R = \underline{A}_S + \underline{A}_T, \tag{8.70}$$

where the subscript T once more designates the torsional component.

It follows from this that the functions $k(t)$ in the time domain are composed of corresponding components:

$$k_R(t) = k_S(t) + k_T(t); \tag{8.71}$$

the same is true of the Green's functions generated by a single-pulse excitation,

$$g_R(t) = g_S(t) + g_T(t). \tag{8.72}$$

Also, both components have the same structure. The functions k_S and g_S remain the same as k and g introduced earlier, since here, too, we normalize the excitation force to the point-impulse impedance for transverse waves, $2Z$, as shown in (8.43). In this way, we retain the same meaning for f.

It is not self-evident that we should do this; the point-impulse impedance for torsional waves, expressed here as F/aw, is not the same [(see (6.19)], but rather

$$\frac{F}{aw} = \frac{1}{a^2}\frac{Fa}{w} = \frac{1}{a^2}2\Theta'c_T = \frac{2Z}{2\zeta}. \tag{8.73}$$

From this, it follows that the first pulse generated at the point of excitation retains its value of $\delta(0)$ in $g_S(t)$; but in $g_T(t)$, this pulse and all succeeding ones are smaller by a factor of 2ζ. It is always true that aw, generated through such excitation, is smaller than v_S by this factor. Besides, the reflection factors for torsional waves are different from those for transverse waves. It should be stressed that there are still no experimental results relating to reflection factors for torsional waves. Furthermore, the density of reflected pulses is greater in proportion to their propagation speed; and finally, their round-trip time $2l/c_T$ is generally unrelated to the period T, which is, in the case of the bowed string, based on the round-trip time of transverse waves.

In earlier sections, we discussed the transformation of transverse waves into torsional waves and vice versa. Taking this into consideration, we might be surprised that the Green's functions are completely independent of one another. But each of them alone incorporates no reflection from the bow, as long as the force at the bow is taken to be determined externally and to be unaffected by the velocities generated in the string.

This independence would not hold if we could establish that reflection of the transverse and torsional waves at the bridge, nut, or finger were accompanied by significant transformation of one type of wave into the other. To be sure, the string lies in a notch at the bridge and the nut, but it can still fit loosely enough to allow a certain amount of rolling motion at these boundaries. The finger, too, cannot be seen as absolutely preventing such motion.

8.7 Integral equation of the bowed string

Only if we regarded the force in the integral (8.44) as being influenced by feedback from the velocity on the left, does our treatment correctly represent the basic attributes of a self-sustained oscillation. We then obtain a Volterra's integral equation

$$v = \int_0^t g(t - t')f(v(t'))dt'. \tag{8.74}$$

We know that the torsional waves are important, in particular, for the fate of the secondary waves; we will therefore incorporate torsional waves in of our analysis from this point on. We will understand v to mean v_R; similarly, we will replace g with g_R, so that

$$v_R = \int_0^t g_R(t - t')f(v_R(t'))dt'. \tag{8.75}$$

Only now can we account for reflections from the bow and for the transformation between the two types of waves; though the individual processes will be much less clear than they were in the signal-flow diagram of figure 8.4.

In the model diagramed in figure 8.4, a group of waves arriving at the bow always produces a new generation of waves leaving the bow. These, modified and delayed by their trip to and from the bridge or nut, return to the bow. But now there is only one series of waves, beginning at $t = 0$ and never interrupted. These waves propagate back and forth between the bridge and the nut; at the bow, they are modified in proportion to the force which they themselves generate. It can even be said that all that matters are the resulting quantities, without any question about their origin or destination.

The equations (8.12), (8.13), (8.29), and (8.30), and their synthesis in

(8.33), conveys this same message. When the normalized force f is incorporated into (8.33), that equation takes on the form

$$v_R = (1 + 2\zeta)\delta f + \sum v_k + a \sum w_k. \tag{8.76}$$

If we compare this equation to (8.75) and note that f and δf take on the same role in both, we see that the first term on the right in (8.76) can be separated in (8.75) as well: we rearrange the Green's functions $g_S(t)$ and $g_T(t)$, which together give $g_R(t)$, to put the first terms $\delta(0)$ and $2\zeta\delta(0)$ at the beginning. We then obtain

$$v_R = (1 + 2\zeta)f + \int_0^t g_R'(t - t')f(v_R(t'))dt', \tag{8.77}$$

in which we introduce the truncated Green's function g':

$$g_R'(t) = g_S'(t) + g_T'(t)$$
$$= [g_S(t) - \delta(0)] + [g_T(t) - 2\zeta\delta(0)]. \tag{8.78}$$

These functions can also be generated using the signal-flow process in figure 8.4. To do this, it is necessary only to introduce a single pulse at the position of the bow in the program, and then to bypass the functions representing the bow. The signal paths for v_S and aw then remain separate until v_R is calculated and displayed. If the initial pulse is also displayed, the result is the original Green's function; otherwise, it is the truncated Green's function.

Since the second term on the right side of (8.77) has the dimension of a velocity, McIntyre and Woodhouse (1979) designated it as

$$v_h = \int_0^t g_R'(t - t')f(v_R(t'))dt', \tag{8.79}$$

because this term is related only to the "past history" of the function of force. This is a better expression than "arriving wave," since this function includes a constant term that cannot correspond to a constant velocity, which cannot exist in a system with a defined rest position.

The form (8.77) of the integral equation allows us to proceed from the physical assumption we made in section 8.3, namely, that it is appropriate to separate the overall force f into a constant term and a time-varying term:

$$f = f_- + \delta f(t). \tag{8.80}$$

(The subscript "−" here means "unidirectional.")

In section 8.3, we took the constant term to be the minimal sliding-friction force, as derived from Δv. This gave us a justification for concerning ourselves only with the additional force of sliding friction and to regard $\delta R = 0$ as the abscissa of all interesting characteristics.

This simplification is permissible only when the system is in a periodic or nearly periodic state. If we are interested in tracing the bowing process starting from the rest condition, f as well as δf is 0 at first. This transition, however, results if we replace $\sum v_k + a \sum w_k$ by the integral at the right in (8.77). If, namely, we insert the sum (8.80) there, we obtain two integrals:

$$f_- \int_0^t g_R'(t - t')dt' + \int_0^t g_R'(t - t')\delta f(v_R(t'))dt', \tag{8.81}$$

The second of these leads to the sum $\sum v_k + a \sum w_k$ mentioned earlier. This represents the incident wave, designated by v_i.

The first integral approaches the value

$$f_- \int_0^\infty g_R(t)dt = f_- \int_0^\infty g_R(t)dt - (1 + 2\zeta)f_-. \tag{8.82}$$

According to (8.48), however, the first term disappears in a system with a defined rest position. Equation (8.77) then takes on the form

$$v_R = (1 + 2\zeta)\delta f + \int_0^t g_R'(t - t')\delta f(v_R(t'))dt'$$
$$= (1 + 2\zeta)\delta f + v_i, \tag{8.83}$$

which we used in section 8.3. There $\delta f = \delta R/2Z$ corresponded to the increase in friction above the minimum value $R_{min}/2Z = f_{min}$ to be expected at the greatest relative velocity between the bow and the string, whose value Δv is given by the bowing speed and the point of bowing. The partition of f into a constant term and an increase, and the validity of the following equations is, however, not dependent on the equality of f_- and f_{min}. It is possible to choose δf such that it represents only a pure oscillating component, and f_- such that it represents the arithmetic average of the periodic function $f(t)$, as is done in Fourier analysis. However, an approximation of f_{min} is given a priori by the Helmholtz motion, while $\overline{f(t)}$ is obtained only from the solution.

8.8 Transients of the bowed string[6]

Up to this point, we have discussed only the periodic solution; first, in terms of the round trip of the rounded corner (section 5.3) and later, in terms of the secondary waves as well (figures 8.2 and 8.4). From the beginning, McIntyre and Woodhouse (1979), on the other hand, incorporated in their analysis the transients which always precede the periodic state, by examining f instead of δf and v_h instead of v_i.

It is well known that the listener's ability to distinguish different musical instruments depends less on the steady-state periodic time dependence of the sound pressure than it does on the transients.

To acquaint ourselves with the specific attributes of the transients occurring in the bowing process, we will look into the simplest possible example: a string supported rigidly at both ends. If we neglect all losses in a forced oscillation with sinusoidal excitation, then the displacement would increase without limit at a resonant frequency. But this does not happen when an oscillator is excited by friction. The restricted velocity during capture prevents the amplitude from exceeding certain limits.

We may further simplify our model, as we did in section 1.1, by assuming that the sliding friction is not only smaller than the maximum friction but also independent of the relative velocity of string and bow.

Friedländer (1953), Keller (1953), and Woodhouse (1977) chose a point of bowing in the middle in order to describe the behavior as easily as possible. Bowing in the middle makes possible a symmetrical periodic solution, in contrast to the Helmholtz motion to be expected in the case of off-center bowing. The symmetrical solution corresponds to the superposition of two symmetrically opposed Helmholtz motions, to which is added the symmetrical, triangle-shaped displacement due to the sliding-friction force.

Motion during the first half-period is the same as if we pluck the string in the middle, with the bow in sliding contact with the string (see figure 3.17, top). The plucked string, however, does not return to the initial

[6] This section, new to the English-language translation of the book, was suggested by correspondence with M. E. McIntyre. He encouraged the author to extend his signal-flow diagrams to transients, corresponding to McIntyre's and Woodhouse's simulation of the bowed string by a computer. Most of the text is adapted from a paper presented by the author at the meeting of the Catgut Acoustical Society, April 1982, De Kalb, Illinois (Cremer 1982).

triangular shape after a full period T. Because of the sliding friction, which always opposes the string's motion, the triangular envelope decreases from period to period. In the case of the bowed string, the frictional force is always exerted in the same direction, and in the periodic case its amount is constant. Consequently, the motion during the interval of sticking friction repeats that of the half-period when the string is sliding, but in the opposite direction and in the opposite sequence.

If we try to obtain this periodic motion by bringing the bow up to a constant velocity v_{b0}, after the string has been at rest at $t = 0$ (see figure 8.7a), the behavior is at first entirely different. At $t = T/8$, the displacement of the string consists of a triangle extending over only half the length l. The ends of the string are at rest as long as the propagating waves generated by the excitation have not yet arrived there. In figure 8.7a, we indicate those parts of the string that are in motion by means of short vertical lines on the side toward which the string is moving. The lengths of the lines are proportional to the magnitude of the velocity.

At $t = T/4$, the waves reach the ends of the string and reflection begins; segments of the string are now at rest (see the drawing for $t = 3T/8$) and form angles of $2v_{b0}/c$ with the abscissa, i.e., the rest position of the string.

At $t = T/2$, the whole string is again at rest, not as a straight line but as a triangle. But this situation occurs once each period in the periodic solution, and could mark the onset of such a solution, under two conditions:

1. We must assume that the normalized frictional force necessary to move the string farther in the direction of the bow, here

$$f_1 = \frac{F_x}{2Z}(6v_{b0}/c) = 3f_0 \qquad (8.84)$$

is larger than the maximum frictional force, so that the bow releases the string.

2. We must assume that, at least after the next half period, the magnitude of the sliding velocity of the string equals the speed of the bow, which is now v_{b1}. The necessary value of this velocity results from the situation during the next $T/4$; the string will approach a situation in which it will have only its static displacement, given by the force of sliding friction, normalized to f_g. We therefore obtain

$$v_{b1} = \frac{2v_{b0} - f_g}{c} \frac{l/2}{T/4} = 2v_{b0} - f_g. \tag{8.85}$$

Since f_g is proportional to the bowing pressure, it is theoretically possible for v_{b1} to equal v_{b0}; but it would be very unlikely for the bowing pressure to have been adapted to the bow's velocity from the onset so that f_g equals v_{b0}. Nonetheless, we have chosen f_g to be equal to $v_{b0}/2$ in figure 8.7a. By changing the bowing speed (in our example $v_{b1} = 3v_{b0}/2$), we obtain a periodic solution after a delay of only one-half period. (Alternatively, we could have adjusted the bowing pressure.)

If we retain the original bowing speed, we get the behavior shown in figure 8.7b, which is extended as far as $t = 9T/4$. Until $t = T$, the situation is the same for all corresponding values of n in $nT/4$. The situation becomes different, however, at subsequent steps of $nT/4$. It is astonishing that the situation at $t = 2T$ is the same as at $t = 0$; this means that a subharmonic of the expected period is generated, as Friedländer (1953) noted. The subharmonic occurs at the very beginning of the bowing process, without any transient.

What we learn from these two simple examples is that the duration of the bowed string's transient depends on changes in bowing speed and bowing pressure, which are up to the player. It is quite normal for the player to increase bowing speed during the transient, though certainly not stepwise by 50% and after only one half-period.

But our examples also give insight into the character of the "past-history" velocity v_h. This is plotted horizontally in figure 8.7, along with the displacement η, the velocity of the string v at the point of bowing, and the normalized friction force f; time increases toward the bottom.

Following (8.77) and the definition of v_h in (8.79), and excluding torsion (i.e., with $\zeta = 0$ and the subscript R deleted), it is easy to see that

$$v_h = v - f. \tag{8.86}$$

The velocity v_h can also be derived from the integral (8.79).

In our simple example in figure 8.7, the string is bowed in the middle and the ends are regarded as ideal reflectors; consequently, it is clear that $g'(t)$ consists of a train of Dirac pulses whose strength is 2, and with a period of $T/2$. Two pulses always arrive simultaneously; the signs alternate, starting with -2. Furthermore, as the dependence of the force on time changes in a known way at intervals of $T/2$, v_h could be

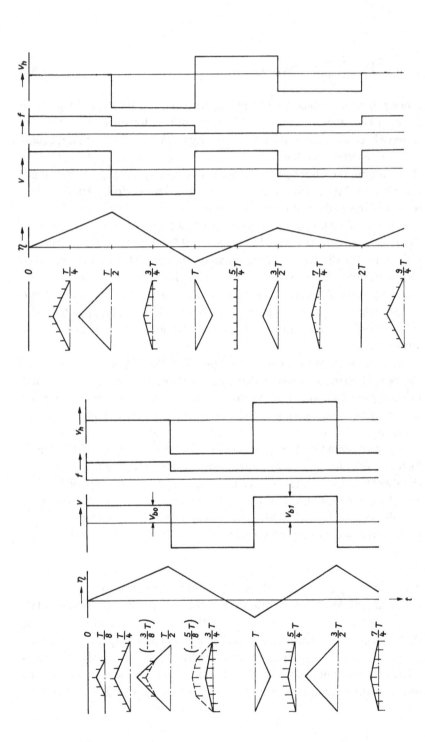

Figure 8.7
Displacement of a string initially at rest at different times during bowing. Time increases downward. At the right are shown functions of displacement, velocity, frictional force and η at $x = x'$ with changing bowing speed (a, left) and with constant bowing speed (b, right).

obtained by a simple summation of all reflections that arrive at the same time each period.

One general attribute of v_h can be seen in figure 8.7a, our simplest example. The velocity of the string has no average value, once the solution has become periodic; otherwise the string would move away from its supports at the ends. In contrast, v_h has an average value, which it needs to compensate the average, i.e., static, value of the force f. This demonstrates that v_h is not a velocity, which may be separated physically and measured; rather, it is an auxiliary term with the dimensions of velocity.

A similar step-by-step procedure can also be used to trace the onset of bowing away from the middle of the string (McIntyre, Schumacher, and Woodhouse 1983). This procedure is much more complicated, however; it can even be seen that the two corners excited by waves propagating in opposite directions from the bow, and then reflected back and forth from the ends of the string, can never lead asymptotically to the one-corner Helmholtz motion without the introduction of losses. Although it would be possible to introduce such losses in the detailed tracing of the individual waves, such calculation would be very difficult, even for the simple bridge model of section 5.4. Again, computer simulation makes it comparatively easy to solve for the transient (McIntyre and Woodhouse 1979).

Figure 8.8 shows the signal-flow diagram for such a procedure. This should be compared with figure 8.2, in which we also neglect the influence of the torsional waves. The boxes at the left and right (this time at the bottom) again represent the reflections at the bridge and the nut or finger. We have changed the convolution integrals by introducing the impulse responses $\varphi_{0,l}$ (replacing the step responses $\psi_{0,l}$) to agree better with McIntyre's and Woodhouse's procedure; this change is not, in fact, necessary. The lower limit of the integrations here designates the moment at which bowing begins.

Correspondingly, the sum of the reflected waves v_{h0} and v_{hl} (instead of v_{k0} and v_{kl}) now is the same as the past-history velocity v_h which enters the box representing the bow (here at the top) and governs the frictional force f. In this box, the switching between sticking and sliding is replaced by Friedländer's (1953) graphic solution for v and f at a given v_h. If we replace δf with f, we can, in any case, no longer approximate the sliding-friction characteristic by a straight line; a straight line could intersect

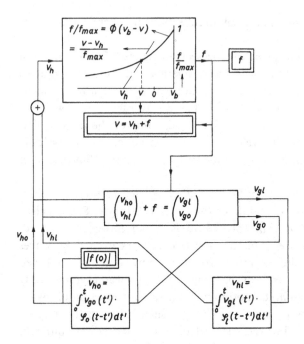

Figure 8.8
Signal-flow diagram for computer simulation including transients of a bowed string not
accounting for torsion, using the procedure of McIntyre and Woodhouse.

the abscissa. Then it would include negative values of sliding friction that
would work in the direction of the relative velocity; such values are
physically wrong. Furthermore, we have to express the characteristic
here by its deviation from the maximum value f_{max}. If, as Friedländer
(1953) does, we normalize the force f once again by dividing by f_{max},

$$f/f_{max} = \Phi(v_b - v) \tag{8.87}$$

and correspondingly, instead of (8.86), use

$$f/f_{max} = (v - v_h)/f_{max}, \tag{8.88}$$

the characteristic is then independent of the bowing pressure. The force
f_{max} is proportional to the bowing pressure, and the change influences
only the slope of the straight line given by (8.88). Consequently, it is
not only possible but rather easy to change the bowing pressure during
the simulation of a transient. The same is true of the bowing speed. All

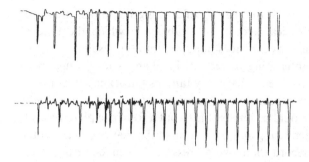

Figure 8.9
Example of the onset of the bowing process (martelé stroke). Top, simulated on a computer, with reflection-free torsional waves; bottom, recorded electrodynamically on an instrument (after McIntyre and Woodhouse).

that is needed is a shift in the origin of the abscissa, since the point $v = v_b$ is moved. This change, too, is easily accomplished during the simulation of a transient.

Again, the easiest condition to define for a transient would be constant bowing speed and bowing pressure. But as McIntyre and Woodhouse (1979) report, several hundred periods may elapse before a periodic solution is approached. They reached this conclusion when they tried to get the same result as Schumacher (1979), who, however, solved the integral equation in another way, restricted to periodic motions.

As McIntyre, Schumacher and Woodhouse (1983) pointed out, these long transients result from neglecting the torsional waves; when transverse waves are transformed into torsional waves, the transients are much shorter.

When these researchers decreased the parameter of bowing pressure over the first dozen periods, they were able to simulate what is called a martelé bowing. Simulated and hand-bowed martelés are compared in figure 8.9. Here, the duration of the transient was only about 12 periods. At the stated frequency of 230 Hz, this corresponds to a transient time of 51 ms, shorter than the shortest duration of a musical note. It is possible that longer transient times occur in legato playing. But it is not only the style of playing that leads to differences; these also depend on the experience of the player, whether master or novice. Since only 1/20th of a second is available for the manipulation of bowing pressure and bowing speed, this can be accomplished only unconsciously. Therefore, it is more difficult to teach the playing of a string instrument than, for example, to teach physical acoustics, a subject in which the time of understanding and reproducing does not enter into the results.

It would be very interesting to record the changes in bowing speed and pressure during transients played by famous musicians. It would be easy to record bowing speed. Bowing pressure has not been successfully recorded thus far. It would be impossible to put a Reinicke piezo receiver (see figure 9.4) between string and bow. It might be that one in the notch of the bridge could record the bowing pressure as a time-average value of the component of force toward the body.

Without such information, and thus without the ability to define a specific adequate playing behavior during the transient, we may not be able to adequately define the "response" of different instruments.

8.9 Periodic integral equation

Just as Volterra's integral equation supplanted pulse synthesis (8.44) as we passed from forced to self-excited oscillation, we now obtain Hammerstein's integral equation

$$v(t) = \int_0^T k(t - t')f(v(t'))dt', \tag{8.89}$$

instead of the periodic pulse synthesis (8.61).

Besides examining Hammerstein's equation thoroughly with regard to the problem of the bowed string, Schumacher has also applied it to organ pipes (1978a) and the clarinet (1978b). Following Schumacher, we use k to represent the kernel of this integral equation.

The problem of self-excited oscillations may be elegantly formulated in this equation, but there is a catch to this elegance. Namely, the period is no longer known *a priori*, but appears as an eigenvalue, which can be determined only by solving the equation. For this reason, and in contrast to the situation with pulse synthesis and forced oscillation, the kernel is known at best as an approximation. Iteration is necessary in order to proceed from this first approximation to one that gives the fundamental frequency of the generated tone with sufficient accuracy. The fundamental frequency, to be sure, is of great musical interest, as the sense of hearing is very sensitive to small changes in frequency.

Above and beyond this, the bowing process presents the unusual difficulty that f is not a function of v in the regime of sticking friction. An unambiguous functional relationship is necessary in order to apply (8.89); Schumacher at first hoped to arrive at such a relationship by accounting for the compliance of the bow hairs, certainly a real attribute of the system, in terms of a longitudinal point impulse impedance (Schumacher 1974); we discussed this in section 6.3. There, we showed that the torsion of the string brings about an effect of the same type, but a far greater one; in this, we were also repeating the findings of Schumacher (1979). Therefore, in what follows, we will be satisfied to account only for torsion; i.e., we discuss (8.89) in the form

$$v_R = \int_0^T k_R(t - t')f(v_R(t'))dt'. \tag{8.90}$$

When, however, we correctly express the friction force physically as a function of the relative velocity at the point at which the bow is in frictional contact with the string, we note again that no such function exists in the regime of sticking friction. Even if we separate v_R into $v_S + aw$ and, similarly, k_R into $k_S + k_T$, as in (8.71), we obtain

$$v_R = v_S + \int_0^T k_T(t - t')f(t')dt'; \tag{8.91}$$

in other words, all of the delayed reflections also must be included. Only if we assume that the reflections of the torsional waves are negligible would we be able to restrict $k_T(t - t')$; we would then retain only the first term of $g_T(t - t')$, which represents the value $2\zeta\delta(0)$. Only then would it be possible to substitute $2\zeta f$ for aw; instead of the relationship $f(v_R)$, we would then obtain another unambiguous relationship, $f(v_S)$. We have already used this in section 6.2, in our investigation of the pulse process at the point of bowing; this approach let us ignore the reflections of the torsional waves, and led to the substitution shown in figure 6.2 of a friction characteristic with a sloping segment representing sticking friction, instead of a vertical segment.

A corresponding assumption about the omission of reflected waves would also have been necessary in our treatment of longitudinal waves in the bow hairs, if these waves had been incorporated in an unambiguous function $f(v_S)$.

Schumacher succeeded in his intentions, however, and without any restriction, by representing the friction force as a function of the velocity

v_h introduced in (8.79). This velocity is better incorporated here through the definition

$$v_h = \int_0^T k_R'(t - t')f(t')dt', \tag{8.92}$$

which is equivalent to (8.79) as we go over to the periodic solution. Here, k_R' differs from k_R in exactly the same way as g_R' differs from g_R, in that the impulse response at the point of excitation, $(1 + 2\zeta)\delta(0)$, is omitted:

$$k_R'(t) = k_R(t) - (1 + 2\zeta)\delta(0). \tag{8.93}$$

Thus there does, in fact, exist between v_R and v_h a relationship in which the only other variable is f:

$$v_R = (1 + 2\zeta)f + v_h \tag{8.94}$$

(We have already discussed this relationship in (8.86) without torsion, i.e., for $\zeta = 0$.) Here, the vertical segment representing sticking friction becomes a straight line of finite slope:

$$f = \frac{v_b - v_h}{1 + 2\zeta}. \tag{8.95}$$

But also, in the regime of sliding friction, it follows from this that we have only to shift a point corresponding to a given $f(v_R)$ by $(1 + 2\zeta)f(v_R)$ to the left in order to respresent f as a function of v_h.

This transformation is illustrated by the change from solid to dashed curves in figure 8.10. Here, we have assumed that more than one value of $f(v_h)$ corresponds to a given bowing pressure; but this problem leads only to the steps of the flattening effect, which appear as vertical, dotted lines if we take v_h as the abscissa.

Our representation here differs in yet another way from the earlier ones. Previously, we always took only the variable part of the frictional force δR as the ordinate. In contrast, $\delta f = 0$ as well as $f = 0$ is included in figure 8.10 as a possible abscissa; and also, the difference between v_h and

$$\sum v_k + a \sum w_k = v_i, \tag{8.96}$$

already described in (8.83), is indicated. Here, we need only apply to k_R' what was said about g_R' in section 8.7. The separation of f into $f_{min} + \delta f$ allows (8.86) to be separated into

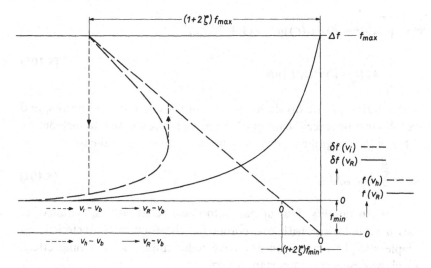

Figure 8.10
Transition from v_h to v_i as abscissae for $f = f_{min} + \delta f$.

$$v_h = f_{min} \int_0^T k'_R(t - t')dt' + \int_0^T k'_R(t - t')\delta f(v_R(t'))dt'. \qquad (8.97)$$

Here the second term corresponds once again to v_i. The first term can, in accord with (8.87), the definition of k'_R, be written as

$$f_{min}\left(\int_0^T k_R(t - t')dt' - (1 + 2\zeta)\right). \qquad (8.98)$$

Here, however, the integral goes to zero; just like the integral in (8.48), it represents a constant velocity. This is, however, impossible in a system with a defined rest position. Therefore, we obtain

$$v_h = v_i - (1 + 2\zeta)f_{min}, \qquad (8.99)$$

as in figure 8.10.

It makes no difference whether we solve the integral equation

$$v_h = \int_0^T k'_R(t - t')f(v_h(t'))dt' \qquad (8.100)$$

or the integral equation

$$v_i = \int_0^T k'_R(t - t')f(v_i(t'))dt' + (1 + 2\zeta)f_{min}$$

$$= \int_0^T k'_R(t - t')\delta f(v_i(t'))dt'; \tag{8.101}$$

both require that we divide the period into N equal time segments, and that N must be twice the order of the partial tone we wish to include.

From the integral equation, we obtain N equations of the form

$$v_m = \sum_{n=1}^{N} k_{(m-n)} f_n(v_n). \tag{8.102}$$

In order to simplify these at least into linear relationships, Schumacher uses a straight line with one corner for the friction characteristic. A simple straight line would not have represented the flattening effect, which was especially important to him.

Nonetheless, the difficulty of programming is definitely greater than in the simulation illustrated in figure 8.4. Furthermore, the form of the equations requires that it be known where capture and release lie. This makes it difficult to accommodate any solutions differing much from Helmholtz motion.

The greatest difficulty, however, is given by the fact already mentioned at the beginning of this section: namely, that the period is not a quantity known a priori, and that T changes the kernel. Therefore we must always start with the best possible approximation and then explore the effects of small changes.

Despite these difficulties, Schumacher succeeded in carrying this method through to the calculation of specific examples, for which he obtained the same results as did McIntyre and Woodhouse with their simulation method.

Figure 8.11 shows two examples from Schumacher (1979). In both cases shown, all conditions except the bowing pressure were the same. He used terminating impedances with complicated frequency dependences to set his boundary conditions; details are in his paper. Schumacher chose 0.2 as the value of 2ζ and $1/(1 + \pi)$ as the value of c_S/c_T. This latter value is intentionally irrational, but may be difficult to realize when 2ζ (which Schumacher called G_r) = 0.2. The value of x_b/l was taken to be 0.18. At the top in figure 8.11 is shown the function v_S over time, which can be derived from v_h and f. At the bottom f/f_{max} is shown. At the left, the bow-pressure parameter is $\beta = 0.058$, barely above the

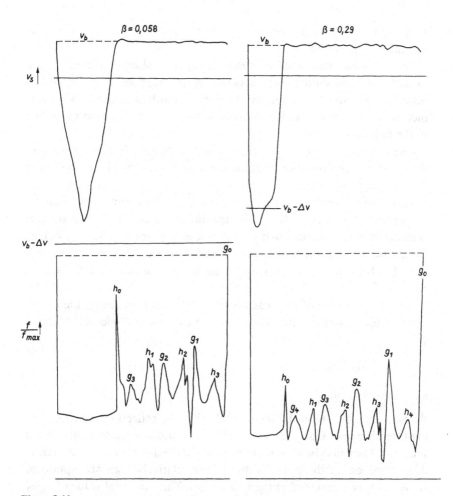

Figure 8.11
Top, velocity of the center of the string at the point of bowing. Bottom, frictional force.
Left, low bowing pressure. Right, high bowing pressure. Examples obtained on a
computer by means of the periodic integral equation (after Schumacher).

minimum. At the right, the value is five times as great, $\beta = 0.29$; this is a rather high value.

On first inspection, v_S show the steepening of the slopes and the decrease in the interval of sliding friction as bowing pressure increases; these two results were also obtained using the pure round trip method. What this method could not lead to was a decrease to below $(v_b - \Delta v)$ in the valley of the function.

The variations of velocity in the interval of sticking friction, increasing with bowing pressure, show that the torsional waves have been accounted for.

It is more interesting to examine the force functions when judging the power of this procedure. The maximum value of sticking friction is attained at release (denoted by g_0), not at capture (denoted by h_0), as the particular friction characteristic used here would seem to indicate. But this behavior is, in fact, even more pronounced at high bowing pressure.

The chosen value of l/x_b was not an integer; consequently, the reflections of the secondary waves generated at capture and release, indicated by

$$h_1, \quad h_2, \quad h_3, \quad h_4,$$

$$g_1, \quad g_2, \quad g_3, \quad g_4,$$

do not coincide. It can be seen clearly that the reflections generated at capture decrease from left to right, while those generated at release increase. The peaks become more rounded in the direction of the decrease. The sharpness of the peaks is due to the relatively high absorption of torsional waves assumed at the boundaries. The irrational value of c_S/c_T also insures that the impulses of torsion will have no systematic relationship to those of the transverse wave. Nonetheless, the torsional waves have a clear effect on the decay of the transverse wave. This effect would be far from attainable with the values at which Schumacher sets their reflection factors for the bridge and the fingers.

If only the bowed string is under consideration, McIntyre's and Woodhouse's simulation procedure is preferable; it is not only easier to program and to grasp, but it can also account for transient processes.

8.10 Possible fluctuations in the length of the period

We have heretofore assumed that a periodic steadly state will eventually
be attained by a self-sustained system. We will now examine this assumption in the case of string instruments. As shown in sections 5.5–5.7, the
length of the period depends on the delay of release, and this depends
on the attainment of maximum sticking friction in each particular case:

$$R_{max} = \mu_{max}F_z;$$
 (8.103)

that is, it depends on the product of the maximum coefficient of sticking
friction and the bowing pressure.

The coefficient μ_{max}, however, depends strongly on the rosining of the
bow, which, despite all effort, is certainly not the same at all points. The
same may be true of the properties of the bow hairs.

As we discussed in section 5.6 and at the end of section 6.3, the bowing
pressure, too, is subject to uncontrollable variations, particularly due to
oscillations of the bow hairs perpendicularly to the string.

All of these effects lead to variations in the delay of release that occurs
with the real, bowed string, but that do not show up in the computer
simulations.

There are other effects, however, which might occur in a computer
simulation in which the threshold of release is always the same. These
are based on the possibility that the secondary waves might arrive at the
bow near the moment of release established by the combination of the
Helmholtz motion and the primary corrective wave. If l/x_b is not a small
integer (less than 10, say), and if c_T/c_S also is not a small integer, or if
it is irrational, the time of arrival of the secondary waves can vary from
period to period.

In order to investigate this question, the present author in 1973 directed
his co-workers Beldie and Hermes in recording the force at the bridge,
and its zero crossings, so as to determine the relative frequency of appearance of various periods. Accuracy was to 1.5 cents (1 cent = 1/100
semitone). A great number of trials were run (Cremer 1974, p. 130).

The result, shown at the top in figure 8.12, showed a width at half the
maximum of approximately 10 cents and a total width of approximately
40 cents. Similar results were also obtained by Cardozo and van Noorden
(1968). These results led the present author to the concept that these
small fluctuations in the period, which occur even when there is no finger
vibrato, contribute to the attractiveness of the sound of string instruments.

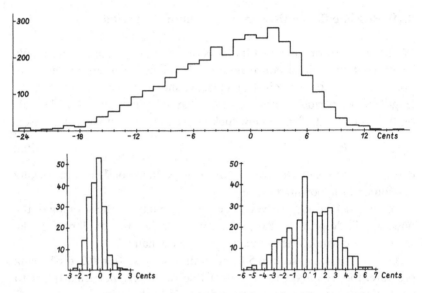

Figure 8.12
Measured frequency distributions of periods. Top, for a mechanically bowed string (after
Beldie and Hermes). Bottom, for a hand-bowed g string. Left, open g string. Right, at a
frequency of 715 Hz on the same string (after McIntyre, Schumacher, and Woodhouse).

McIntyre, Schumacher, and Woodhouse (1981) have contradicted this
generalization, which was derived from one example that established only
the fact of variations in the period. They obtained the much smaller
deviations in the period shown in figure 8.12, bottom. The violin string
was bowed by hand; the graph at the left represents an open g string,
and the one at the right a 715-Hz tone, fingered on the same string near
the end of the fingerboard. The explanation of the difference between the
two graphs is that the duration of the ramps should be the same in both
cases; but the duration of the period was less in the latter case. The
graphs, however, relate the variations in cents to the duration of the
period.

   We (the present author, Beldie and Hermes) believed it was necessary
to remove any individual influence from the variations by bowing the
string mechanically. Beldie (personal correspondence) now does not
exclude the possibility that the larger variations he observed might have
been due to irregular oscillations of the moving band that bowed the
string.

   On the other hand, McIntyre, Schumacher, and Woodhouse attempted to replace the time functions of force at the bridge around the zero points with straight lines, using the method of least squares. It is not certain that they eliminated only random errors.

But even if certain variations in the period are an essential part of the bowing process, it remains appropriate to seek an "average" periodic solution.

Looking back over this chapter, we can in any case state with satisfaction that it is possible today to deal mathematically with the difficult problem of the bowed string. Our mathematical tools are at this moment more highly developed than our physical knowledge of the boundary conditions at the nut, finger, and bridge. In order to develop such knowledge, we must now turn to a study of the bridge and of the body of the violin.

References

Cardozo, B. L. and L. P. A. S. van Noorden, 1968. *3rd Ann. Progr. Rep. Inst. v. Perceptie Onderzoek*, Eindhoven.

Cremer, L., 1974. *Acustica* **30**, 119.

Cremer, L., 1975. *Vorlesungen über Technische Akustik* [*Lectures on Technical Acoustics*]. Berlin: Springer.

Cremer, L., 1982. *Catgut Acoust. Soc. Newsletter* no. 38 (November).

Friedländer, F. G., 1953. *Proc. Cambridge Phil. Soc.* **49**, 516.

Hancock, M., 1977. *Catgut Acoust. Soc. Newsletter* no. 28 (November), 14.

Keller, J. B., 1953, *Comm. Pure Applied Math.* **6**, 483.

Lazarus, H., 1972. Dissertation, Technical University of Berlin.

McIntyre, M. E., R. T. Schumacher, and J. Woodhouse, 1981. *Acustica* **49**, 13 *and* **50**, 294.

McIntyre, M. E., R. T. Schumacher, and J. Woodhouse, 1983. *J. Acoust. Soc. Amer.* **74**, 1325.

McIntyre, M. E., and J. Woodhouse, 1979. *Acustica* **43**, 93.

Schelleng, J., 1963. *J. Acoust. Soc. Amer.* **35**, 328.

Schelleng, J., 1973. *J. Acoust. Soc. Amer.* **53**, 26.

Schumacher, R. T., 1974. Unpublished lecture, convention of the Catgut Acoustical Society, Mittenwald, July.

Schumacher, R. T., 1978a. *Acustica* **39**, 225.

Schumacher, R. T., 1978b. *Acustica* **40**, 298.

Schumacher, R. T., 1979. *Acustica* **43**, 109.

Wagner, K. W., 1919. *Arch. Elektrotechn.* **8**, 61.

Woodhouse, J., 1977. Dissertation, Cambridge University.

II THE BODY OF THE INSTRUMENT

9 The Bridge

9.1 Function and form of the bridge

The string must be regarded as thin in comparison with even the shortest wavelengths of interest to us, which amount to a few centimeters. For this reason, and as we will show in part III, the string is ineffective as a radiator of sound, and is therefore connected to a larger radiating surface, to which its oscillations are conducted. Even a tuning fork, much thicker than a string, sounds stronger, as is well known, when its foot is placed on a tabletop. This "reinforcement" is based on the extraction of a larger amount of power from the oscillating system. Energy from an outside source is, then, not "valved," as is the case, for example, with electronic amplifiers. The oscillating energy in the tuning fork is so great compared to the energy radiated in each cycle that the increased withdrawal of energy by the tabletop has hardly any effect on the oscillatory process. This is an advantage of low mechanicoacoustical efficiency.

In the same way, we can explain the ratio of power radiated from a violin to the mechanical-power input at the bow. Sivian, Dunn and White (1931) have given 6 mW as the maximum radiated power of the violin. In contrast, with a bowing pressure of 6×10^4 dyne, leading to an average frictional force F_{yb} of approximately 1.5×10^4 dyne, and at a high bowing speed of 100 cm s^{-1}, the power input to the bow is approximately 150 mW. These figures suggest an efficiency of only 4%. As we showed in section 3.6, the total power appearing in the denominator of our efficiency equations represents only about 10% of the total oscillating energy per cycle. This indicates that in each period only 0.4% of the oscillating energy is released as the power of interest, the radiated sound. However, since additional losses occur in the body of the violin, the total energy extracted from the string through the bridge is greater.

The bridge defines a boundary of the oscillating part of the string and transfers the transverse force of the string from one of the points where this force is at a maximum.

The shape of the bridge results from the requirement that the radiating surface not interfere with the motion of the bow; the radiating surface must, then, be more or less parallel to the bow. Since the bow must excite four different strings, this requirement cannot be fulfilled without restrictions. Bowing on the outer strings dictates the well-known waist-like narrowing of the body near the bridge (see figure 9.1). The particular

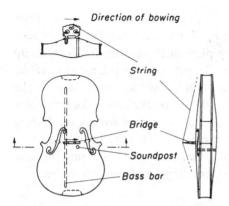

Figure 9.1
The bridge and the body of the instrument.

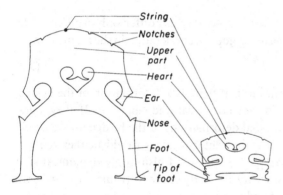

Figure 9.2
The cello bridge and the violin bridge.

shape, like almost all details of a string instrument, results not only from functional considerations, but also from esthetic ones.

Both functional and esthetic considerations contribute to the shape of the bridge (figure 9.2). The traditional proportions of the violin bridge and the cello bridge are very different; yet both bridges have the same task. The input force to the bridge F_{0y} and the corresponding velocity v_{0y} are essentially parallel to the top plate of the instrument; the bridge must turn these around to accomplish a power transfer perpendicular to the top plate (in the z-direction). This change can be accomplished only through rotational motions and moments of a body that can rotate in the y, z-plane. Conditions are favorable for the generation of motions in the z-direction when the moment $F_{0y} \cdot h$, where h is the distance between the notch of the bridge and the top plate of the instrument, is taken up by two feet separated by the distance $2a$, with a being as large as possible. These feet also serve to support the bridge. The "legs" are short on the violin (like those of a dachshund) and long on the cello (like those of a Great Dane). The difference arises, certainly, in part for reasons having to do with playing technique: the two instruments are held differently. On the violin, a higher bridge would bring the bow farther into the field of view of the eyes; on the cello, a lower bridge would restrict the freedom of bowing on the outer strings.

Statically, the forces on the two feet are equal and opposite:

$$F_{1z} = -F_{2z} = -F_{0y}\frac{h}{2a}. \tag{9.1}$$

But dynamically as well, their amplitudes can differ only insofar as the center of mass of the bridge undergoes an acceleration in the z-direction:

$$F_{1z} + F_{2z} = j\omega m v_{Sz}. \tag{9.2}$$

Such an acceleration can occur if there is a deformation of the bridge, because v_{1z} and v_{2z} are in fact not equal and opposite. If they were, and if we could neglect internal deformations of the bridge (as we surely can at low frequencies), then the center of mass would neither rise nor fall, and the relationship $F_{1z} = -F_{2z}$ would also apply dynamically. It would, however, be necessary for the body of the instrument to load both feet with the same input impedances.

The body of every stringed instrument appears outwardly to be symmetrical (see figure 9.1). This appearance is certainly intentional. However, as soon as we look underneath the top plate of the instrument, we observe that a wooden reinforcing bow, parallel to the long axis of the body, lies nearly under the left foot of the bridge (as seen from the position of the player of the violin). This is called the *bass bar*, probably because the lower-pitched strings are on the same side. Near the right foot, a short pillar leads from the top plate to the back plate; this is called the *soundpost*, probably because small changes in its position significantly affect the sound of the instrument.

Both the bass bar and the soundpost serve structural purposes as well, and were perhaps first introduced for structural reasons. The stretched strings exert a force on the top plate of the instrument of approximately 10 kp;[1] approximately half the free baggage allowance for an air traveler. This load is spread out over the top plate by the bass bar and by the sound-post to the back plate. In addition, however, these elements have an important acoustical purpose. The bass bar leads to in-phase excitation of the largest possible surface area of the top plate of the instrument; the soundpost, along with the *ribs* and the enclosed air, transmits oscillations to the back plate of the instrument.

Based on static considerations, it can be hypothesized that a higher impedance opposes the motion of the foot of the bridge near the sound-post. Helmholtz (1877, p. 147) even assumed that if "is the other foot alone which ... sets the elastic wooden plate into vibration." On account of Helmholtz's authority, this assertion has been repeated again and again by others. According to recent measurements by Moral and Jansson (1982), the velocity v_1 of the left foot is indeed greater than that of the right, v_2, at low frequencies. These authors have also verified, at higher frequencies (> 500 Hz, for the violin), the careful measurements of the point admittances at the feet of the bridge (see below, figure 9.17, bottom) that had been earlier carried out by several researchers (Eggers, 1959, Reinicke and Cremer 1970, Reinicke 1973). These measurements show that equal excitation forces can lead to velocities of the same order of magnitude at both feet: in other words, the body of an instrument is excited by both feet. The measurements also show that the velocities of the two feet need by no means be equal. The functions of frequency are very complicated; at some frequencies the motion of the left foot pre-dominates; at others, the motion of the right foot.

For this reason, the bridge cannot be considered to have only one input, the string notch, and one output at one of the feet or divided equally between the feet. Rather, the bridge must be regarded as having two outputs (see figure 9.3).

In communications engineering, the expressions "three-port" or "six-pole" are used to designate systems of this type. The meaning of the latter expression is, however, not intuitively clear when applied to mechanical systems.

The relationship between the three forces F_{0y} ($\equiv F_y$), F_{1z} ($\equiv F_1$), and

---

[1] The kilopond (kp) is the unit for kilogram force, in contrast to kilogram mass.

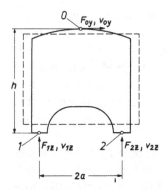

Figure 9.3
Input and output variables of the bridge considered as a three-port device.

F_{2z} ($\equiv F_2$) and the corresponding velocities v_y, v_1, and v_2 is linear. This relationship can be described simply by an impedance matrix linking a force vector to a velocity vector:

$$\begin{pmatrix} \underline{F}_y \\ \underline{F}_1 \\ \underline{F}_2 \end{pmatrix} = \begin{pmatrix} \underline{Z}_{yy} & \underline{Z}_{y1} & \underline{Z}_{y2} \\ \underline{Z}_{1y} & \underline{Z}_{11} & \underline{Z}_{12} \\ \underline{Z}_{2y} & \underline{Z}_{21} & \underline{Z}_{22} \end{pmatrix} \begin{pmatrix} \underline{v}_y \\ \underline{v}_1 \\ \underline{v}_2 \end{pmatrix}$$

$$= \begin{pmatrix} \underline{Z}_{yy} & -\underline{Z}_{y2} & \underline{Z}_{y2} \\ -\underline{Z}_{y2} & \underline{Z}_{11} & \underline{Z}_{12} \\ \underline{Z}_{y2} & \underline{Z}_{12} & \underline{Z}_{11} \end{pmatrix} \begin{pmatrix} \underline{v}_y \\ \underline{v}_1 \\ \underline{v}_2 \end{pmatrix}. \tag{9.3}$$

In the upper matrix, we have written out the double subscripts of the elements formally, so they appear in nine different ways. In the lower matrix, however, we have undertaken a reduction to only five separate elements (with the possibility of positive and negative signs), based on the physical properties of the bridge. This reduction follows from two general properties, without our having to know anything about the particular form of these five elements.

Because the system is linear, and no energy is introduced from inside it, we can write

$$\underline{Z}_{1y} = \underline{Z}_{y1}, \quad \underline{Z}_{2y} = \underline{Z}_{y2}, \quad \underline{Z}_{12} = \underline{Z}_{21}. \tag{9.4a}$$

The law that leads to this symmetry with respect to the principal diagonal is called the law of reciprocity.[2] It states, for example, that a forced motion at the point 0 with the velocity v_y will generate a force at the point 1 given by

$$\underline{F}_1 = \underline{Z}_{1y}\underline{v}_y \qquad (9.4\text{b})$$

if the feet of the bridge are supported rigidly (i.e., if $v_1 = v_2 = 0$). The amplitude and phase of this force are the same as those of the force

$$\underline{F}_y = \underline{Z}_{y1}\underline{v}_1 \qquad (9.4\text{c})$$

that would appear at the point 0 if we were to set $v_y = v_2 = 0$ and introduce an excitation at 1 with $\underline{v}_1 = \underline{v}_y$.

We can undertake a second simplification because we can consider the bridge to be symmetrical. This condition leads to the relationship

$$\underline{Z}_{11} = \left(\frac{\underline{F}_1}{\underline{v}_1}\right)_{v_y=v_2=0} = \underline{Z}_{22} = \left(\frac{\underline{F}_2}{\underline{v}_2}\right)_{v_y=v_1=0}. \qquad (9.4\text{d})$$

But also:

$$\underline{Z}_{y1} = \underline{Z}_{1y} = \left(\frac{\underline{F}_1}{\underline{v}_y}\right)_{v_1=v_2=0} = -\underline{Z}_{y2} = -\underline{Z}_{2y} = -\left(\frac{\underline{F}_2}{\underline{v}_y}\right)_{v_1=v_2=0}; \qquad (9.4\text{e})$$

in other words, equal and opposite forces must result at points 1 and 2 if we force a velocity v_y at the input to the bridge, as long as the feet are supported rigidly ($v_1 = v_2 = 0$). As (9.4e) shows, a positive $\underline{v}_y$ generates a force in the negative direction at point 1 and one in the positive direction at point 2, when these are rigidly supported. (The sign of $\underline{Z}_{12}$ cannot be determined without further assumptions.)

9.2 Measurement of the input impedance and the force transfer factor of a rigidly supported bridge

With the aid of circuit diagrams, it is usually possible to determine the frequency dependence of the individual elements of an impedance matrix in a multi-port electronic device. It is also possible, however, to determine the frequency dependence by measurement, using a given excitation for each of the field variables in turn, and examining the behavior of the other

---

[2] A treatment of the law of reciprocity based on energy considerations can be found in Cremer (1976).

variables; it must also be known whether the ports are open or short-circuited.

The field variables for other conditions of loading can be determined without reference to the physical basis for the measured functions of frequency. The bridge can be considered to be a black box (shown dashed in figure 9.3). Only the three ports at which power enters or leaves the bridge need be accessible.

In considering physical problems relative to the violin, the main goal of theoretical speculations and calculation is not, however, to arrive at precise predictions, but rather to demonstrate physical relationships. Consequently, we will not be satisfied with a black-box procedure. Rather, as soon as measurements at the ports pose questions, we will turn to the study of processes within the bridge. By studying the details of the bridge, we will also spare ourselves the trouble of measuring six different functions of frequency under boundary conditions some of which are very difficult to realize.

To the best of the present author's knowledge, Steinkopf (1963) was the first to study a bridge—the cello bridge—as a multi-port device. (Steinkopf made the assumption that the bridge could be regarded as a three-port device.)

Steinkopf is also the only researcher who suspended the cello bridge at its center of gravity, and horizontally. Using this approach, he could measure vanishingly small forces at its ports. He attached a coil of wire to one of the feet, and placed this in the magnetic field of an electro-dynamic transducer. He thus chose F_2 as the excitation; he kept the current constant. He measured v_y, by attaching a wire to the middle of the upper edge of the bridge and placing this in a magnetic field. The induced voltage, read on a vacuum-tube voltmeter, represented the magnitude of the admittance element a_{y2}. This appears in the admittance matrix derived by inversion of (9.3), giving

$$\begin{pmatrix} v_y \\ v_1 \\ v_2 \end{pmatrix} = \begin{pmatrix} a_{yy} & a_{y1} & a_{y3} \\ a_{1y} & a_{11} & a_{12} \\ a_{2y} & a_{21} & a_{22} \end{pmatrix} \begin{pmatrix} F_y \\ F_1 \\ F_2 \end{pmatrix},$$

$$= \begin{pmatrix} a_{yy} & a_{y2} & -a_{y2} \\ a_{y2} & a_{11} & -a_{12} \\ -a_{y2} & -a_{12} & a_{11} \end{pmatrix} \begin{pmatrix} F_y \\ F_1 \\ F_2 \end{pmatrix}. \tag{9.5}$$

To obtain the matrix in the second equation, we have once more made use of the law of reciprocity, the symmetrical structure of the bridge, and the physically appropriate signs for the elements.

It is physically clear that this admittance is determined at low frequencies by a combination of translational and rotational inertias; at low frequencies, it must first fall as ω^{-1}. What could not be forseen was that the admittance exhibits a maximum at 4,000 Hz, indicating the possibility of a natural resonance at that frequency. 4,000 Hz is, however, a frequency which may be regarded as in the upper range of hearing; this frequency is as far above the fundamentals of the cello as a frequency of 12,000 Hz would be above those of the violin.

The case in which the feet of the bridge are supported by a very massive body is more suitable for measurement: in this case, it is possible to assume that, approximately, $v_1 = v_2 = 0$. Steinkopf investigated the input impedance $\underline{Z}_{yy}$ under these conditions. He attached the top edge of the bridge to one end of a rod at the other end of which a coil of wire in the field of a magnet generated the excitation force, just as before. An accelerometer was attached to the rod in order to measure velocity. This apparatus had already been used by Eggers (1959) to measure the magnitudes and phases of various impedances of the body of the instrument. It was necessary for Steinkopf to determine the internal impedance of the generating device in order to separate it mathematically from that of the bridge. He was able to establish that the bridge worked as a stiff spring at low frequencies, with a reactance varying as ω^{-1}. A resonance occurred at higher frequencies in this case as well, as a minimum of Z_{yy}; with different bridges, this resonance occurred at 1,000, 1,500, and, in two cases, at 1,250 Hz.

The same resonance was later confirmed by Reinecke (1973). Figure 9.4 shows the measuring apparatus he used. He, too, excited the bridge electrodynamically, but he measured the forces in the string notch and under the feet by means of a piezoelectric sensor that he devised. He sensed transverse displacement at the upper part of the bridge using a capacitive probe to frequency-modulate a carrier wave above the audible range (see section 2.2). With this apparatus, he was able to conduct measurements at frequencies up to 10,000 Hz (see figure 9.5, top). He discovered another resonance at 2,000 Hz and the necessary antiresonance between the two resonances at approximately 1,500 Hz. When a resonance

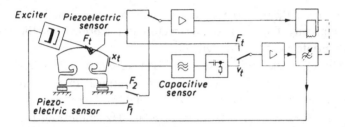

Figure 9.4
Experimental arrangement for measuring the input impedance and the force transfer
factor of a rigidly supported bridge (after Reinicke).

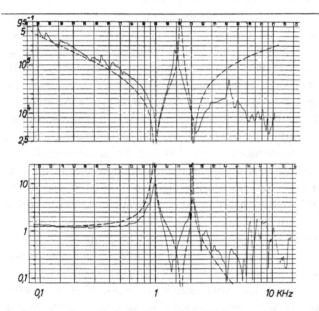

Figure 9.5
Measured (solid line) and calculated (dashed line) frequency curves. Top, magnitude of
the input impedance. Bottom, magnitude of the force transfer factor of a cello bridge
(after Reinicke).

is followed by an antiresonance and the force is given, the motion at the point of excitation must change sign. Therefore, if no losses are present, the motion must go to zero at some particular frequency. When there are losses, on the other hand, the phase changes continuously, and there is no frequency at which motion ceases entirely.

The frequency minima in $\underline{Z}_{yy}$ may be of importance in the oscillation of the strings. However, it is more important, in connection with the task of the bridge—the transfer of force from the string to the body—that the resonances express themselves as maxima of the force transfer factor

$$\ddot{u} = \left| \frac{F_1}{F_t} \right| = \left| \frac{F_2}{F_t} \right|, \tag{9.6}$$

which, as a measured function of frequency, is reproduced in figure 9.5, bottom. [The distinction between F_y and F_t, the force tangential to the bridge, will be made later in (9.25).]

It should be expressly mentioned, even at this early stage of the discussion, that the output forces at the feet when the supports are rigid are different from those when the bridge rests on the body of an actual instrument. Nonetheless, it is possible, even in the latter case, for bridges to act as filters, emphasizing certain high frequency ranges. This effect counteracts the monotonic decrease in the input force to the bridge resulting from Helmholtz motion.

Reinicke ascertained the same properties in his examination of the violin bridge (see figure 9.6). The lower resonance of the violin bridge is at 3,000 Hz and the upper one is at 6,000 Hz. The resonances of the cello and violin bridges seem to differ exactly in correspondence with the fundamentals of the open strings, which are an octave and a fifth (a duodecimo) higher on the violin—the frequencies, in other words, being three times as great. The proportions of the violin and cello bridges differ, however, so this correspondence is by no means the result of a simple law of similarity. On the other hand, the correspondence can hardly be regarded as an accident. It is entirely probable that the empirical development of the shapes of the bridges tended to bring the filtering effects of the bridges to bear on partial tones of the same order. This is all the more surprising because, as we shall see, the deformations of the two bridges are entirely different.

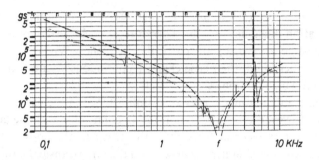

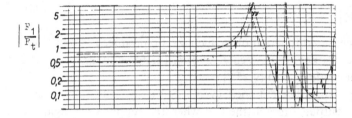

Figure 9.6
Measured (solid line) and calculated (dashed line) frequency curves. Top, magnitude of
the input impedance. Bottom, magnitude of the force transfer factor of a violin bridge
(after Reinicke).

9.3 Holography applied to the study of the oscillatory modes of the bridge

The resonances discussed in the preceding section certainly include natural
oscillatory modes of the rigidly supported bridge. It is riskly, however,
to try to draw conclusions only from the frequencies at which the resonances occur.

Direct observation of the deformation of the bridge, even if only at a
few points, is much better.

Minnaert and Vlam (1937) attempted—long before the technique of
acoustical measurement at the ports was developed—to examine the
motions of a bridge, installed in the normal way on an instrument, by
gluing mirrors to its broad and narrow sides. Through telescopes, they
observed points of light reflected from these mirrors, which moved as a
result of the oscillations of the bridge. It was impossible with their apparatus to distinguish resonances of the bridge from those of the body;

furthermore the excitation of the string was by bowing, and blurred all of the resonance peaks.

These authors were able to establish, nevertheless, that the bridge oscillates not only in the y, z-plane but also perpendicularly to this plane, bending around the y-axis and twisting around the x-axis. This conclusion is still of interest today.

Minnaert's and Vlam's observations led them to the conclusion that these bending and twisting motions did not contribute in any significant way to the motion of the body of the instrument. The torsional motions must contend with the great impedance of the front of the instrument to tangential motions; but even if such motions occur, they contribute nothing to the radiated sound. The bending motions lead only to very small lever arms for the moments transmitted to the front of the instrument, and must attempt to bend it in the direction of the grain of the wood.

With the invention of holography, point-by-point sensing of the oscillatory modes has become unnecessary; these modes can now be observed in their entirety. We will make extensive use of holography in our study of the natural modes of the body of the instrument in section 12.2; so we will now briefly discuss the principle of holography and the interpretation of holographic images.[3]

When a light wave from an illuminated object O arrives at a photographic plate at an angle of incidence ϑ, its phase is different at different points along the x-axis in the plane of incidence:

$$\underline{E}_0(x) = E_0 e^{-ik\sin\vartheta x}. \tag{9.7}$$

These phase differences have no effect on the darkening of the plate; this depends only on the exposure of the plate. The exposure, in turn, is proportional to the intensity of the wave, given by

$$I_0 = |\underline{E}_0(x)|^2 = E_0^2. \tag{9.8}$$

In figure 9.7, left, showing the maxima of the wave arriving from the illuminated object at $t = 0$, the exposure is represented by the evenly-spaced vertical shading lines in the plate.

However, a reference wave from the same source may be superimposed

[3] Readers wishing to study this subject thoroughly might read Brown. Grant, and Stroke (1966).

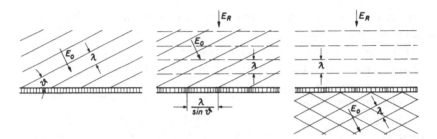

Figure 9.7
The principle of holography. Left, object wave alone, even darkening of the photographic
plate. Center, the object wave and reference wave generate a periodic pattern of changes
in darkening. Right, the reference wave, striking the developed hologram, generates the
original object wave as a first-order diffracted wave.

on the object wave. For simplicity's sake, we show the reference wave
arriving perpendicularly to the plate (figure 9.7, middle). The intensity,
and so the darkening of the plate, then varies periodically along the
x-dimension, with a "trace wavelength" of $\lambda/\sin\vartheta$:

$$I_{O+R} = \left| E_O e^{-ik\sin\vartheta x} + E_R \right|^2$$

$$= E_O^2 + E_R^2 + 2E_O E_R \cos(k\sin\vartheta x). \qquad (9.9)$$

In this equation, the two waves are in phase at $x = 0$. If we change the
position of the illuminated object so that an increase Δl occurs in the
length of the light path, the periodic variations in exposure will also change
position by $\Delta l/\sin\vartheta$. Unlike normal photography, the technique using
a superimposed reference wave, called holography, records the phase of
the light wave, producing a photograph called a *hologram*. It is, to be
sure, necessary for the emulsion on the plate to be able to resolve variations in darkening separated by the distance $\lambda/2\sin\vartheta$. This task becomes
easier as the difference between the angle of the object wave and that of
the reference wave becomes smaller.

It is useful to store information only if there is a way to reproduce it;
in this case, a means of reproducing the original object wave. After
photographic development, reproduction is achieved by allowing the
reference wave alone to fall on the hologram, whose darkened parts act
like a grating (figure 9.7, right). Among the waves it generates are two
first-order diffracted waves. The direction and phase of one of these are
identical with those of the original object wave; the diffracted wave is

shifted according to the position of the grating, just as when the holographic pattern on the plate was generated.

The phase information is contained in the fine structure of the intensity information, the only information used in ordinary photography. The holographic method works, however, even if the fine structure to be resolved is of the order of magnitude of the spacing of the elements of the grating (Gabor 1948, 1949). Everywhere on the developed plate, the reference wave generates a wave like that which arrived from the object; since the phase differences, i.e., the differences in distance, were stored, a person looking toward the former position of the object has the surprising impression of seeing it in three dimensions behind the plate.

These concepts were first expressed by Gabor, but they could not be applied to the study of oscillating objects until a frequency-stable light source, the laser, was developed. Even when the laser is used, the path-length difference between the object wave and the reference wave must not be too great.

What happens, then, when we generate a hologram by illuminating an object as its shape changes, during many periods of oscillation? In connection with figure 9.7, we have already mentioned that the periodic darkening changes position as the path to and from the object changes. Let us assume that the object (for example the top or back plate of a violin) oscillates with a peak-to-peak amplitude of $2\hat{\zeta}$ in the z-direction, that an illuminating wave falls on the object in the x, y-plane at an angle of incidence α, and that the wave proceeds from the object to the photographic plate at the angle β; then the change in path length is given by

$$\Delta l = 2\hat{\zeta}(\cos \alpha + \cos \beta). \tag{9.10a}$$

The exposure of the hologram oscillates correspondingly, cosinusoidally with respect to time, back and forth over a distance $\Delta l/\sin \vartheta$.

If we now illuminate the object stroboscopically only at its extreme positions—as has been done successfully (Archbold and Ennos 1968)—we would obtain a double exposure. The simple image of the object would exhibit greater brightness where the oscillating object remains at rest, that is, at the nodes of its natural oscillatory modes. The same would, however, also be the case where

$$\Delta l = n\lambda, \quad n = 1, 2, 3, \ldots, \tag{9.10b}$$

since the periodicity of the second grating coincides with that of the first at these points.

In between, however, there occur points where

$$\Delta l = (2n + 1)\frac{\lambda}{2}, \quad n = 1, 2, 3, \ldots; \tag{9.10c}$$

at these intermediate points the peaks of one exposure coincide with the valleys of the other. The result is a constant darkening as in figure 9.7, left. The diffracted wave needed to reconstruct the object wave (in figure 9.7, right) disappears. The term *holographic interference* is consequently used. (Δl, to be sure, should not be so great that the image regions on the plate no longer coincide for the most part. For this reason, n is restricted to $n \lesssim 10$.) The dark lines corresponding to the values given in (9.10c) may be recorded on a photograph of the hologram as illuminated by the reference beam. We will call this a *holographic photograph*. It indicates the amount of displacement like the contour lines on a map.

Such contour lines, however, are still produced if we use a constant, rather than a stroboscopic, illumination in generating the hologram. An object oscillating sinusoidally spends more time near the extreme values than near the zero crossings. However, the differences in contrast decrease with increasing n. But, even so, constant illumination has an advantage: the greatest brightness occurs in the areas at rest, and increasing oscillation can be recognized by decreasing contrast. The first contour lines do not correspond to the Δl given in (9.10b and c), but rather occur at somewhat lower values. This, however, is usually of little importance; though (9.10b and c) may not hold exactly, the holographic photograph nonetheless still represents the oscillation clearly.

The apparatus used by Reinicke (1973, fig. 7) to apply these principles of *time-average holography* is shown in figure 12.1 below. There, the procedure is discussed in connection with the natural oscillations of the body of violins; Reinicke was first to apply holography to them.

In attempting to apply this method to the study of the oscillations of the bridge, however, Reinicke faced an additional, particularly difficult, problem; namely, he had to record the distribution of the modes of oscillations in the y, z-plane; the interesting displacements η and ζ lay in this plane. In this case, α and β do not represent the angle with respect to a line perpendicular to the illuminated surface, but rather an angle complementary to this. The illumination must arrive at the surface at a nearly grazing angle. A low value of β was most appropriate also for the light reflected from the object.

On the other hand, the surface under observation had to be projected

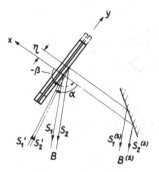

Figure 9.8
An overhead view of the bridge and the control mirror (after Reinicke). Note: The angle
β here is not the same as that in (9.10a).

Figure 9.9
Time-average holographic photographs of a rigidly supported violin bridge at its lowest
natural frequency (3,000 Hz). Left, η-component. Right, ζ-component (after Reinicke).

onto a large enough area of the holographic photograph so that the details
of the oscillation could easily be observed.

If the values of α and β were large, the danger arose that displacements
ξ in the x-direction, perpendicular to the plane of the bridge, might be
recorded. These were the motions noted by Minaert and Vlam. Reinicke
used a trick to separate these from the displacements η and ζ: he placed
two holograms side by side, one of which was generated using a mirror
at an angle of approximately 45° to the bridge (see figure 9.8). If no
displacement ξ occurred along with the motions of interest, the holographic photographs should differ only in that the paths including a
reflection from the mirror S would lead to a greater—nearly doubled—
separation between the contour lines, since $\sin\beta_S \approx 0$. Fortunately, this
proved to be the case.

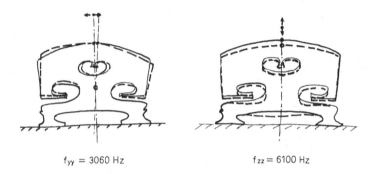

$$f_{yy} = 3060 \text{ Hz} \qquad\qquad f_{zz} = 6100 \text{ Hz}$$

Figure 9.10
Natural modes of a rigidly supported violin bridge (solid line, rest position; dashed line, displaced position). (after Reinicke).

This can be seen clearly in the holographic photographs in figure 9.9, which correspond to the lowest oscillatory modes of the bridge. In figure 9.9, left, the laser light came from the y-direction, and revealed the displacement η, parallel to the front of the instrument. In figure 9.9, right, the light came from the $-z$-direction, and revealed the displacement ζ, perpendicular to the front of the instrument. In both cases, the interference stripes are parallel and equidstant—indicating a linear increase away from the white area, which is at rest. Figure 9.9, left, shows that $\eta = 0$ in the part of the bridge below the slits; figure 9.9, right, shows that the vertical axis of the bridge is at rest with respect to ζ. The combination of both motions is a rocking oscillation of the upper part of the bridge with respect to the lower part, as shown in figure 9.10, left.

Using the same method, the oscillation at 6,100 Hz, shown in figure 9.10, right, was revealed. In this oscillation, the upper part of the bridge moves vertically, as the lower transverse beams bend up and down.

Neither of these oscillatory modes would be possible without the slits at the sides of the bridge. The additional material removed, beyond that necessary to make the slits, may also have an effect on the resonant frequencies. The details of the design, for example the "heart" in the middle, are probably based, however, on purely esthetic considerations.

It makes sense to assume that the indentations in the cello bridge are also of importance. This is shown in the mode in figure 9.11, right, which was also observed holographically. This, however, corresponds to the higher resonant frequency, 2,100 Hz. The mode in figure 9.11, left, at 985 Hz, corresponds to an almost purely translational motion of the upper

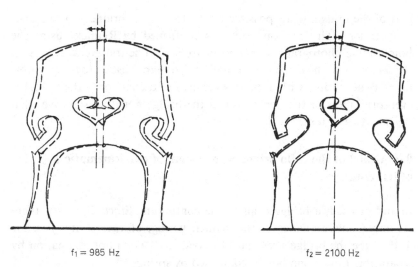

$f_1 = 985$ Hz $f_2 = 2100$ Hz

Figure 9.11
Natural modes of a rigidly supported cello bridge (solid line, rest position; dashed line,
displaced position) (after Reinicke).

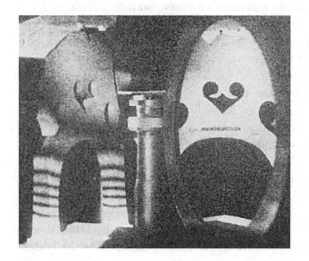

Figure 9.12
Time-average holographic photograph of a rigidly supported cello bridge at its lowest
natural frequency (985 Hz), η-component (after Reinicke).

part of the bridge, made possible by the feet of the bridge—in this case, actually legs. This motion, too, was confirmed by Reinicke using the holographic photograph shown in figure 9.12. The transverse stripes on the legs clearly show their deformation. Though these stripes are unfortunately not visible within the viewing area of the mirror at the right, the even brightness of the upper part of the bridge nonetheless proves that this moves as a unit.

9.4   Model of the cello bridge as a system of two kinematic degrees of freedom

In and of itself, a bridge is an elastic continuum. Since, however, there are only a few resonances in the frequency range of interest, it is possible to describe the bridge's resonant behavior to a good approximation by separating it into rigid bodies connected by springs.

It is possible to represent the measured frequency response as a fraction with a polynomial in ω in the numerator and another in the denominator. These polymonials could be realized using mechanical or, preferably, electrical circuits. For the electrical engineer, who is accustomed to attributing frequency responses to circuits, this approach helps in understanding the behavior of the bridge. This representation is, however, not only difficult; it is also ambiguous, since different circuits may be hypothesized for the same function.

Knowledge of the way the bridge is separated into parts not only simplifies the analysis, but also leads immediately to a model corresponding to the structure of the bridge.

This method was first applied to the larger bridge, that of the cello, by Zimmermann (1967). Holographic photographs were not yet available; so he based his analysis on static deformations measured by Bathe (1966), using a rigidly supported bridge, an optical strain gage, and an inductive displacement sensor. Figure 9.13 shows the bridge, preloaded vertically as it would be by the stretched strings, and deformed by a horizontal force. The trapezoid drawn in the upper part of the body of the bridge rotates around a point in the lower part. This point itself moves only horizontally, as a result of the deformation of the legs, as shown by the rectangular frame in the lower part of the bridge.

These measurements led Zimmermann to develop the model of the bridge shown in figure 9.14, top. The deformations shown at the right

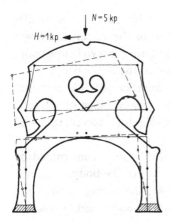

Figure 9.13
Change in shape of a bridge, preloaded with the vertical force N, due to the horizontal force H, measured with an inductive displacement sensor (after Bathe).

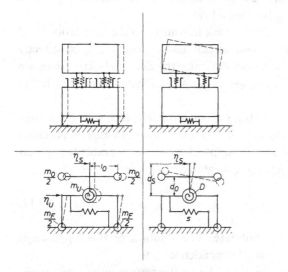

Figure 9.14
Mechanical model of a rigidly supported cello bridge. Top, after Zimmermann, with three degrees of freedom. Bottom, after Reinicke, with two degrees of freedom. Left, the lower natural mode (approximated). Right, the higher natural mode (approximated).

and at the left make it clear that model can exhibit the resonant modes
sketched in figure 9.11: the horizontal motion of the upper and lower
parts of the body with respect to the horizontal spring shown below them,
and the rocking of the upper body with respect to the lower body.

Zimmermann's model, however, includes another possible resonant
mode which has not been observed in the frequency range of interest:
a vertical up-and-down motion. The model is, therefore, too complicated
by one degree of freedom.

It is also unusual that the upper part of the body is constrained by
vertical springs to move along with the lower part of the body.

Reinicke, therefore, uses a special symbol for the spring between the
upper and lower parts of the body. This spring allows no vertical oscillations, and transfers horizontal motions as a stiff link (see figure 9.14,
bottom). Reinicke represents the bodies, including the rotating part of
the upper body, by point masses and rigid, massless levers; these symbols
are like those of mechanical circuit analogs. This representation proves
advantageous in modeling the violin bridge.

It is, however, always a difficult task to transform the translations and
rotations of rigid bodies into a circuit model which can be directly
translated into an electrical circuit diagram; the electrical circuit diagram
is limited in that it can encompass only one kinematic form, translation
or rotation.

In deriving the system equations of such models, too, it is a generally
applicable principle to start with the expressions for kinetic and potential
energy, W_k and W_p. The relationships of forces and moments are best
derived from these energies, using Lagrange's equation

$$\frac{d}{dt}\left(\frac{\partial W_k}{\partial \dot{q}_i}\right) + \frac{\partial W_p}{\partial q_i} = F_i, \tag{9.11}$$

where q_i is the ith local coordinate. The solution of this equation is
simplified by ignoring the spatial dependence of W_k.

We will go only so far here as to derive the system equations for the
rigidly-supported bridge. We choose the horizontal translation of the top
of the bridge, η_S ($\equiv \eta_0$), as one local coordinate q_i; this is the input
displacement that we are trying to find. Our other coordinate is the inner
horizontal displacement η_U of the mass of the lower part of the body
m_U, concentrated at a point. Furthermore, we designate the distance of
this mass from the top of the bridge by d_S; and its distance from the

center of the upper mass by d_O. The upper mass m_O is divided into two partial masses; its radius of inertia is designated by i_O, the horizontal stiffness by s, and the torsional stiffness by D. Then the expressions for kinetic and potential energy are

$$W_k = \frac{m_U}{2}\dot{\eta}_U^2 + \frac{m_O}{2}\left(\dot{\eta}_U + \frac{d_O}{d_S}(\dot{\eta}_S - \dot{\eta}_U)\right)^2 + \frac{m_O}{2}\left(\frac{i_O}{d_S}\right)^2(\dot{\eta}_S - \dot{\eta}_U)^2,$$

$$W_p = \frac{s}{2}\eta_U^2 + \frac{D}{2d_S^2}(\eta_S - \eta_U)^2. \tag{9.12}$$

If we substitute these into (9.11) and replace the displacement vectors η by the corresponding velocities $\underline{v}$, we obtain two system equations of the form

$$Z_{SS}\underline{v}_S - Z_{SU}\underline{v}_U = F_S \quad (\equiv F_0), \tag{9.13a}$$

$$-Z_{US}\underline{v}_S + Z_{UU}\underline{v}_U = 0. \tag{9.13b}$$

The input impedance of the bridge may be obtained by solving these equations, whence we obtain

$$Z_{00} = \frac{F_0}{\underline{v}_0} \equiv \frac{F_S}{\underline{v}_S} = \frac{Z_{SS}Z_{UU} - Z_{SU}^2}{Z_{UU}}. \tag{9.14}$$

(This equation makes it clear why we must replace the subscript 0 with the new subscript S in the system equations (9.13).

The zeros of the numerator represent the resonant frequencies, and those of the denominator represent the antiresonant frequencies.

Reinicke attempted to construct the simplest possible model, and we have restricted ourselves to an examination of the rigidly-supported bridge. Nonetheless, the equation which determines the resonant frequencies is not simple, as all of the impedances in (9.13) include terms representing masses and springs.

We begin with the discussion of

$$Z_{US} = Z_{SU}. \tag{9.15}$$

Adjacent, nondiagonal elements of the impedance matrix, by pairs, are based on common mixed products $\eta_U\eta_S$ and $\dot{\eta}_U\dot{\eta}_S$ from (9.12). In this way, the derivation from the expressions representing energy fulfills the requirement for reciprocity. We obtain

$$\underline{Z}_{SU} = j\omega m_O \left[\left(\frac{d_O}{d_S}\right)^2 + \left(\frac{i_O}{d_S}\right)^2 - \left(\frac{d_O}{d_S}\right) \right] + \frac{D}{j\omega d_S^2}. \tag{9.16}$$

This quantity, representing coupling, may disappear as well at a particular frequency. According to (9.13b), however, this means that finite values of v_S are possible without any motion of the lower part of the body. We have foreseen this possibility in the representation of the second resonant mode shown in figure 9.14, right. According to (9.13a), however, this possibility exists only with an excitation force $\underline{Z}_{SS}\underline{v}_S$ at the top of the bridge; it therefore only approximately fulfills the requirement for a natural frequency, namely, that $F_S = 0$.

It is appropriate, therefore, to write Z_{SS} and Z_{UU} in the form

$$\underline{Z}_{SS} = \underline{Z}_{SU} + \delta\underline{Z}_{SS}, \tag{9.17a}$$

$$\underline{Z}_{UU} = \underline{Z}_{SU} + \delta\underline{Z}_{UU}. \tag{9.18a}$$

Consequently

$$\underline{Z}_{SS} = \underline{Z}_{SU} + j\omega m_O \frac{d_O}{d_S}, \tag{9.17b}$$

and

$$\underline{Z}_{UU} = \underline{Z}_{SU} + j\omega m_U + j\omega m_O \frac{d_S - d_O}{d_S} + \frac{s}{j\omega}, \tag{9.18b}$$

since the product $\underline{Z}_{SU}^2$ vanishes in the numerator of $\underline{Z}_{00}$, and the condition for the resonance consequently simplifies to

$$\underline{Z}_{SU}(\delta\underline{Z}_{SS} + \delta\underline{Z}_{UU}) + \delta\underline{Z}_{SS}\delta\underline{Z}_{UU}$$

$$= \left[j\omega m_O \frac{i_O^2 + d_O^2 - d_S d_O}{d_S^2} + \frac{D}{j\omega d_S^2} \right]\left[j\omega(m_O + m_U) + \frac{s}{j\omega} \right]$$

$$+ j\omega m_O \frac{d_O}{d_S}\left[j\omega m_U + j\omega m_O \frac{d_S - d_O}{d_S} + \frac{s}{j\omega} \right]$$

$$= \left[j\omega m_O \frac{i_O^2 + d_O^2}{d_S^2} + \frac{D}{j\omega d_S^2} \right]\left[j\omega(m_O + m_U) + \frac{s}{j\omega} \right]$$

$$+ \left(\omega m_O \frac{d_O}{d_S} \right)^2 = 0. \tag{9.19}$$

If the last term were absent from this equation, a resonance would exist where either of the preceding expressions in brackets, one multiplied by the other, go to zero. The resonant frequencies would be those of the simplified resonant modes assumed in figure 9.14:

$$\omega_{\mathrm{IO}} = \sqrt{\frac{s}{m_{\mathrm{O}} + m_{\mathrm{U}}}}, \qquad\qquad\qquad (9.20_{\mathrm{I}})$$

$$\omega_{\mathrm{IIO}} = \sqrt{\frac{D}{m_{\mathrm{O}}(i_{\mathrm{O}}^2 + d_{\mathrm{O}}^2)}}. \qquad\qquad\qquad (9.20_{\mathrm{II}})$$

Physically, however, it can be seen that these simple modes cannot exist in this form without an external force F_y. If m_{O} is to be moved back and forth, an accelerating force is necessary. This force can be obtained only by deformation of the torsion springs if $F_0 = 0$. Also, a rotation of the upper body with relation to the lower part of the body is bound up with a translational motion of the center of mass of the upper part of the body. The upper part of the body is supported by the lower part of the body, and must therefore move the lower part of the body as well. In agreement with these considerations, the term that complicates (9.19) vanishes when d_{O} goes to zero. If the centers of both masses lie on the same horizontal axis, the modes in equations (9.20) would be independent of one another. On the other hand, the deformations of the upper part of the body are clearly small in the first case, as are the deformations of the lower part of the body in the second case. The formulas in (9.20) can, then, be regarded as good approximations. We see this also if we substitute

$$\omega_{\mathrm{I}} = \omega_{\mathrm{IO}} + \Delta\omega_{\mathrm{I}}, \qquad\qquad\qquad (9.21_{\mathrm{I}})$$

$$\omega_{\mathrm{II}} = \omega_{\mathrm{IIO}} + \Delta\omega_{\mathrm{II}}, \qquad\qquad\qquad (9.21_{\mathrm{II}})$$

for the actual natural frequencies. Then, if we assume that in both equations $\Delta\omega \ll \omega_0$, we can rewrite (9.19) as

$$\frac{2\Delta\omega_{\mathrm{I}}}{\omega_{\mathrm{IO}}}\left[1 - \left(\frac{\omega_{\mathrm{IIO}}}{\omega_{\mathrm{IO}}}\right)^2\right] = \frac{m_{\mathrm{O}}}{m_{\mathrm{O}} + m_{\mathrm{U}}}\frac{d_{\mathrm{O}}^2}{i_{\mathrm{O}}^2 + d_{\mathrm{O}}^2}, \qquad (9.22_{\mathrm{I}})$$

$$\frac{2\Delta\omega_{\mathrm{II}}}{\omega_{\mathrm{IIO}}}\left[1 - \left(\frac{\omega_{\mathrm{IO}}}{\omega_{\mathrm{IIO}}}\right)^2\right] = \frac{m_{\mathrm{O}}}{m_{\mathrm{O}} + m_{\mathrm{U}}}\frac{d_{\mathrm{O}}^2}{i_{\mathrm{O}}^2 + d_{\mathrm{O}}^2}. \qquad (9.22_{\mathrm{II}})$$

Recalling that $\omega_{\mathrm{II}} > \omega_{\mathrm{I}}$, then we see that the relative frequency shift $\Delta\omega_{\mathrm{I}}/\omega_{\mathrm{IO}}$ (detuning) is negative, while the relative frequency shift $\Delta\omega_{\mathrm{II}}/\omega_{\mathrm{IIO}}$

is positive and significantly greater. In both cases, however, they are small enough to justify our approximation, as shown by the data which Reinicke obtained by averaging values obtained by measuring two cello bridges:

m_O	m_U	d_O	i_O	ω_{II}/ω_I
5.3 g	4.45 g	1 cm	2.45 cm	2.

The value of the expression on the right sides in (9.22) is then 0.08, and so

$$\frac{\Delta\omega_I}{\omega_{I0}} = -0.013; \quad \frac{\Delta\omega_{II}}{\omega_{II0}} = 0.052.$$

Additionally, using data derived from the dimensions of bridges and the natural frequencies,

d_S	D	s
3.05 cm	6.3×10^9 g cm^{-2}	3.6×10^8 g s^{-2},

Reinicke obtained $|Z_{00}|$ as a function of frequency and of the quantity $|F_1/F_0|$, shown as dashed lines in figure 9.5. Both quantities could also be derived from his model.

If we no longer restrict ourselves to the rigidly supported bridge—including the free bridge as a limiting case—it is certainly reasonable to represent the inertia of the feet as two point masses m_F. (Reinicke provides this extension of the system equations.) If the feet are unsupported, another oscillatory mode is possible: the rotation of the upper body around the center of mass of the lower body, which we have already noted, can work against a rotation of the two masses $m_F/2$, with their radii of inertia. It can be shown that the resonant frequency corresponding to this mode is much higher than $f_{II} = \omega_{II}/2\pi$. This result agrees with Steinkopf's observation of this resonance at 4,000 Hz.

9.5 A corresponding model of the violin bridge

From the evidence of the holographic photographs, we know that one of the natural modes of the violin bridge, too, consists of a rotation of the upper part of the body with respect to the lower part of the body. In constructing a model, we can, then, as we did with the cello bridge, represent the upper part of the body as two masses $m_O/2$, rigidly connected by a rod. We can represent the lower part of the body, which does not rotate, by a single mass, and we can connect the two parts of the body

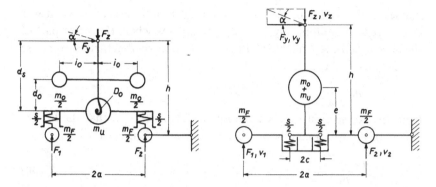

Figure 9.15
Mechanical model of a violin bridge. Left, as in figure 9.14; right, simplified, after Reinicke.

with a torsion spring D_0. Reinicke's model shown in figure 9.15, embodies these criteria.

This model once again allows rotational motion. The motion, however, corresponds to the lower of the two natural modes of the violin bridge. The translational motion observed in the cello bridge does not occur in the violin bridge, because of its short legs; in other words, the body of the violin bridge cannot move in the y-direction.

Instead, a vertical oscillation occurs at the upper end of the frequency range. The compliance allowing this oscillation is found mostly in the horizontal beams of the lower part of the bridge, and to a lesser degree in the elasticity of the feet. In Reinicke's model, the springs embodying this compliance are placed at the right and the left over the circles representing the masses of the feet.

These springs could also influence the rotation of the upper part of the bridge with respect to the lower part. In the case of the violin bridge, however, it is appropriate to avoid the influence of several springs. The vertical housings around these springs are intended to indicate, as before, that the springs are guided so that they can carry out only a vertical motion.

Rather than complicate the model, we will simplify it further, in a way suggested by Reinicke himself. This is shown in figure 9.15, right. The simplified model allows the essential rotational and vertical motions. However, the center of rotation is moved to the middle, between the masses of the feet. This model does not correspond exactly to what was discovered holographically, shown in figure 9.9, where the lower part of the body and the feet remain white; but it is possible nonetheless to choose the distance e of the mass

$$m = m_O + m_U \tag{9.23a}$$

from the connection between the masses of the feet so that the kinetic energy of the rotational motion is the same as in figure 9.15:

$$e = \sqrt{\frac{m_O}{m_O + m_U}(d_O^2 + i_O^2)}. \tag{9.23b}$$

If s is the combined stiffness of the two springs with respect to vertical motion, and if we define the new torsional stiffness as

$$sc^2 = D \tag{9.23c}$$

then the formulas for the lower natural frequency,

$$\omega_I = \sqrt{\frac{D}{me^2}} \tag{9.24_I}$$

and for the upper one,

$$\omega_{II} = \sqrt{\frac{s}{m}} \tag{9.24_{II}}$$

can be read directly from this model.

A force exactly in the y-direction could not excite the higher mode. This mode is evident, however, in figure 9.6. Its presence is easy to explain: the excitation force F_t and the recorded velocity v_t are not in the y-direction, but rather are in a direction tangential to the upper edge of the bridge. This direction is at a slight angle α to the y-direction. The same situation is to be expected for actual bowing, particularly on the outer strings. We must expect an excitation in the y-direction, with

$$F_y = F_t \cos \alpha \approx F_t \tag{9.25_y}$$

and a simultaneous one in the z-direction, with

$$F_z = F_t \sin \alpha \approx F_t \alpha. \tag{9.25_z}$$

These relationships also are valid for the cello bridge, which, however, does not exhibit any resonant reinforcement of z-components in the frequency range of interest; rather, F_z is divided evenly between the two feet. For this reason, we did not examine the small effect of F_z in our analysis of the cello bridge.

Here, however, we obtain both

$$\underline{v}_y \approx \frac{h^2}{(D/j\omega) + j\omega me^2} F_t \tag{9.26$_y$}$$

and

$$\underline{v}_z \approx \frac{\alpha}{(s/j\omega) + j\omega m} F_t. \tag{9.26$_z$}$$

In connection with the reaction on the string, the velocity in the tangential direction generated at the input to the bridge

$$\underline{v}_t = \underline{v}_y \cos \alpha + \underline{v}_z \sin \alpha \approx \underline{v}_y + \alpha \underline{v}_z \tag{9.27}$$

is of interest, and was examined by Reinicke. For the input impedance of the bridge, which was the variable to be found, he obtained

$$\underline{Z}_{tt} = \frac{F_t}{\underline{v}_t} = 1/(\underline{v}_y/\underline{F}_t + \alpha \underline{v}_z/\underline{F}_t)$$

$$\approx \frac{\left(\dfrac{D}{j\omega} + j\omega me^2\right)\left(\dfrac{s}{j\omega} + j\omega m\right)\Big/h^2}{\dfrac{1}{j\omega}\left(s + \dfrac{\alpha^2}{h^2}D\right) + j\omega m\left(1 + \dfrac{\alpha^2}{h^2}e^2\right)}. \tag{9.28}$$

The measured function of frequency of this impedance was reproduced in figure 9.6, top. The function includes not only the two resonances indicated by the factors in the numerator, but also an antiresonant frequency, a zero in the denominator between ω_I and ω_{II}. At this frequency, $\underline{v}_t$ goes to zero because the resultant velocity composed of v_y and v_z is perpendicular to the tangential direction. The values of the spring stiffness and mass differ from those for ω_{II} only by terms which are multiplied by α^2; so it can be understood that this pole lies only slightly below ω_{II}. Nonetheless, it is clearly present in the measurements. (In figure 9.6, which was made using excitation at the d-string notch, α is approximately 12°.)

Accounting for F_z and v_z, as is done here, makes it possible to retain the three-port conception introduced in section 9.1, but with F_t and v_t as input quantities. However, we must substitute different equations for each of the strings. In order to avoid this problem, we may separate the excitation force into a y- and a z-component. If we neglect the differences in position of the string notches, which are not so important, we may combine the equations which we obtain for F_y and F_z as excitation forces, according to the current value of the bow angle α.

In doing this, we can retain the three-port equation (9.3); we need only expand it with a corresponding three-port equation for $\underline{F}_{z0}$, $\underline{v}_{z0}$ (instead of $\underline{F}_{y0}$, $\underline{v}_{y0}$):

$$
\begin{pmatrix} \underline{F}_{z0} \\ \underline{F}_1 \\ \underline{F}_2 \end{pmatrix} = \begin{pmatrix} \underline{Z}_{zz} & \underline{Z}_{z1} & \underline{Z}_{z1} \\ \underline{Z}_{z1} & \underline{Z}_{11} & \underline{Z}_{12} \\ \underline{Z}_{z1} & \underline{Z}_{12} & \underline{Z}_{11} \end{pmatrix} \begin{pmatrix} \underline{v}_{z0} \\ \underline{v}_1 \\ \underline{v}_2 \end{pmatrix}.
\tag{9.29}
$$

Here, even the change of sign in the elements drops out, due to symmetry around the z-axis.

After we have crossed out the first row and the first column, the remaining matrix, which relates the ports at the feet, is the same as before. For this reason, it is possible, as Reinicke has suggested, to combine (9.3) and (9.29) into an equation in four-dimensional vectors and the corresponding matrix:

$$
\begin{pmatrix} \underline{F}_{y0} \\ \underline{F}_{z0} \\ \underline{F}_1 \\ \underline{F}_2 \end{pmatrix} = \begin{pmatrix} \underline{Z}_{yy} & 0 & -\underline{Z}_{y2} & \underline{Z}_{y2} \\ 0 & \underline{Z}_{zz} & \underline{Z}_{z1} & \underline{Z}_{z1} \\ -\underline{Z}_{y2} & \underline{Z}_{z1} & \underline{Z}_{11} & \underline{Z}_{12} \\ \underline{Z}_{y2} & \underline{Z}_{z1} & \underline{Z}_{12} & \underline{Z}_{11} \end{pmatrix} \begin{pmatrix} \underline{v}_{y0} \\ \underline{v}_{z0} \\ \underline{v}_1 \\ \underline{v}_2 \end{pmatrix}.
\tag{9.30}
$$

This corresponds to the physically appropriate concept that the bridge has not only two outputs, but also two independent inputs, but without noting that these two inputs are at the same point.

Finally, the symmetry of the bridge also makes it possible to reduce the two three-port equations (9.3) and (9.29) to two two-port equations, by combining the second and third rows, and, similarly, the second and third columns:

$$
\begin{pmatrix} \underline{F}_{y0} \\ \dfrac{\underline{F}_1 - \underline{F}_2}{2} \end{pmatrix} = \begin{pmatrix} \underline{Z}_{yy} & -\underline{Z}_{y2} \\ -\underline{Z}_{y2} & \tfrac{1}{2}(\underline{Z}_{11} - \underline{Z}_{12}) \end{pmatrix} \begin{pmatrix} \underline{v}_{y0} \\ \underline{v}_1 - \underline{v}_2 \end{pmatrix},
\tag{9.31$_y$}
$$

$$
\begin{pmatrix} \underline{F}_{z0} \\ \dfrac{\underline{F}_1 + \underline{F}_2}{2} \end{pmatrix} = \begin{pmatrix} \underline{Z}_{zz} & \underline{Z}_{z1} \\ \underline{Z}_{z1} & \tfrac{1}{2}(\underline{Z}_{11} + \underline{Z}_{12}) \end{pmatrix} \begin{pmatrix} \underline{v}_{z0} \\ \underline{v}_1 + \underline{v}_2 \end{pmatrix}.
\tag{9.31$_z$}
$$

With the equations factored in this way, torsional and vertical oscillations appear separately. These equations are, then, especially well-suited to the discussion of physical problems and to the construction of circuit models (Reinicke 1973, p. 32).

However, as soon as the value of the output load is allowed to be chosen freely, it makes no difference which of these representations is used. All include the six elements:

$$\underline{Z}_{yy}, \quad \underline{Z}_{zz}, \quad \underline{Z}_{11}, \quad \underline{Z}_{y2}, \quad \underline{Z}_{z1}, \quad \underline{Z}_{12},$$

and all must lead to the same relationships for the input impedances and the force-transfer factors.

These six elements may be derived from the expressions for energy by means of Lagrange's equations, an in section 9.14. They may also be derived from direct observation of figure 9.15, with three ports held rigidly while the fourth is displaced forcibly and the forces at the ports are noted. We have already derived two of these elements in this way in (9.26), using the variable e from (9.23b):

$$\underline{Z}_{yy} = \left(\frac{F_y}{v_y}\right)_{v_z=v_1=v_2=0} = \left[\frac{D}{j\omega} + j\omega m e^2\right]\Big/ h^2, \tag{9.32a}$$

$$\underline{Z}_{zz} = \left(\frac{F_z}{v_z}\right)_{v_y=v_1=v_2=0} = \frac{s}{j\omega} + j\omega m. \tag{9.32b}$$

For the remaining elements, we obtain

$$\underline{Z}_{11} = \left(\frac{F_1}{v_1}\right)_{v_y=v_z=v_2=0} = \frac{s}{4j\omega} + \frac{D}{4j\omega a^2} + \frac{j\omega m_F}{2}, \tag{9.32c}$$

$$\underline{Z}_{y2} = \left(\frac{F_2}{v_y}\right)_{v_z=v_1=v_2 \ 0} = \frac{D}{2j\omega a h}, \tag{9.32d}$$

$$\underline{Z}_{z1} = \left(\frac{F_1}{v_z}\right)_{v_y=v_1=v_2=0} = \frac{s/2}{j\omega}, \tag{9.32e}$$

$$\underline{Z}_{12} = \left(\frac{F_1}{v_2}\right)_{v_y=v_z=v_1=0} = \frac{s}{4j\omega} - \frac{D}{4j\omega a^2}. \tag{9.32f}$$

Reinicke gives the following values for the structural variables in these equations: $m = 1.77$ g, $m_F = 0.38$ g, $e = -1.3$ cm. The measured natural frequencies $f_I = 3,000$ Hz and $f_{II} = 6,000$ Hz were used in deriving the values of D and s. The curves shown with dashed lines in figure 9.6 were calculated using these values.

A musical fifth below f_1, in other words, below approximately 2,000 Hz, all of the elements appear as pure spring reactances. This fact leads most significantly to the relationship

$$\lim_{\omega \to 0} \left(\frac{\underline{F}_z}{\underline{F}_y} \right)_{v_z = v_1 = v_2 = 0} = \lim_{\omega \to 0} \frac{\underline{Z}_{y2}}{\underline{Z}_{yy}} = \frac{h}{2a}, \tag{9.33}$$

which we were able to derive in section 9.1 based on purely static considerations. The slight compliance of the rigidly supported bridge makes no difference to the frequency-independent force transfer factor at these frequencies.

On the contrary, the bridge, freely movable at the feet, acts like a rigid plate with inertia. This transition is easiest to see if we set F_1 [of (9.31_z)] = $F_2 = 0$. We can then eliminate $(v_1 + v_2)$, and we obtain

$$\lim_{\omega \to 0} \left(\frac{\underline{F}_z}{\underline{v}_z} \right)_{F_1 = F_2 = 0} = j\omega(m + m_F), \tag{9.34a}$$

which is what we would expect for the equal acceleration of all parts of the total mass in the z-direction.

If, on the other hand, the bridge is rigidly supported, so that $v_1 = v_2 = 0$, we obtain

$$\lim_{\omega \to 0} \left(\frac{\underline{F}_z}{\underline{v}_z} \right)_{v_1 = v_2 = 0} = \frac{s}{j\omega}. \tag{9.34b}$$

This example should make it clear that the boundary conditions at the feet of the bridge are very important if we are to make any assertions about its behavior on the instrument.

9.6   The bridge on the instrument

We now turn to the question of whether the behavior of the bridge on the instrument is closer to the limiting case in which the feet are supported rigidly, or to that in which they are unconstrained.

We are equally interested in the question as to whether the bridge loads the body of the instrument heavily enough to alter the body's characteristics significantly.

The two questions are related, as we shall first demonstrate in connection with the simple case in which contact at only one point is considered. This case occurs when sound-bearing structures such as the bridge or the body of an instrument are excited by a transducer, or when we sense their motion using a transducer.

Electrodynamic transducers, used again and again to generate the excitation in experiments on string instruments, provide a particularly simple example. With a given magnetic induction B, length of conductor l, and alternating current I, the internal force $\underline{F}_i$ is given by

$$\underline{F}_i = Bl\underline{I}. \tag{9.35}$$

This force is, however, not equal to the external force $\underline{F}_e$ exerted on the surrounding continuum, which presents an external impedance Z_e:

$$\underline{F}_e = \underline{Z}_e \underline{v}, \tag{9.36}$$

The discrepancy exists because the given internal force must also overcome the inertia of the mass of the transducer coil, $j\omega m_{coil}$:

$$\underline{F}_i = (j\omega m_{coil} + \underline{Z}_e)\underline{v}. \tag{9.37}$$

A similar relationship holds for every transducer. Consequently, we introduce, in place of $j\omega m_{coil}$, an *internal impedance* $\underline{Z}_i$ and obtain, in place of (9.37), the equation

$$\underline{F}_i = (\underline{Z}_i + \underline{Z}_e)\underline{v}. \tag{9.38}$$

It follows that F_e and F_i satisfy

$$\frac{\underline{F}_e}{\underline{F}_i} = \frac{\underline{Z}_e}{\underline{Z}_i + \underline{Z}_e}; \tag{9.39}$$

in other words, the more it is true that

$$|\underline{Z}_i| \ll |\underline{Z}_e|, \tag{9.40}$$

the less F_e and F_i differ. If we set the transducer on a rigid surface, as we did the bridge, we can determine the force F_i at the point of contact using a piezoelectric force sensor.

Furthermore, Z_i can be measured at the mechanical output of the transducer, when the electrical excitation—or any other one—is stopped.

It is even expedient to base the definitions of F_i and Z_i on these experimental conditions. Then we can make use of (9.36) and (9.38) to describe the case where the internal system between the source proper and the point of excitation is a two-port.

This case pertains where there is a thin rod between the coil of the transducer and the object of being excited or monitored. If this rod is of stiffness s, then the so-defined Z_i is the inverse of the sum of the admittances $1/(j\omega m)$ and $j\omega/s$. This internal impedance would become infinite at the resonant frequency $(s/m)^{1/2}$.

In order to fulfill the condition (9.38) and to guarantee a small loading of the object by the inserted two-pole, this frequency must lie far above the frequency range of interest. This will be guaranteed by an adequately high stiffness s; then Z_i again approaches $j\omega m$.

The other requirement, that the velocity excited at the input of the object, v_e, should be influenced as little as possible by the two-port device, is fulfilled by (9.38). If we regard F_i as the given force, that would excite an object with a totally rigid surface, the real velocity v_e follows again from (9.36), wherein Z_i again means the impedance of the two-port element on its output side. The influence of the two-port device is proportional to the ratio $|Z_i/Z_e|$.

There are two points of contact between the bridge and the body of the violin. This means not only that we must go through the same calculations for each point, but also that the force vector consisting of $\underline{F}_{e1}$ and $\underline{F}_{e2}$ is related to the velocity vector consisting of $\underline{v}_1$ and $\underline{v}_2$, through an impedance matrix:

$$\begin{pmatrix} \underline{F}_{e1} \\ \underline{F}_{e2} \end{pmatrix} = \begin{pmatrix} \underline{W}_{11} & \underline{W}_{12} \\ \underline{W}_{12} & \underline{W}_{22} \end{pmatrix} \begin{pmatrix} -\underline{v}_1 \\ -\underline{v}_2 \end{pmatrix}. \tag{9.41}$$

(The change of sign in v accounts for the sign chosen for motions at the bridge, positive moving toward the bridge. Also, we choose the symbol W to indicate the impedances of the body, in order to avoid having to introduce an additional subscript.)

We begin again with the question as to when we have to consider the body significantly loaded by the bridge: in other words, with the question as to how much an internal force vector consisting of F_{i1} and F_{i2} differs from that in (9.41), if the same velocity is to be generated. In this cases the impedance matrix of the body must be added to that of the bridge, excited at its feet, with F_y and F_z set equal to zero. The elements, then, are different from those in (9.32a–f). Consequently, we designate them by the symbol Z':

$$\begin{pmatrix} F_{i1} \\ F_{i2} \end{pmatrix} = \left[\begin{pmatrix} W_{11} & W_{12} \\ W_{12} & W_{22} \end{pmatrix} + \begin{pmatrix} Z'_{11} & Z'_{12} \\ Z'_{12} & Z'_{22} \end{pmatrix} \right] \begin{pmatrix} -v_1 \\ -v_2 \end{pmatrix} \tag{9.42}$$

$$= \begin{pmatrix} W_{11} + Z'_{11} & W_{12} + Z'_{12} \\ W_{12} + Z'_{12} & W_{22} + Z'_{22} \end{pmatrix} \begin{pmatrix} -v_1 \\ -v_2 \end{pmatrix}. \tag{9.42}$$

In this case, a slight discrepancy between the values of F_e and F_i is related not only to

$$|Z'_{11}| \ll |W_{11}| \tag{9.43a}$$

and to

$$|Z'_{22}| \ll |W_{22}|; \tag{9.43b}$$

in addition, there is a third condition,

$$|Z'_{12}| \ll |W_{12}|. \tag{9.43c}$$

This last condition becomes unimportant when the coupling elements W_{12} and Z'_{12} and their sum are much smaller than those of the principal diagonal. In this case, (9.43a–c) can be replaced by

$$|W_{12}| \ll |W_{11}|, \quad |W_{22}| \tag{9.43d}$$

and

$$|W_{12} + Z'_{12}| \ll |W_{11} + Z'_{11}|, \quad |W_{22} + Z'_{22}|. \tag{9.43e}$$

If this is so, both cases represent pairs of one-port devices that are for all practical purposes independent.

The issues we have just raised about the coupling between the bridge and the body were first investigated by Eggers (1959) for the cello bridge. Figure 9.16 shows $|W_{11}|$ (top) and $|W_{22}|$ (bottom) as functions of frequency, with (a) and without (b) the bridge in place on the instrument. The curves differ only by amounts within the limits of the accuracy of measurement; we must take into account that F_i can excite the body only in the immediate vicinity of the bridge. The upper curves are for the side with the bass bar; the lower ones, for the side with the soundpost; this is evident from the curves labeled (c), giving measurements with the bridge and soundpost removed from the instrument. The effect on the

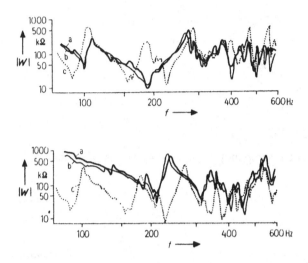

Figure 9.16
Measured input impedances of the body of a cello: a, with the bridge; b, without the
bridge; c, without the bridge and the soundpost. Top, $|W_{11}|$; bottom, $|W_{22}|$ (after
Eggers).

bass-bar side is a shift in the position of the peaks and valleys; but on
the soundpost side, the average impedance decreases significantly. We
may also gather from this data that $|W_{22}|$ surpasses $|W_{11}|$ only at low
frequencies; consequently, the motion of the bridge is principally a
rocking one as in figure 5.11. The difference between F_e and F_i, then, is
small.

The small difference between F_e and F_i in the problem of the loading
of the body by the bridge corresponds to a small difference in the problem
of the excitation of the body by the bridge. In this problem, F_i again
represents the forces under the bridge when its feet are rigidly supported.
F_e, on the other hand, represents the forces with which the body is excited.
We can conclude from figure 9.16 that these two forces are very nearly
the same. The body appears to the bridge as a rigid support, at least in
the measured frequency range; but this frequency range unfortunately
lies well below the first resonance of the rigidly supported bridge, which
occurs at 1,000 Hz. So, unfortunately, the effect of this resonance when the
bridge is in place on the instrument cannot be determined from the
results of Eggers's experiment.

The experiments carried out by Reinicke (1973, figs, 54–57) on the

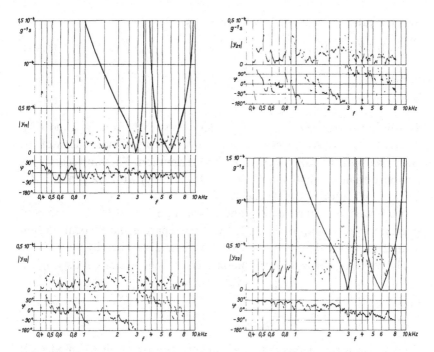

Figure 9.17
Measured frequency functions of the magnitudes and phases of the input admittance
matrices (after Reinicke). The heavy curves show the input admittance of the
unconstrained bridge excited at one foot.

violin give us a better look at this question. Figure 9.17 shows the admittance elements of the body obtained by inverting (9.41):

$$
\begin{pmatrix} -\underline{v}_1 \\ -\underline{v}_2 \end{pmatrix} = \begin{pmatrix} \underline{Y}_{11} & \underline{Y}_{12} \\ \underline{Y}_{12} & \underline{Y}_{22} \end{pmatrix} \begin{pmatrix} \underline{F}_{e1} \\ \underline{F}_{e2} \end{pmatrix}
$$

$$
= \frac{1}{\underline{W}_{11}\underline{W}_{22} - \underline{W}_{12}^2} \begin{pmatrix} \underline{W}_{22} & -\underline{W}_{12} \\ -\underline{W}_{12} & \underline{W}_{11} \end{pmatrix} \begin{pmatrix} \underline{F}_{e1} \\ \underline{F}_{e2} \end{pmatrix}. \tag{9.44}
$$

The measured results are shown in figure 9.17 in positions corresponding
to those in the admittance matrix.

The graphs for the coupling elements should in theory be identical.
In fact, they differ only slightly, indicating that the accuracy of measurement is satisfactory.

Though there may be frequencies at which $Y_{12} > Y_{11}$ or Y_{22}, the average value, as evaluated by Reinicke, is

$$Y_{12} = 0.3 Y_{22} = 0.6 Y_{11}, \tag{9.45a}$$

and so

$$Y_{12}^2 = 0.18 Y_{11} Y_{22}. \tag{9.45b}$$

This indicates that $1/Y_{11}$ and $1/Y_{22}$ differ only slightly from W_{11} and W_{22}. Since there is no available measurement of the impedance matrix of the body, we must be satisfied with an approximation. We can regard it as an appropriate estimate if we compare Y_{11} and Y_{22} with $1/Z_{11}'$ and $1/Z_{22}'$ as independent one-port elements.

We obtain the values of Z_{11}' and Z_{22}' using figure 9.15, through considerations in which, as in (9.31), we first separate the vertical motion from the rotational motion of the bridge when it is excited by the body.

The resulting vertical force generates an average velocity satisfying

$$\underline{F}_1 + \underline{F}_2 = \left[j\omega m_F + \cfrac{1}{\cfrac{j\omega}{s} + \cfrac{1}{j\omega m}} \right] \frac{\underline{v}_1 + \underline{v}_2}{2}. \tag{9.46a}$$

The chosen form of this equation is characteristic of circuit models. It shows that the inertial reactance of the mass of the feet is "in series" with the "paralleled" forces due to the spring reactance and the inertial reactance of the remaining mass (see the force-voltage analogy in section 8.1).

In the same way, we obtain the relationships between the moments,

$$\underline{F}_1 - \underline{F}_2 = \left[j\omega m_F + \cfrac{1/a^2}{\cfrac{j\omega}{D} + \cfrac{1}{j\omega m e^2}} \right] \frac{\underline{v}_1 - \underline{v}_2}{2}. \tag{9.46b}$$

From the sum of the two preceding relationships, we obtain

$$\underline{F}_1 = \left[j\omega m_F/2 + \frac{j\omega m/4}{1 - \omega^2/\omega_{II}^2} + \frac{j\omega m e^2/4a^2}{1 - \omega^2/\omega_I^2} \right] \underline{v}_1$$
$$+ \left[\frac{j\omega m/4}{1 - \omega^2/\omega_{II}^2} - \frac{j\omega m e^2/4a^2}{1 - \omega^2/\omega_I^2} \right] \underline{v}_2 = \underline{Z}_{11}' \underline{v}_1 + \underline{Z}_{12}' \underline{v}_2. \tag{9.47}$$

A plot of $1/Z_{11}'$ ($= 1/Z_{22}'$, by symmetry) as a function of frequency, based on Reinicke's data, is included in the diagrams for Y_{11} and Y_{22}

in figure 9.17. In the low-frequency range, $1/Z'_{11}$ lies far above the measured values of Y_{11} and Y_{22}, indicating that the same force is transmitted to the input of the body at these frequencies as would be transmitted to a rigid plate.

But at the frequency $f = f_1$, in which we are most interested, this conclusion is not possible. On the countrary, here $1/Z'_{11}$ vanishes, as can easily be derived from (9.44). It is also physically clear that the body is extremely highly loaded at the natural frequency that produces a filtering of the input forces by resonance. This does not exclude a filtering effect at either a lower or a higher frequency.

On the other hand, $1/Z'_{11}$ vanishes only if there are no losses at the bridge. Here we must consider at least the power transfer in the strings; the oscillation of the bridge excites transverse waves, and since at 3,000 Hz their wavelengths are short, even for the string segments between the bridge and tailpiece, we can estimate these losses by loading the summit of the bridge in the y and z directions with a resistance r that is the sum of the point-impulse impedances of all four strings, i.e., by $2(Z_g + Z_d + Z_a + Z_e)$ $= 2(375 + 280 + 230 + 175) = 2,120$ g s^{-1}. If we compare this resistance to the mass reactance $j\omega m$ to which it must be added, i.e., to 32,000 g s^{-1} at $m = 1.77$ g and 3,000 Hz, we obtain a ratio of 0.066—a remarkably high damping. For the rocking motion of the bridge we might expect the ratio to be four times as great, because the resistance lies at a doubled lever-arm from the center of rotation. But calculation shows the corresponding value of the magnitude of $1/Z'_{11}$ to be only about 0.046×10^{-4} g^{-1} s, by no means surpassing the value of Y_{11} in figure 9.17, not to mention Y_{22}, in which some resonance peaks occur near 3,000 Hz.

Nevertheless we here anticipate measurements of bridge-input admittances on the instrument by Moral and Jansson (1982) (see section 12.4, esp. figure 12.9), which show a resonance peak at 3 kHz with a broad half-power bandwidth.

9.7 The effect of the mute

We have been able to show by measurement that the bridge must have some filtering effect at high frequencies, and we can make this effect understandable by analyzing models of the bridge. One possible approach is to shift the frequency response for force by altering the bridge; however,

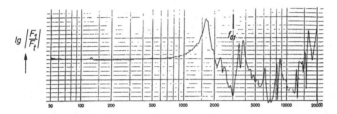

Figure 9.18
Measured frequency function of the force transfer factor of a rigidly supported violin
bridge with a "mute" mass of 1.5 g attached to its top (after Reinicke).

a very precise knowledge of the input impedances of the instrument body
would be necessary in order to predict the effect in any specific case.

On the other hand, we have established that the cello bridge is seen by
the string as a rather stiff lever with a constant force-transmission factor
at frequencies below 700 Hz, and that the violin bridge has the same
effect below 2,000 Hz. At these frequencies, the bridge is seen by the body
of the instrument as a free rigid plate whose inertia does not have any
appreciable loading effect. We should perhaps state these as the principal
results of our investigations.

In this frequency range, slight alterations in the shape of the bridge
have no effect.

There is only one exception to this rule: placing the mute on the bridge
to play *con sordino*. Here the effect is considerable, but so is the alteration
to the bridge. It can be seen immediately that the partial mass m_0 is
substantially increased, and that all of the natural frequencies in which
it plays a part must be lowered.

As Reinicke (1973, p. 46) showed, an additional mass at the top of
the violin bridge of only 1.5 g lowers the lower natural frequency from
3,000 Hz to 1,700 Hz (see figure 9.18). The mute has a mass of approximately 4 g and so the frequency is lowered even more. The effect of the
mute is not only in the reinforcement of the frequency range of the
resonance, but perhaps subjectively even more in the weakening of the
frequency range above the resonance.

This effect can be seen especially vividly in the recordings of force that
Reinicke (1965) carried out using piezoelectric sensors in the string notch
and under the feet of the bridge with mute attached (see figure 9.19).
Without the mute, it would be expected that the forces at the feet would

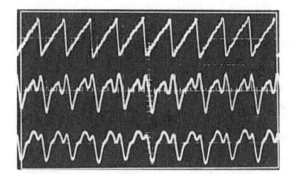

Figure 9.19
Piezoelectrically recorded time functions of force with a mute on the bridge. Top, in the
string notch. Center, under the foot of the bridge near the bass bar. Bottom, under the
foot of the bridge near the soundpost (after Reinicke).

retain their sawtooth shape, with rounded corners and superimposed
secondary waves, up to approximately the tenth partial. Instead, we obtain
force functions that hardly show partials above the third or fourth at all.
The differences between the shapes of the functions at the feet over the
bass bar and soundpost (with the sign of F_1 reversed in the oscillograms)
are less remarkable than their similarities. This behavior shows that the
inertial forces resulting from a vertical motion of the bridge are small in
comparison with $F_2(\approx -F_1)$ at these lower frequencies (see section 9.1).

References

Archbold, E., and A. E. Ennos, 1968. *Nature* **217**, no. 5132, p. 942.

Bathe, K., 1966. Unpublished student work, Institute for Technical Acoustics, Technical
University of Berlin.

Brown, G. M., R. M. Grant, and G. W. Stroke, 1966, Theory of Holographic Interferometry,
J. Acoust. Soc. Amer. **45**, 1166.

Cremer, L., 1976. *Ing. Arch.* **45**, 371.

Eggers, F., 1959. *Acustica* **9**, 453.

Gabor, D., 1948. *Nature* **161**, 777.

Gabor, D., 1949. *Proc. Roy. Soc.* (London) A **197**, 454.

Helmholtz, H. v., 1877. *Lehre von den Tonempfindungen* [*On the Sensations of Tone*], 4th
edition. Braunschweig: Vieweg.

Minnaert, M., and C. C. Vlam, 1937. *Physica* **4**, 361.

Moral, J. A., and E. V. Jansson, 1982. *Acustica* **50**, 329.

Reinicke, W., 1965. Unpublished thesis work, Institute for Technical Acoustics, Technical University of Berlin.

Reinicke, W., 1973. Dissertation, Institute for Technical Acoustics, Technical University of Berlin.

Reinicke, W., and L. Cremer, 1970. *J. Acoust. Soc. Amer.* **48**, 988.

Sivian, L. J., H. K. Dunn, and D. S. White, 1931. *J. Acoust. Soc. Amer.* **2**, 330.

Steinkopf, G., 1963. Unpublished thesis work, Institute for Technical Acoustics, Technical University of Berlin.

Zimmermann, P., 1967. *Acustica* **18**, 287.

10 The Body of the Instrument as a System of Few Degrees of Freedom

10.1 Choice of the number of partial masses of the model

It was not an easy task to analyze the bridge in terms of masses linked by springs. The model had to have the correct number of natural oscillatory modes in the frequency range of interest, and also to reproduce the observed details of the motion. It was also important to avoid over-complicated formulas. The much larger body of the instrument certainly must pose even greater difficulties; every peak in the curves of admittance against frequency in figure 9.17 represents a natural frequency.

The plots in figure 9.17 begin at 400 Hz, due to limitations in the measuring technique; the bridge contributes no peaks below 3,000 Hz. Consequently, it is better to base the following discussion on the curves shown in figure 10.1, measured by Beldie (1975).[1] These show the input admittances A_{0y} of the bridge as it rests on the body. Six different violins were measured, so these curves give an idea of the differences and similarities one might expect from different instruments.

The peaks appear to be closer to one another at higher frequencies; this is due primarily to the logarithmic frequency scale. This scale is commonly used in acoustics; it conforms to the musical keyboard and is justifiable on grounds related to sensory psychology. On a logarithmic scale, the density of peaks with equal frequency spacing would double with every octave. But even on a linear frequency scale the peaks in the low-frequency range of the violin would appear to be especially well-separated, on account of the lesser overlapping accompanying lesser damping. Sound production and reaction effects on the bowing process in the low-frequency range are of particular importance, too, since this range contains the fundamentals in the first (lowest) hand position.

If we limit ourselves to consideration of the first three or four peaks, we also keep the number of independent degrees of freedom and hence the model masses within tolerable limits. However, we still have to include in our model the more or less pronounced peak and following dip near 300 Hz and consequently the velocity of the air oscillating in the f-holes. This velocity represents the kinetic energy of a natural oscillatory mode whose potential energy is primarily contained in the compressed or rarefied air in the cavity of the instrument. In the literature, the corre-

---

[1] Similar bridge input admittances as functions of frequency have been measured by Eggers (1959) for the cello and by Reinicke (Reinicke and Cremer 1970) for the violin.

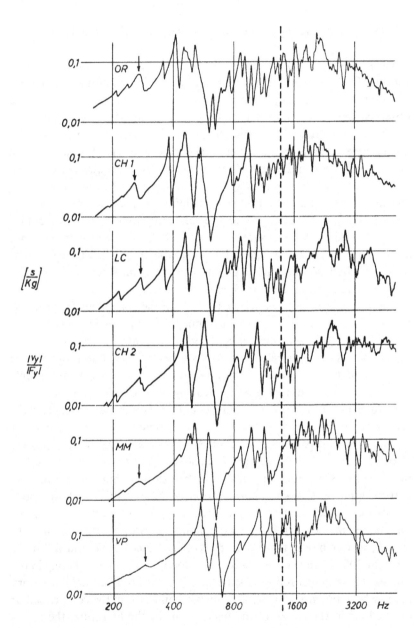

Figure 10.1
Measured admittances at the input of the bridge for six different instruments (after Beldie).

sponding resonance is often called the "air resonance" or "air cavity resonance." At higher frequencies, however, other resonances occur in the cavity, with the air oscillating as a continuum. Consequently, we will designate the lowest air resonance by the term usual in the acoustical literature, *f-hole resonance.*

The motion of the top plate results from the pair of forces transferred from the feet of the bridge. Consequently, the top plate, if it were as symmetrical as it appears from the outside, could be deformed only anti-symmetrically around its longitudinal axis. Such a deformation would clearly result in motion of the air within cavity, but not to compression.

In our discussion of research on the bridge, however, we have already mentioned the asymmetry introduced by the soundpost and the bass bar. The soundpost is most important to compression in the cavity, since it transfers part of the force F_2 from the right foot to the back plate of the instrument. This single force on the back plate bends it in one direction, and would lead to compression of the air in the cavity even if two equal and opposite deformations of the top plate canceled one another out. The motions of the top plate do not cancel, however; this is primarily due to the bass bar, which makes the effects of the two f-holes different from one another, though they are the same in appearance and have the same effect on airborne sound. Their elongated shape introduces free boundaries for the center section of the top plate, boundaries that are capable of restricting the out-of-phase motions proceeding from the bridge to the region between the longitudinal axis and the f-holes. This effect is particularly significant on the right side; here, there occurs an "island." The mass m_{2D} of this island, oscillating in phase with the back plate and out of phase with the rest of top plate, and its area S_{2D}, can then be much smaller than those of the remainder of the top plate, m_{1D} and S_{1D}. (See figure 10.2, where the part S_{2D} of the top plate is hatched from the upper left to the lower right). On the side with the bass bar, on the other hand, a longitudinally much larger part is excited in phase. This part can in turn set the area behind the f-hole on its side into oscillation in phase, as well as areas on the right side including some behind its f-hole. In figure 10.2, these parts are consequently hatched the same way (upper right to lower left) as the left side of the top plate. The different effect of the two f-holes is symbolized by the interrupted upper and lower bound-aries of the right f-hole, which forms the boundary of the out-of-phase areas. The boundaries of the left f-hole are shown as solid lines, since it

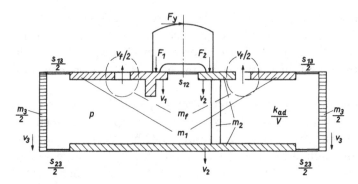

Figure 10.2
Model of the body of the instrument as four independent partial masses (simplified,
after a model by Beldie).

lies between in-phase areas. Since the area S_{2D} of the island is smaller than
that of the rest of the top plate S_{1D}, the top plate can contribute to the
compression of the air in the cavity. (We note here that a string instrument
called the "chrotta," developed in the seventh century, had the its right
foot of its bridge connected directly to the back plate through a circular
hole in the top plate.[2])

The two out-of-phase parts of the top plate are coupled by a small
section between the feet of the bridge, which is deformed in bending. It is
modeled by a leaf spring of stiffness s_{12}. Note the other leaf springs
shown in figure 10.2; The springs are shown as pairs of parallel solid
lines.

The springs s_{12} and s_{23} couple the main mass of the top plate, m_1
($\equiv m_{1D}$), and of the back plate, m_2 (including the mass of the soundpost
and the island), to the ribs. In the present model, the ribs are represented
as an additional mass m_3 oscillating as a unit.

In his dissertation, Beldie considered the ribs to be a coupling mass
between the top and back plates, and the bridge to be a three-port device.
Figure 10.2 is largely based on Beldie's model, though he regarded the
island as being separated from the back plate by a spring. This approach
could not have be motivated by a consideration of the elastic properties
of the soundpost, since it is so short in comparison with the length of
longitudinal waves in the wood of which it is made that it may be regarded

---

[2] This instrument is again being built by the luthier Weidler in Nuremburg.

as rigid even at high frequencies. The model can account, however, for a certain amount of elasticity of the top plate between the foot of the bridge and the upper end of the soundpost. In this way, it might be possible to model the effect of the precise position of the soundpost. This effect certainly exists, though likely in frequency ranges in which the top and back plates must be subdivided into greater numbers of out-of-phase regions.

In any case, we want our model to be simple and easy to understand, so we will regard the mass of the island as one with that of the soundpost and back plate. Our system, then, is comprised of the springs s_{12}, s_{13}, and s_{23}; the air cavity V, of adiabatic modulus of compression K_{ad}; and the four masses m_1, m_2, m_3, and m_f. There are consequently four natural frequencies. One of them, however, is zero; in other words, it corresponds to motion of the entire body as a unit. The three others, however, correspond to the first three peaks of the admittance function. One is near the f-hole resonance, and the others are resonances of the body, in which, as we shall see, the top plate, the back plate, and the ribs play a part.

It must, however, be emphasized that the partial masses in figure 10.2, shown with rectangular outlines, account only for vertical translational motions relative to each other. This approach is in contrast to that in our earlier models of the bridge. It would be incorrect, for example, to introduce rotational motions of the top and back plates with respect to the ribs, flexing the springs at the right and left in opposite directions. Such motions would be possible only if the top and back plates were rectangular and were supported by the ribs only along their longitudinal edges. Rather, such rotations of the top and back plates are strongly resisted by the appreciably thick strip of glue between each plate and the ribs, and by the strongly curved boundaries between these parts. Rotational motions, then, cannot occur in the frequency range in which our model is useful.

10.2 Motions of the body as a unit

The actual instrument as held by the player has a defined rest position. If, on the other hand, we study the behavior of the model shown in figure 10.2 beginning at low frequencies, we immediately see that it has no rest position.

Still, the player's chin and shoulder do not rigidly clamp the violin, and the player's hand does not rigidly hold the violin's neck. The parts of the player's body holding the violin may be characterized as dynamically compliant and lossy springs. We could account for them in figure 10.2 as a combination of springs and dashpots installed between the ribs at one side of the model and a rigid body; this would define a rest position and raise the lowest natural frequency of the model above zero.

But this approach would place another limitation on the validity of our model: the frequency range in which measurements are to be made would have to lie sufficiently far above this resonance.

This requirement is certainly fulfilled in the case of the measurements reproduced in figure 10.1; if it were not, the admittance curve would fall as it moved out of the lowest range of frequencies. In contrast, the magnitude of the admittance of an unconstrained mass set into a motion at a distance d from its center is given by

$$|\underline{F}_y| = \omega m |v_{y1}| = \omega m \frac{i^2}{d^2}|v_{y2}|, \tag{10.1}$$

in which i is the radius of gyration. This leads to

$$\left|\frac{v_y}{\underline{F}_y}\right| = \left|\frac{v_{y1} + v_{y2}}{\underline{F}_y}\right| = \frac{i^2 + d^2}{i^2} \cdot \frac{1}{\omega m}. \tag{10.2}$$

If we insert the values $m = 0.45$ kg, $i^2 = 25$ cm^2, and $d^2 = 25$ cm^2, all representative of a violin, the admittance at the lowest frequency of interest is 0.0035 s kg^{-1}. This lies below the range of ordinates in the logarithmic scale in figure 10.1. The violin is, then, prevented dynamically by its own inertia from oscillating as a unit.

It should be mentioned that the violin was rotated 90° around the x-axis in Beldie's experiments. The endpin was suspended from a 60-cm cord and the scroll rested on a foam-rubber cushion. Translational motion (v_{y1}), then, was constrained more than rotational motion; but the effect was only to make the admittance smaller.

A more pertinent objection to Beldie's experimental arrangement and his model is that the inertial mass of the chinrest does not constrain the motion of the ribs. Moreover, Beldie does not account for the constraint against motion of the ribs posed by the mass of the instrument's neck; rather, in his model (figure 10.2), he assumes that all parts of the ribs move equally. It cannot be excluded that an additional natural resonance within the range to be measured might result from this spatial clamping due to the neck. To evaluate this possibility, we must further examine the mechanical properties of the ribs in their position between the top and back plates (see also section 10.5).

The clamping, too, is in a dimension not represented in our one-dimensional model. Such a model might be fundamentally unable to account for it.

However, if we restrict our examination to purely translational motions in the z-direction, then unconstrained ribs with inertial mass as in Beldie's model certainly reflect actual behavior better than would rigidly clamped ribs.

In contrast to Beldie's model of the violin, Eggers's (1959) measurements of the cello show that rotational motions play an important role. Support by the peg and possibly the different ratios of the variables probably are factors leading to this observation. Eggers gave the instrument under measurement a defined rest position (see figure 10.3) by pressing it against foam rubber blocks at the upper block and ribs "where the instrument normally rests lightly on the chest and legs of the player."

Eggers excited the instrument using an electrodynamic transducer driving a rod attached to the d-string near the bridge. All strings were stretched, but were damped with foam rubber.

He sensed the motion of the top and back plates capacitively; as Backhaus (1930) had done previously, he used an FM capacitive microphone. In this microphone, the change in capacitance between two electrodes, one attached to the body under measurement and the other close to it but fixed, brings about frequency modulation of a high-frequency carrier. A receiver detuned by half the width of its selectivity curve converts the frequency modulation to amplitude modulation; then, as in every radio receiver, the modulation is rectified and filtered so as to appear alone at the output. The advantage of sensing in this way is that the high frequencies to be amplified in the input stages are far from the mechanically excited changes in displacement at low frequencies, avoiding any effect of the strong electrical field from the exciter on the sensor. The disadvantage is that the rest distance between the two electrodes affects the results, and must be adjusted again and again to be the same. This adjustment is especially difficult when the object to be measured has curved surfaces, such as those of the top and back plates of stringed instruments.

Eggers sensed the motion along linear strips, unlike Backhaus, who had

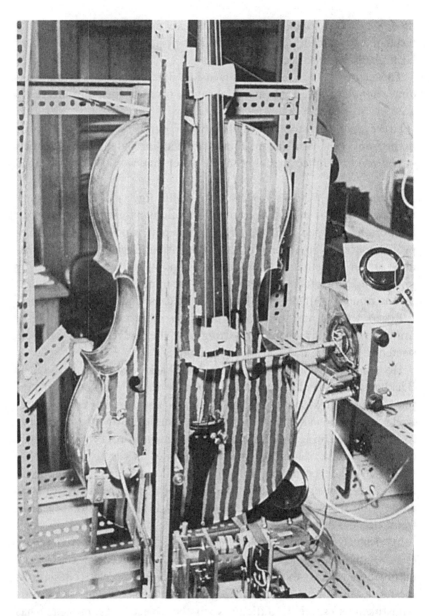

Figure 10.3
A cello painted with strips of conductive paint, for capacitively sensing displacement, with electrodynamic excitation at the bridge (after Eggers).

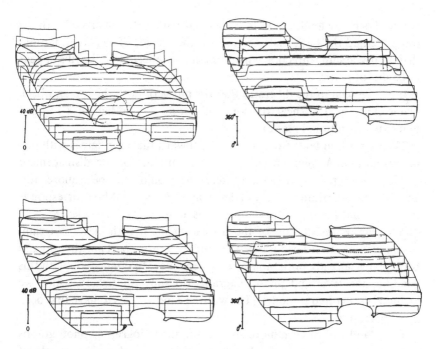

Figure 10.4
Level (left) and phase (right) of the displacement of the top plate (top) and back plate (bottom) of a cello, shown in axonometric relief (after Eggers).

sensed it point by point. He painted strips of conductive silver on the instrument (the light strips in figure 10.3). With two electrodes placed one above the other in front of it, a strip formed the third plate of a three-plate variable capacitor.

Eggers moved the electrodes along the strips, adjusting the rest distance manually, using a potentiometric null technique. He was capable of continuously recording the levels corresponding to the displacement and the phase differences with respect to the current exciting the instrument. He assembled the curves corresponding to the different strips into axonometric relief diagrams. Figure 10.4 gives the results for the lowest frequency he studied, 82 Hz. (This is below the f-hole resonance.) The top and back plates are shown horizontally, with the peg at the left and the scroll at the right. The outside of both the top plate and the back plate face the reader. The lower edge of the top plate lies directly above the upper edge of the back plate, as they are shown in figure 10.4. Also, Eggers defines his phase angles so as to coincide when displacements of both the top plate and the back plate are toward the outside. The actual motion, then, is in the same direction when the edges connected by the ribs appear to be out of phase.

It can be seen in the diagrams that the phase of both the top and back plates shifts by 180° near the longitudinal axis. This behavior in itself points to a rotational motion.

The increase in level from the middle toward the edge agrees with this interpretation. A diagram showing the amplitude of the displacement would, however, have defined the zero and would have been more suitable. In the logarithmic scale of a level diagram, a nodal line corresponds to $-\infty$. What actually appears instead is an undefined noise level.

A linear representation of the displacement, like the level diagram, would also have revealed the bending of the central parts of the top and back plates, already noticeable at this frequency. In connection with this bending, the level and phase diagrams do show the island which we included earlier in our model. (The word "island" seems to have first been used by Eggers.)

The island will be even more apparent in the holographic photographs of the violin we will study in chapter 12. No holographic photographs of the cello are yet available, probably due to its size. Capacitive sensing is indeed more cumbersome, and axonometric reliefs are more difficult to read, but they do show phase relationships; holographic photographs do not. This more complicated technique will always be necessary in order to arrive at a complete description of the instrument's motions.

10.3   Helmholtz resonance and f-hole resonance

As the curves of figure 10.1 show, the f-hole resonance, indicated by the arrows, corresponds to the lowest of the natural frequencies represented in the model in figure 10.2. This frequency lies near the fundamental of the second-lowest open string, and may be found by sliding the finger up and down the lowest string. Like any other resonance, it makes itself evident not as much by an increase in loudness—since the other partials also contribute to the loudness—as by a tone color in which the fundamental if especially predominant. If one f-hole is covered with adhesive tape (of a type that does not damage the instrument's varnish), the resonance is lowered by a tritone: on the violin, to a_0-flat.

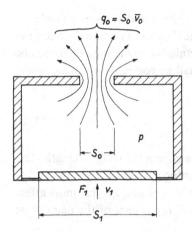

Figure 10.5
A simple example of an f-hole resonator amenable to calculation, excited by pistonlike
motion of one wall. (In the violin, the f-holes are located in one of the oscillating walls,
but this makes no difference for any practical purpose, as $v_1/v_0 \ll 1$.)

The most vivid demonstration of the f-hole resonance is to bow the note
corresponding to it while a second person removes tape from both f-holes
at once.

Before we study the excitation of the f-hole resonance in our model, let
us look at the basic phenomenon as shown in figure 10.5. A volume sur-
rounded by an enclosure, simulating a violin body, has an opening in the
front which comprises only a small part of its surface area, thus simulating the f-holes. The elastically supported back plate moves as a piston with
the velocity v_1.

At frequencies far below the f-hole resonance, the sound flux generated
by the back plate

$$q_1 = S_1 v_1, \tag{10.3}$$

where S_1 is the area of the back plate set into motion, is forced through the
upper opening. It is inevitable that the air is accelerated, since a significantly greater velocity v_0 must be present in the small opening S_0:

$$q_1 = q_0 = S_0 \bar{v}_0. \tag{10.4}$$

The bar over v_0 indicates that it represents an average value; the velocity
is generally anything but equal over the surface area.

In order to bring about this acceleration, the sound pressure p inside the enclosure must be greater than that outside; beyond a certain distance, the latter can be set equal to zero. We then obtain for an opening of any shape —given only that its dimensions are small in comparison to the wavelength—the relationship

$$\underline{p} = j\omega\varrho\frac{l}{S_0}\underline{q}_0 = j\omega\varrho l \bar{v}_0. \tag{10.5a}$$

The dimensions in the equation show the constant l to be a length; this constant depends on the shape of the opening.

If the plate containing the opening has a thickness l_0, this must affect the characteristic length l, since the mass $S_0 l_0 \varrho$ is part of the mass to be accelerated by the force difference $S_0 \Delta p$.

The kinetic energy, however, is not contained only in the opening. Air particles converging in front of it and diverging behind it also make a contribution.

If l_0 predominates, then the difference $(l - l_0)$, which may be divided equally between the two sides, is taken as the *aperture correction* $2\Delta l$. It might be noted that a fall in pressure would also be necessary in order to bring about an acceleration if the plate with the opening were infinitesimally thin: what is called a "sheet" in the language of physics. In this case, l would reduce to two equal aperture corrections, one for the front of the sheet, and one for the back:

$$\underline{p} = j\omega\varrho\frac{2\Delta l}{S_0}\underline{q}_0 = j\omega\varrho(2\Delta l)\bar{v}_0. \tag{10.5b}$$

This limiting case may be treated as a potential-field problem. Helmholtz solved this problem for elliptical openings of any given eccentricity:

$$\varepsilon = \sqrt{1 - (b/a)^2}, \tag{10.6}$$

where b is the minor diameter and a the major diameter of the ellipse (see Rayleigh 1878, II, sec. 306). Helmholtz obtained

$$\frac{2\Delta l}{S_0} = \frac{1}{\pi a}\int_0^{\pi/2} \frac{d\varphi}{\sqrt{1 - \varepsilon^2 \cos^2 \varphi}}, \tag{10.7}$$

and

$$2\Delta l = b\int_0^{\pi/2} \frac{d\varphi}{\sqrt{1 - \varepsilon^2 \cos^2 \varphi}}. \tag{10.8a}$$

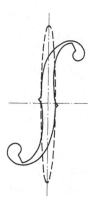

Figure 10.6
f-hole of a violin and a substitute ellipse.

The elliptic integral that appears here takes on the value $\pi/2$ for a circular opening, and so

$$2\Delta l = \frac{\pi}{2}b = \frac{\pi}{2}a; \tag{10.8b}$$

that is, the obstruction is least, for a given area, when the opening is circular. The greater the eccentricity, the greater Δl becomes. For this reason, it would be incorrect to substitute a circle of the same area for the elongated f-hole.

In figure 10.6, an f-hole traced from a violin is superimposed on an ellipse with the same maximum width (0.8 cm) and approximately the same length (9.2 cm) and area (5.78 cm^2). Here, ε is 0.996 and the elliptical integral takes on the value 3.83 (see, e.g., Jahnke and Emde 1945). We consequently obtain

$$2\Delta l = 1.53 \text{ cm}. \tag{10.8c}$$

Since there are two f-holes working in parallel, the total area is 11.56 cm^2. The value of $2\Delta l/S_0$ to be substituted into (10.5b) is, then, 0.132 cm^{-1}. Though the shape is quite different from that of an f-hole, this substitution results in a good approximation to the actual resonant frequency.[3]

(Since a potential equation must be solved here, it would also be possible to determine the exact value of $2\Delta l/S_0$ for an f-hole by means of an

---

[3] The substitution of an ellipse for the f-hole was also made by Itakawa and Kumagai (1952).

electrical analogy. One approach, for example, would be to saw an f-hole in an insulating plate as thick as the front of a violin and to place this in a trough filled with a conductive fluid, between two plane-parallel electrodes. The value of l/S_0 could be calculated from the increase in resistance.)

As the frequency approaches the f-hole resonance, the air volume between the opening and the oscillating plate is compressed adiabatically. If only the geometric volume V of the enclosure, independent of its shape, determines the compression, we may call the system a "Helmholtz resonator": not because Helmholtz attached particular importance to this type of oscillation as it applies to the violin, but because he used resonators employing this principle in his acoustical analyses. The particular resonators he used were hollow spheres with one small opening presented to the sound field and another, even smaller, presented to the ear canal. Helmholtz's resonators were excited by the sound field, through the equivalent mass of the opening. In the violin, excitation is through the motions of the top and back plates, which enclose the volume V. In calculating the effective stiffness of a Helmholtz resonator, we can neglect the effect of the openings, though the air near them is compressed somewhat less. If we introduce the adiabatic modulus of compression

$$K = \varkappa P, \tag{10.9}$$

where P (atmospheric pressure) $\approx 10^5 \mathrm{Pa}$,[4] and $\varkappa = 1.4$, we can obtain the increase in pressure,

$$\underline{p} = \frac{K}{j\omega V}(S_1 \underline{v}_1 - \underline{q}_0). \tag{10.10}$$

In combination with (10.5a), this leads to

$$\left(\frac{j\omega \varrho l}{S_0} + \frac{K}{j\omega V}\right) \underline{q}_0 - \frac{KS_1}{j\omega V} \underline{v}_1 = 0, \tag{10.11a}$$

and to

$$\underline{q}_0 = \frac{K/j\omega V}{K/j\omega V + j\omega \rho l/S_0} S_1 \underline{v}_1. \tag{10.11b}$$

[4] Pascal (Pa) = Newton/m^2 (International Standards Organization).

At the natural frequency of the Helmholtz resonator, the reactances of the volume stiffness and of the mass inertia of the f-hole cancel out:

$$\omega_{\mathrm{H}} = \sqrt{\frac{K}{\varrho}} \sqrt{\frac{S_0}{lV}} = c \sqrt{\frac{S_0}{lV}}, \tag{10.12}$$

where c is the propagation speed of sound waves in air; these are longitudinal waves just as in section 6.3, except that the adiabatic volume stiffness K takes the place of the modulus of elasticity E.

We use the value of c corresponding to a room temperature of 18° C, 34,200 cm s^{-1}; we also take $V = 2{,}220$ cm^3, and $l/S_0 = 2\Delta l/S_0 = 0.132$ cm^{-1}, corresponding to our substitute ellipse. The result is

$$f_{\mathrm{H}} = \frac{34{,}200}{2\pi} \sqrt{\frac{1}{2{,}220 \cdot 0.132}} = 318 \text{ Hz}, \tag{10.13a}$$

This corresponds approximately to d_1-sharp, near the frequency of the open d-string.

If we include $l_0 = 0.25$ cm in l, the frequency sinks further, to:

$$f_{\mathrm{H}} = 318 \sqrt{\frac{2\Delta l}{2\Delta l + l_0}} = 318 \sqrt{\frac{1.53}{1.53 + 0.25}} = 295 \text{ Hz}, \tag{10.13b}$$

corresponding to d_1.

In contrast to the situation for a Helmholtz resonator, the increased pressure in the cavity of an f-hole resonator acts on the surface S_1, that is, against the stiffness s_1 of the spring. A force is generated which increases as S_1, and the resulting velocity leads to a pressure which also increases as S_1. Consequently, the spring admittance given by q/p in (10.10), $j\omega V/K$, is increased by $j\omega S_1^2/s_1$. The resulting stiffness of the resonator is lowered, and the natural frequency is brought down to its actual value, ω_0.

The resonance appears, however, only if we regard as a given, not v_1, but rather the force F_1 exerted against the plate at the back. We must then supplement (10.11a) by

$$\underline{F}_1 = \left[j\omega m_1 + \frac{s_1}{j\omega} + \frac{KS_1^2}{j\omega V} \right] v_1 - \frac{KS_1}{j\omega V} \underline{q}_0. \tag{10.14}$$

If we assume that the natural frequency of the back plate, $\omega_1 = \sqrt{s_1/m_1}$, is far higher than ω_0 and ω_{H}, we can neglect the first term. We will also neglect it later in our model of the violin even though the interval is only a fifth; accounting for it would only complicate the calculations without effecting any basic changes near the f-hole resonance; if we neglect it, the resonance is shifted only slightly.

As is always the case, the natural frequency and the resonance of the system appear at the frequency at which the determinants incorporating the coefficients of the system equations (10.14) and (10.11a) go to zero; here:

$$
\begin{vmatrix}
\left(\dfrac{s_1}{j\omega} + \dfrac{KS_1^2}{j\omega V}\right) & \left(-\dfrac{KS_1}{j\omega V}\right) \\[2ex]
\left(-\dfrac{KS_1}{j\omega V}\right) & \left(\dfrac{K}{j\omega V} + \dfrac{j\omega\varrho l}{S_0}\right)
\end{vmatrix} = 0.
\tag{10.15}
$$

It is possible to write this condition as

$$
-\frac{s_1 K}{\omega^2 V}\left[1 - \omega^2 \frac{l\varrho}{S_0}\left(\frac{V}{K} + \frac{S_1^2}{s_1}\right)\right] = -\frac{s_1 K}{\omega^2 V}\left[1 - \frac{\omega^2}{\omega_0^2}\right],
\tag{10.16}
$$

with the frequency of the f-hole resonance to be found:

$$
\omega_0 = c\sqrt{\frac{S_0}{lV}\Big/\left(1 + \frac{KS_1^2}{Vs_1}\right)} < \omega_{\mathrm{H}}.
\tag{10.17}
$$

This determines the resonant peak q_0,

$$
\underline{q}_0 = \frac{j\omega S_1/s_1}{1 - \omega^2/\omega_0^2}\underline{F}_1,
\tag{10.18}
$$

and also the peak in the input admittance,

$$
\frac{\underline{v}_1}{\underline{F}_1} = \frac{j\omega}{s_1}\frac{1 - \omega^2/\omega_{\mathrm{H}}^2}{1 - \omega^2/\omega_0^2}.
\tag{10.19}
$$

A zero occurs very close to this peak, at $\omega = \omega_{\mathrm{H}}$. We have already observed a curve consisting of a neighboring peak and dip in the range of the f-hole resonance in the measured frequency curves of the input admittance of the bridge. This behavior can be derived even from the simplified model in figure 10.5.

More generally, the relationships obtained using our four-mass model of the violin (figure 10.2) take on exactly the same form as (10.11a) and (10.14)–(10.19) if we ignore all of the inertial masses except that within the f-holes. If we do this, the ribs joining the top and back plates represent only an additional spring with the reciprocal stiffness $(1/s_{13} + 1/s_{23})$. Along with the stiffness s_{12}, this results in the value

$$s_{res} = s_{12} + \frac{1}{1/s_{13} + 1/s_{23}}. \qquad (10.20)$$

Based on this relationship, we obtain

$$\underline{F}_1 = \frac{1}{j\omega}\left[s_{res}(\underline{v}_1 - \underline{v}_2) + \frac{K}{V}S_{1D}[S_{1D}\underline{v}_1 - (S_2 - S_{2D})\underline{v}_2 - \underline{q}_0] \right] \qquad (10.21)$$

as the balance of forces in the top plate.

We can simplify this as well, by neglecting the very small area of the f-holes in the geometrical relationship

$$S_2 = S_{1D} + S_{2D} + S_0. \qquad (10.22)$$

If we do this, v_1 and v_2 appear in (10.21) only as a difference:

$$\underline{F}_1 = \frac{1}{j\omega}\left[s_{res} + \frac{KS_{1D}^2}{V} \right](\underline{v}_1 - \underline{v}_2) - \frac{KS_{1D}}{j\omega V}\underline{q}_0. \qquad (10.23)$$

Similarly, for the back plate,

$$\underline{F}_2 = \frac{1}{j\omega}\left[s_{res} + \frac{KS_{1D}^2}{V} \right](\underline{v}_2 - \underline{v}_1) + \frac{KS_{1D}}{j\omega V}\underline{q}_0. \qquad (10.24)$$

From these two equations, we see immediately that

$$F_2 = -F_1 \quad \left(= \frac{F_y h}{2a} \right), \qquad (10.25)$$

as we would expect from the negligible vertical acceleration of the bridge. This relationship, however, cannot hold strictly, since the air in and around the f-holes is accelerated vertically; there must be an opposite acceleration of the entire body of the violin. However, its 450-g mass compensates for the acceleration of an air mass of only $(\varrho/S_0) = 0.025$ g. If we neglect this motion of the entire body, then

$$\text{sign } v_2 = -\text{sign } v_1. \qquad (10.26)$$

These relationships express in an ideal way the excitation of the air oscillating in the f-holes by the lever mechanism of the bridge; the front and back compress and rarefy the air in the cavity equally and in phase.

Now in (10.23) and (10.24) $v_2 - v_1$ can be replaced by

$$v_2 - v_1 = v_y \frac{2a}{h}. \tag{10.27}$$

Using this relationship, we can express the difference (10.24) − (10.23) as

$$\underline{F}_y \frac{h}{2a} = \left[\frac{1}{j\omega} \left(s_{res} + \frac{KS_{1D}^2}{V} \right) \right] \frac{2a}{h} \underline{v}_y + \frac{KS_{1D}}{j\omega V} \underline{q}_0. \tag{10.28}$$

Correspondingly, (10.11a) must be changed to

$$\frac{KS_{1D}}{j\omega V} \frac{2a}{h} \underline{v}_y + \left[\frac{K}{j\omega V} + \frac{j\omega \varrho l}{S_0} \right] \underline{q}_0 = 0. \tag{10.29}$$

We must, then, also change (10.18) to

$$-\underline{q}_0 = \frac{j\omega S_{1D}/s_{res}}{1 - \omega^2/\omega_0^2} \frac{\underline{F}_y h}{2a}, \tag{10.30}$$

and now, instead of (10.19), and taking (10.27) into account, we obtain

$$\underline{v}_y \left(\frac{2a}{h} \right) = \underline{v}_2 - \underline{v}_1 = \frac{j\omega}{s_{res}} \frac{1 - \omega^2/\omega_H^2}{1 - \omega^2/\omega_0^2} \frac{\underline{F}_y h}{2a}. \tag{10.31}$$

From both of these formulas, it follows that

$$\frac{S_{1D}(\underline{v}_1 - \underline{v}_2)}{\underline{q}_0} = 1 - \left(\frac{\omega}{\omega_H} \right)^2. \tag{10.32a}$$

Especially at the f-hole resonance $\omega = \omega_0$, this means that

$$\frac{S_{1D}(\underline{v}_1 - \underline{v}_2)}{\underline{q}_0} = 1 - \left(\frac{\omega_0}{\omega_H} \right)^2 = \frac{K/V}{\frac{s_{res}}{S_{1D}^2} + \frac{K}{V}}. \tag{10.32b}$$

As we will show in section 13.2, the radiated sound in this low-frequency range depends on the total flux, given here by

$$\sum \underline{q} = \underline{q}_0 + S_{1D}(\underline{v}_2 - \underline{v}_1) = \underline{q}_0 \left(\frac{\omega}{\omega_H} \right)^2 \tag{10.32c}$$

This formula contains the limit

$$\lim_{\omega \to 0} \sum \underline{q} = 0, \tag{10.32d}$$

and so the resulting flux must go to zero at very low frequencies, as emphasized by Benade (1976, p. 532). At the f-hole resonance, particularly, we get

$$\sum \underline{q} = \underline{q}_0(\omega_0/\omega_H)^2. \tag{10.32e}$$

When f_H is 300 Hz and f_0, as determined by measurement, is 280 Hz, then $\sum \underline{q} = 0.87 \underline{q}_0$; this means that radiation from the f-holes predominates.

The quick succession of a resonant peak and an antiresonant dip appears as only a moderate hill and valley in figure 10.1; this can be due only to relatively great losses.

One of these is the radiation loss, the one of most interest as regards the violin. It is easy to calculate this, as we shall show in section 13.2. The flux $\sum \underline{q}$ can be expressed by means of the radiation impedance, which amounts to

$$\frac{\underline{p}}{\sum \underline{q}} = \varrho c \frac{\pi}{\lambda_0^2} = \frac{\pi \varrho}{c} f_0^2 \tag{10.33}$$

at the resonant frequency. In order to account for this loss in (10.31), we divide (10.33) by the mass reactance to which we add it, and which appears in the denominator of $1/\omega_0^2$ or $1/\omega_H^2$. In terms of (10.5a), the reactance is

$$\frac{\underline{p}}{\underline{q}_0} = j2\pi f_0 \varrho \frac{l}{S_0}. \tag{10.34}$$

The quotient is

$$-j\frac{f_0}{2c}\frac{S_0}{l}\frac{\sum q}{\underline{q}_0} = -0.023j, \tag{10.35}$$

if we substitute the numerical values we have been using.

A second type of loss is due to the viscosity of air. This always occurs when, as is the case here, air must move through an opening with a small cross section. The only available solutions are for cylindrical holes; these include aperture corrections (see Cremer and Müller 1982, vol. 2, p. 201). The aperture correction for resistance proves to be about the same as for mass. We have no alternative other than to apply corresponding values to our elliptical opening. We must, then, extend (10.35) by

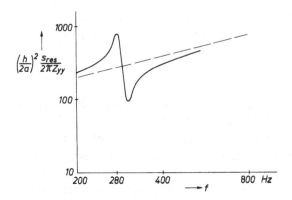

Figure 10.7
Calculated input admittance of the bridge near the f-hole and Helmholtz resonances, taking radiational and viscous losses into account.

$$-j\frac{U_0\sqrt{\omega_0\varrho\eta/2}}{S_0\omega_0\varrho}. \tag{10.36}$$

Here U_0 is the circumference of the opening (for both f-holes this is about $8a$) and η is the "dynamic viscosity" (1.8×10^{-4} g cm^{-1} s^{-1}). Using our numerical values, we obtain $-0.02j$ for this term. The losses due to viscosity are, then, of the same order of magnitude as those due to radiation.

Figure 10.7 shows the curve of bridge input admittance calculated on this basis and normalized to $(2\pi/s_{res})\,(2a/h)^2$; comparison of figure 10.7 and the corresponding parts of the curves in figure 10.1 shows that our formulas do not yet account for all of the losses in an actual string instrument.

10.4   Coupled oscillations of the top and back plates

We earlier neglected the inertias of the masses of the body in our treatment of the f-hole resonance; now, similarly, we assume that we are examining a frequency high enough above this resonance that it no longer qualitatively influences the characteristics of the oscillations in which the three masses of the body, m_1, m_2, and m_3, participate. We may, then, set

$$v_0(\equiv v_f) = 0, \tag{10.37}$$

and we must now regard the velocity of the ribs, v_3, as an additional field quantity.

Equation (10.37) does not imply that we also ignore the spring effect of the cushion of air inside the cavity of the instrument. Its stiffness, connecting the back plate (including the island) to the top plate, must be added to the spring stiffness s_{12} just as before; we therefore introduce the stiffness

$$s_{12}' = s_{12} + \frac{KS_1^2}{V} \tag{10.38}$$

If, as in the previous section, we also assume that

$$F_2 = -F_1 = F_y \frac{h}{2a}$$

[see (10.25)], then we can embrace the three system equations in matrix form:

$$
\begin{pmatrix} -j\omega \dfrac{h}{2a} F_y \\[2mm] +j\omega \dfrac{h}{2a} F_y \\[2mm] 0 \end{pmatrix} =
$$

$$
\begin{pmatrix} s_{12}' + s_{13} - \omega^2 m_1 & -s_{12}' & -s_{13} \\ -s_{12}' & s_{12}' + s_{23} - \omega^2 m_2 & -s_{23} \\ -s_{13} & -s_{23} & s_{13} + s_{23} - \omega^2 m_3 \end{pmatrix} \begin{pmatrix} v_1 \\ v_2 \\ v_3 \end{pmatrix}. \tag{10.39}
$$

Once more, the natural frequency is obtained by setting the determinant equal to zero. The corresponding equation, called the *characteristic equation* in the theory of vibration, is of the third order in ω^2:

$$\Delta = a_0 \omega^6 + a_1 \omega^4 + a_2 \omega^2 + a_3 = 0. \tag{10.40}$$

From the s-terms in (10.39), we see that

$$a_3 = 0; \tag{10.41}$$

in other words, a root $\omega^2 = 0$ can be factored out multiplicatively. This result was to be expected; it has to do with the undefined rest position and a possible constant transverse motion of the whole system. In accord with this, the sum of the three system equations gives the result

$$m_1 v_1 + m_2 v_2 + m_3 v_3 = 0; \tag{10.42}$$

that is, the center of mass remains at rest if the excitation is as we assumed. This condition can now be fulfilled strictly. We obtain as the three other constants

$$a_0 = -m_1 m_2 m_3, \tag{10.43a}$$

$$a_1 = (s'_{12} + s_{13})m_2 m_3 + (s_{23} + s'_{12})m_3 m_1 + (s_{13} + s_{23})m_1 m_2, \tag{10.43b}$$

$$a_2 = -(m_1 + m_2 + m_3)(s'_{12}s_{13} + s_{23}s'_{12} + s_{13}s_{23}). \tag{10.43c}$$

(The validity of such formulas can be checked by substituting the sub-scripts cyclically; only s_{12} need have the prime, so as to correlate correctly with the stiffness of the air cavity.)

Reducing the number of natural frequencies to two has the advantage that we obtain simple, explicit formulas for both ω_I and ω_II. These formulas allow us to discuss the influences of the individual constructive constants:

$$\omega^2_{\mathrm{I},\mathrm{II}} = -\frac{a_1}{2a_0} \pm \sqrt{\left(\frac{a_1}{2a_0}\right)^2 - \frac{a_2}{a_0}}$$

$$= \frac{1}{2}\left(\frac{s'_{12} + s_{13}}{m_1} + \frac{s_{23} + s'_{12}}{m_2} + \frac{s_{13} + s_{23}}{m_3}\right)(1 \pm \sqrt{1 - \gamma}). \tag{10.44}$$

Here, γ represents the discriminant

$$\gamma = \frac{4a_2 a_0}{a_1^2}, \tag{10.45a}$$

which takes on the value

$$\gamma = \frac{4(s'_{12}s_{13} + s_{23}s'_{12} + s_{13}s_{23})\dfrac{m_1 + m_2 + m_3}{m_1 m_2 m_3}}{\left(\dfrac{s'_{12} + s_{13}}{m_1} + \dfrac{s_{23} + s'_{12}}{m_2} + \dfrac{s_{13} + s_{23}}{m_3}\right)^2}. \tag{10.45b}$$

This quantity cannot exceed 1 here; if it did, it would lead to complex natural frequencies, and these cannot occur in a system consisting only of masses and springs.

Due to the absence of energy-absorbing components, the amplitudes of velocity given by (10.44) with sinusoidal excitation go to infinity. In order to prevent this, losses must be introduced. It is here physically appropriate and mathematically simple to introduce them by replacing the spring stiffnesses with complex ones:

$$\underline{s}_{12} = s_{12}(1 + j\eta),$$

$$\underline{s}_{13} = s_{13}(1 + j\eta),$$

$$\underline{s}_{23} = s_{23}(1 + j\eta). \tag{10.46}$$

These represent phase shifts between forces and displacements and so make power losses possible. The quantity η, resulting from deformation losses in the wood, is called the loss factor. In his calculations, Beldie sets $\eta = 0.02$, obtaining resonant peaks whose half-power bandwidth is close to the measured values.

Between these resonances, there can be an antiresonance at which the observed velocity goes to zero, or, if losses are taken into account, at which there is a deep dip in amplitude. Whether this antiresonance occurs, and at what frequency, depends on which velocity is examined.

We first examine the velocity $\underline{v}_1$ of the top plate, which we obtain from (10.39):

$$\underline{v}_1 = -\frac{j\omega h}{2a\Delta}\underline{F}_y[(s'_{12} + s_{23} - \omega^2 m_2)(s_{13} + s_{23} - \omega^2 m_2)$$

$$- s_{23}^2 - s'_{12}(s_{13} + s_{23} - \omega^2 m_2) - s_{12}s_{23}]. \tag{10.47a}$$

This can be simplified to

$$\underline{v}_1 = -\frac{j\omega h}{2a\Delta}\underline{F}_y[\omega^4 m_2 m_3 - \omega^2((s_{13} + s_{23})m_2 + s_{23}m_3)]. \tag{10.47b}$$

The constant term, then, goes to zero here as well, and so a factor of ω^2 can be extracted and canceled by the corresponding one in Δ. The square of the radian frequency of the zero of v_1, designated as o_1^2, is then obtained from a formula surprisingly simple in comparison with (10.44):

$$o_1^2 = \frac{s_{13} + s_{23}}{m_3} + \frac{s_{23}}{m_2}. \tag{10.48}$$

This represents an absence of motion of m_1; only m_2 and m_3 participate in determining the result. It also at first seems surprising that s_{12} is absent from (10.48). The reason for its absence first becomes apparent if we derive (10.48) from the balance of forces on the mass m_3; the spring s_{12} does not play a part in this. If $v_1 = 0$, then

$$s_{13}\underline{v}_3 + s_{23}(\underline{v}_3 - \underline{v}_2) = o_1^2 m_3 \underline{v}_3. \qquad (10.49)$$

The requirement that the center of mass remain at rest reduces to

$$m_2 v_2 + m_3 v_3 = 0, \qquad (10.50)$$

and allows us to express v_2 in (10.49) in terms of v_3; the condition (10.48) follows from this derivation.

Similarly, we obtain v_2 from (10.39):

$$\underline{v}_2 = \frac{j\omega h}{2a\Delta} F_y[\omega^4 m_1 m_3 - \omega^2[(s_{13} + s_{23})m_1 - s_{13}m_3]], \qquad (10.51a)$$

with the zero condition

$$o_2^2 = \frac{s_{13} + s_{23}}{m_3} + \frac{s_{13}}{m_1}; \qquad (10.51b)$$

and by combining (10.47a) and (10.51a), we obtain

$$\underline{v}_y = (\underline{v}_2 - \underline{v}_1)\frac{h}{2a} \qquad (10.52a)$$

$$= \frac{j\omega h^2}{4a^2\Delta} F_y[\omega^4(m_1 + m_2)m_3 - \omega^2(s_{13} + s_{23})(m_1 + m_2 + m_3)],$$

with the zero condition

$$o_y^2 = \frac{s_{13} + s_{23}}{m_3} + \frac{s_{13} + s_{23}}{m_1 + m_2}. \qquad (10.52b)$$

Finally, we are interested also in the behavior of the velocity v_3 of the mass of the ribs m_3. We obtain this velocity from (10.39):

$$-\underline{v}_3 = \frac{j\omega h}{2a\Delta} F_y[(s_{23}m_1 - s_{13}m_2)\omega^2]. \qquad (10.53)$$

Here, too, the constant term goes to zero. This time, however, only a term in ω^2 remains. Once more, it can be combined with the corresponding factor in the denominator. The numerator consequently becomes frequency-independent. The velocity v_3, then, has no zero between ω_I and ω_{II}.

These formulas would make it possible to predict the positions of the resonances and antiresonances if it were possible to measure the three masses and three stiffnesses sufficiently accurately before assembly of the instrument.

It is easiest to measure masses which can be determined by weighing. Since the back plate is thicker than the top plate, it is understandable that its mass m_B is greater than that of the top plate m_D. According to information provided by the Mittenwald violin-builders' school,[5] m_B averages 115 g and m_D averages 70 g, including the bass bar, which amounts to only 5 to 10 g. It must, however, be noted that individual deviations from these averages can be considerable.

The masses m_B and m_D, however, are not the same as m_2 and m_1. The mass of the island has to be subtracted from the top plate and added to the back plate. Also, the areas near the edges must be considered to be springs, not masses; the groove in which the purfling around the edges is inlaid adds to this effect. More generally, the effective mass depends on the distribution of velocity in the top and back plates.

These difficulties are alleviated if we seek only to determine the ratio m_2/m_1. We may normalize all masses to a reference mass, for example, m_1, and all stiffnesses to a reference stiffness, for example, s_{13}; but if we do this, we can determine only the ratios between the frequencies of the resonances and the antiresonances. If we account for the corrections we have discussed, the ratio comes out as approximately $m_2/m_1 = 2$.

The ribs have the smallest mass. Their curvature prevents them from bending, though they are only approximately 1 mm thick. They are not, however, homogeneous. Stiffening blocks are glued to them at the corners near the waist, as well as the upper and lower end blocks (terms defined as if the body were held vertically). The upper end block is attached to the neck, whose much greater mass restricts oscillation in the frequency range of interest. Consequently, the upper end block should not be considered a part of the effective mass. On the other hand, the mass of the overhanging rims at the edges of the top and back plates must be added to that of the ribs. As an approximate value, we then obtain $m_3/m_1 = 0.6$.

It is much more difficult to evaluate the stiffnesses or even their ratios. It is even riskier than it was for the masses to try to determine the stiffnesses

[5] For this data the author is indebted to H. A. Müller.

through static experiments, since forces exerted at different points result in different distributions of deformations, which are in turn different from those occurring in the oscillating instrument.

For this reason, it is better to determine the stiffnesses from measured natural frequencies, based on assumed values for the masses.

In section 12.2, we will discuss thoroughly the investigations conducted by Jansson et al. (1970) using back and top plates resting on rigid ribs. (These cases are analogous to that of a bridge resting on a rigid top plate.) The stiffness of the cushion of air was absent in the experiments of Jansson et al., since the rigid plate of the experimental apparatus was open underneath. The natural frequency of 465 Hz for the top plate obtained with this arrangement corresponds to

$$465 \text{ Hz} = \frac{1}{2\pi}\sqrt{\frac{s_{13}}{m_D}}, \tag{10.54}$$

if we use the symbols we have previously introduced. The work of Jansson et al. does not give the mass of the plate; but from the given thicknesses and their distribution, a value of approximately 119 g with the bass bar can be assigned. This is much greater than the average value given above. From the estimated mass, s_{13} follows:

$$s_{13} = 119(2\pi \times 465)^2 \approx 10^9 \text{ g s}^{-2}. \tag{10.55}$$

The authors repeated their measurement with the soundpost in place between the top plate and a rigid support; as would be expected, they obtained the higher frequency of 540 Hz. If we next assume that the island, comprising approximately 10% of the area of the top plate, is at rest, it follows from the frequency ratio that

$$\frac{540}{465} = \sqrt{\frac{s_{12} + s_{13}}{0.9 s_{13}}}, \quad \text{or} \quad \frac{s_{12}}{s_{13}} = 0.21. \tag{10.56}$$

Since s'_{12}/s_{13} plays a part in our formulas, we must also determine the stiffness of the cushion of air. This is the only stiffness that we can determine directly; here again, we must diminish the effective area of the top plate by a factor of 0.9. We obtain

$$K(0.9 S_D)^2/V = 1.4 \times 10^6 \times 495^2/2{,}200 = 156 \times 10^6 \text{ g s}^{-2}. \tag{10.57}$$

If we divide this by the value of s_{13} calculated above, we see that the stiffness of the cushion of air is of the same order of magnitude as s_{12}—in no way negligible. The parameter to be found amounts, then, to $s'_{12}/s_{13} \approx 0.4$.

Jansson et al. determined the natural frequency of the back plate, supported in the same way, to be 490 Hz, and were able to conclude from the given thicknesses that the mass ratio $m_B/m_D = 1.2$; the ratio s_{23}/s_{13} comes out in this case to be approximately 1.35.

Given the ratio of masses mentioned above, $115/70 = 1.65$, the ratio of stiffnesses must, however, be much greater. If the top and back plates were of the same construction, s_{23}/s_{13} would in fact be proportional to the cube of the masses, or $s_{23}/s_{13} = 4.4$.

The value of s'_{12}/s_{13} also must be assumed to be greater, given that $m_2/m_1 = 2$, if a ratio of f_{II}/f_I is to be obtained which is not too great in comparison with the observed one.

In summary, then, more research is necessary in order to determine stiffnesses with an accuracy sufficient for this model. Nonetheless, the present author believes that determining the natural frequencies of the top and back plates resting on rigid ribs can be a very valuable aid in violin building.

For a physicist, then, it is very surprising that violin builders instead tap the unsupported top and back plates, holding them between the thumb and index finger at a vibrational node.

Since they excite several natural frequencies, it is difficult to state the composition of the "bell tone" that they hear. To substitute objective measurements for the subjective impression, Hutchins, Stetson, and Taylor (1971) have electrodynamically excited the "free" top and back in the middle and have made holographic photographs of their oscillations. The photographs show that one mode is particularly pronounced and particularly efficient at generating sound pressure. In this mode, the edge oscillates as a unit, but out of phase with the middle; they call this the *ring mode*. They found that the frequencies of the top and back plates of a well-tuned instrument in this mode are in the ratio $f_{B,R} : f_{D,R} = 349 : 307$, approximately a whole tone apart.

It is conceivable, however, that the properties of the two plates are determined by the ratio of the ring-mode frequencies in such a way as to allow evaluation of the ratio of natural frequencies even of the supported plates. (If the plates were of similar construction, this would be expected.) If the volumn of the air cavity is given, even the ratio of natural frequencies of the assembled instrument could be determined approx-

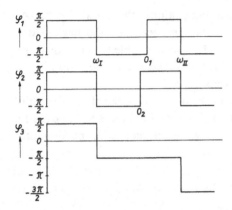

Figure 10.8
Phase functions $\varphi_1 = \mathrm{arc}\,(-\underline{v}_1/\underline{F}_y)$, $\varphi_2 = \mathrm{arc}\,(\underline{v}_2/\underline{F}_y)$, $\varphi_3 = \mathrm{arc}\,(\underline{v}_3/\underline{F}_y)$ for the model shown in figure 10.2.

imately. This conclusion, however, is based on risky assumptions, and has not yet been proven.

This holds even more for the comparison of the tap tones. It therefore appears likely that violin builders also—or primarily—judge the decay time of the resonances; this, too, can be influenced by the shaping of the plates.

Even, however, if the three-mass model discussed here proves too unreliable in determining f_I and f_{II}, it provides a physical insight into the way in which the corresponding oscillatory modes arise. Figure 10.8 shows the function of the phase angle described by

$$\varphi_1 = \mathrm{arc}\,(-\underline{v}_1/\underline{F}_y), \quad \varphi_2 = \mathrm{arc}\,(\underline{v}_2/\underline{F}_y), \quad \varphi_3 = \mathrm{arc}\,(\underline{v}_3/\underline{F}_y). \tag{10.58}$$

It is assumed here that

$$\omega_I < o_2 < o_1 < \omega_{II}. \tag{10.59}$$

If φ_1 and φ_2 are equal, the body of the instrument behaves like a breathing body: that is, the radiating surfaces move inward and outward in phase.

In this sense, all three masses oscillate in phase at low frequencies. At the first resonance (ω_I), all three phase angles jump 180° with respect to the excitation force. The same occurs once again at ω_{II}. At the anti-resonances (o_1 and o_2), φ_1 and φ_2 jump back +180°. Only in the small range o_1 to o_2 do the top and back plates oscillate in the same direction.

The ribs, however, do not participate in this phase reversal. For this reason, they are spatially in phase with the back plate at ω_I and with the top plate at ω_{II}; or vice versa, depending on whether s_{13}/m_1 is smaller or larger than s_{23}/m_3. Additionally, the modes differ in the distribution of amplitudes. If $s_{13}/m_1 = s_{23}/m_2$, the ribs remain at rest, and thus the only difference is in the amplitude ratio v_D/v_B; the back oscillates more strongly in one mode, and the top in the other.

10.5 Comparison of the model and measurements on an instrument

Beldie (1975) himself has furnished the most convincing experimental evidence that his model is largely valid at low frequencies. He excited the bridge in the y-direction and measured the frequency curves of velocity for the top plate, the back plate, and the ribs. The magnitude and phase of all measured quantities was referred to the input force at the bridge. Figure 10.9 gives an example.[6]

The recorded velocities are designated as D (top plate), B (back plate), Z_1, Z_2 (ribs at the right and left) and S (the island region of the top plate, which was treated as part of the back plate in our simplified model).

The validity of the model up to approximately 600 Hz, past the second resonance peak of the body, is shown in the comparison of the frequency curves of φ_S and φ_B. These differ by a constant 180°, with the reference motion directed outwards, since the island and the back plate cannot both move outwards at the same time. Beldie's signs are opposite those in the previous section. The level curves are still parallel to one another, though the amplitudes differ. Their difference does not indicate that different coordinates are necessary for the island region; after all, the motions are not those of pure pistons. Rather, the amplitude of oscillation within the patterns varies smoothly from point to point. Agreement of the amplitudes is to be expected only directly above and below the soundpost.

Otherwise, the phase curves in figure 10.8 correspond to those in figure 10.9 in the region of the resonances of the body, as long as we account for the different phase reference. At ω_I ($\equiv \omega_2$) and ω_{II} ($\equiv \omega_3$), the phases jump 180° downward; between these two frequencies, at the antiresonances, they jump upward; φ_Z, however, jumps only downward,

---

[6] A violin built by Carleen M. Hutchins was lent by her for these investigations.

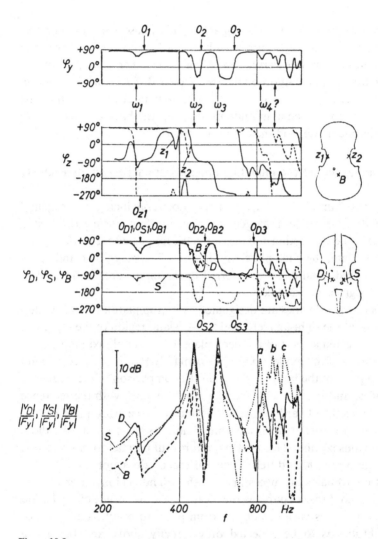

Figure 10.9
Measured velocity phases vs. frequency (three upper diagrams) and velocity-levels (bottom) normalized to the driving force, at the input of the bridge v_Y; the ribs at two points Z_1 and Z_2, v_Z; the top plate at D, v_D; the top plate at S (island region), v_S; and the back plate at B, v_B (after Beldie).

at the frequencies of the two resonances. These measurements prove that the model correctly incorporates the dynamic behavior of the ribs as a coupling element.

The first jump does not occur at exactly the same frequencies at the points Z_1 and Z_2, possibly due to rotation or torsion of the body. At very low frequencies, Z_1 and Z_2 are out of phase, seeming to indicate that the violin rotates around its longitudinal axis as a unit, as Eggers demonstrated for the cello. The considerable differences in the narrow band at the f-hole resonance seem rather to indicate a torsional mode. The model, then, cannot describe all phenomena even in the low-frequency range. Failure of the model beyond the second resonance peak was to be expected.

What casts the most suspicion on the model is that only the three instruments at the bottom of figure 10.1, out of six, exhibit only two body resonance peaks; the remaining instruments exhibit additional complications in this region, which might be approximated as three peaks.

The model could, however, exhibit a third peak only if the mass of the ribs were linked by a spring to a fixed point or a large mass. This could possibly be the mass of the neck, which "fixes" the ribs dynamically at one point. However, this hypothesis is questionable, as the third peak or separation of peaks does not occur in all instruments.

These phenomena must, then, be based on a coupling difficult to comprehend, which depends on small differences between instruments. One possible explanation is that the lowest natural frequency of the air cavity taken as a continuum (see sections 11.5, 11.6) falls within the frequency range we are considering. Such resonances cannot be incorporated in Beldie's model, with its exclusively vertical motions.

10.6 The wolf tone

The frequency curves of admittance in figure 10.1 and the model discussed in this chapter show two natural oscillatory modes of the body of equal importance. However, a phenomenon, particularly prevalent with the cello, that makes the player's task more difficult occurs at only a single "main resonance." At this *wolf tone*, it is difficult to produce a constant sound; rather, the tone tends to vary strongly and harshly at a frequency of approximately 5 Hz.

There is hardly any phenomenon related to the playing of string instruments which has brought forward as many attempts at explanation as this pathological process. It would take not a section but an entire chapter to discuss all of these explanations.

As different as these may be, they all agree that the wolf tone arises at the consonance of the fundamental frequency of the string with a very poorly damped natural frequency of the body of the instrument. (For this reason, there is no improvement between the low range of the d-string or higher positions on the g- or c-string when playing at the pitch of the wolf tone.)

A weakly damped oscillation of the body, however, means that the input impedance Z_0 of the bridge is no longer large in comparison with the characteristic impedance $m'c$ of the string. The bridge impedance may in fact be equal to that of the string, or even lower. In either of these cases, it would be impossible to excite oscillations at fundamental frequencies approximating $f_1 = c/(2l)$. Either the transverse velocity of the reflected wave at the bridge would be in the same phase as that of the incident wave, or the incident wave would be totally absorbed.

But even in these limiting cases, this would be true only once the natural oscillations of the body had built up to their full strength. If we think of the nut as being at an infinite distance from the bridge, and if we abruptly switch on a sinusoidal wave propagating toward the bridge, this is fully reflected at first. As the body begins to oscillate, however, the reflected wave diminishes; if the velocities of the incident wave and that of the bridge become equal, it diminishes to zero.

This elementary reflection process makes it plausible that, even when a string of finite length is in self-sustained oscillation, the body, at first oscillating weakly, conforms to the condition

$$m'c \ll |Z_0|, \tag{10.60}$$

which is necessary in order for a fundamental to be generated. Once this process is underway, however, the oscillations of the body build up; but the the "momentary impedance mismatch" is disturbed to such a degree that the generation of the fundamental collapses. The condition (10.60) then exists once again, and the fundamental is excited once more. The change in tone color due to this periodic process recalls that which occurs when organ pipes are overblown. (Whether wolves actually produce such sounds is unknown to the author.)

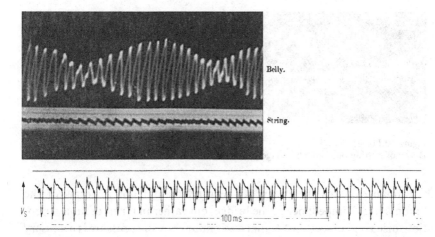

Figure 10.10
Oscillograms of time functions of a wolf tones. Top, displacement of the top plate; center, displacement of the string (after Raman). Bottom, velocity of the string (after Firth).

Impedance comparisons such as those we have mentioned are helpful even in explaining why wolf tones are much more common with the cello than with the violin. As early as 1963, Schelleng was able to use the law of similarlity to show that the ratio of the characteristic impedance of the strings to that of the resulting input impedance of the body under the feet of the bridge is 1.7 times as great for the cello as for the violin. Besides this, the mechanical properties of the bridge, acting as a lever system, reduce the impedance at the input to the bridge by a factor of the square of the ratio of the separation of the feet to the height of the bridge. Consequently, the much higher cello bridge (see figure 9.2) contributes a factor of 2.4 to the ratio of the characteristic string impedance the input impedance of the cello bridge; generation of wolf tones in the cello is, then, made easier by a factor of 3.4.

The classic oscillogram by Raman (1916) reproduced in figure 10.10, top, shows the alternation between oscillation of the string and of the top plate of the instrument. The oscillation of the top plate reaches its maximum amplitude when the fundamental of the string collapses.

Raman gives the displacement of the string at the point of bowing; we add an oscillogram from Firth (1978), obtained using an electrodynamic transducer, which shows velocity (figure 10.10, bottom), revealing the same periodic alternation in the waveform.

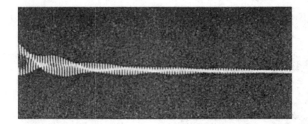

Figure 10.11
Decay of a string plucked while fingered at the position of the wolf tone (after Meamari).

Raman explains the decay of the fundamental not in terms of the "momentary" impedance of the bridge, but rather in terms of his assumption, discussed in section 4.6, that a minimum bowing pressure is necessary in order to generate Helmholtz motion and that this bowing pressure increases along with losses at the bridge. Losses, however, increase along with the oscillation of the body; with a given bowing pressure, a condition is attained in which Helmholtz motion with the fundamental frequency can no longer be sustained. This explanation incorporating the bowing pressure certainly corresponds much better to the frictional excitation mechanism than does an explanation based on linear processes related to impedances.

Raman's explanation is supported by the players' ability to suppress the wolf tone by increasing bowing pressure, at least at certain points of bowing (Meamari 1978, Güth 1978).

Explaining that wolf tones are always associated with two neighboring natural frequencies that can generate beats clearly does not account for this ability.

The concept of beats nonetheless explains very well the linear oscillations occurring in the plucked string when it is stopped at the wolf tone. Figure 10.11 gives an example, an oscillogram from Meamari (1978). (It is appropriate here to use the phrase "beats at the fundamental"; the term "wolf" should be reserved for situations in which the tone color changes periodically becuase the fundamental is periodically absent.)

The two natural oscillations that superimpose to produce beats in the plucked response may be obtained from the impedance condition at the input to the bridge:

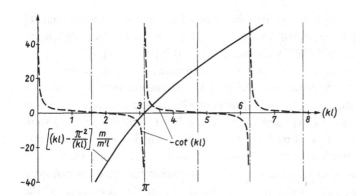

Figure 10.12
Graphic construction of the natural frequencies of a string terminated in a simple
oscillator (after Wagner).

$$\frac{\bar{F}}{\underline{v}} = j\omega m_0 + \frac{s_0}{j\omega} = -\frac{m'c}{j}\cot(kl). \tag{10.61}$$

(See also Wagner 1947, p. 225.) Here, we describe the input impedance
of the bridge by the equivalent mass m_0 and the equivalent stiffness s_0.
The input impedance for a string of wave number $k = \omega/c$ and length l
is shown at the right. The sign is negative, since this impedance corresponds to $-\bar{F}/\underline{v}$. If we express the natural frequency of the body in terms
of its wave number k_0,

$$k_0 = \frac{1}{c}\sqrt{\frac{s_0}{m_0}}, \tag{10.62}$$

then we can write (10.61) as

$$\frac{m_0}{m'l}\left(kl - \frac{(k_0l)^2}{kl}\right) = -\cot(kl). \tag{10.63}$$

Figure 10.12 shows the graphic solution of this equation for the case in
which the natural frequency of a string stretched between rigid supports
exactly matches that of the free oscillations of the body. In this case,
$k_0l = \pi$; we have chosen the value 10 for $m_0/m'l$. It can be seen that two
intersections occur near $kl = \pi$; also, that the intersections corresponding
to higher kl and so to higher partials occur very near $n\pi$ ($n = 2, 3, 4, \ldots$)
and are hardly altered.

The graphic construction also shows that beats can occur at any partial if they string is tensioned so slackly that $k_0 l = n\pi$. Raman and, more recently, Dünnwald (1979) also pointed out this possibility.

If—as is the case with some wind instruments and most electronic oscillators—the mechanism underlying the excitation were only slightly nonlinear, it would be possible to explain the basic characteristics of the wolf tone in terms of the self-excitation of beats. Schelleng (1963) took this approach, followed by Firth (1978) and Güth (1978). However, due to the strong nonlinearity of frictional excitation in bowing, the present author regards this approach as so questionable that it does not bear further examination. (The nonlinearity is not in the curvature of the friction characteristic, but rather in the alternation of sticking and sliding friction.) Besides, the simulation of the bowed string developed by McIntyre and Woodhouse (1979) and described in sections 8.2, 8.3, and 8.8 permits a treatment of the wolf tone incorporating the effect of frictional excitation properly.

Figure 10.13 shows oscillograms of wolf tones simulated on a computer by these authors with R. T. Schumacher (McIntyre, Schumacher, and Woodhouse 1983); in each pair of oscillograms, velocity at the point of bowing is shown at the top and that at the bridge, at the bottom. While the lower oscillograms show only beatlike functions, the upper ones show the typical change in waveform from one period of the fundamental to the next that we saw in the oscillogram of a cello in figure 10.10. Additionally, the oscillograms show the influence of bowing pressure on the waveform and the period. Bowing pressure was twice as great in the lower pair of oscillograms.

It would certainly be possible to examine the influence of the point of bowing systematically, and so to account for all of the nonperiodic manifestations of the wolf tone recorded by Meamari and by Güth.

The only question for the player, as Schelleng correctly emphasizes, is how to suppress the wolf tone. There have been many suggestions as to how to increase damping of the oscillatory mode of the body which the wolf tone excites. Selective damping has been suggested, so as not to damp other natural modes as well.

One possibility would be to couple the body of the instrument to another oscillator, turned to the wolf tone but more heavily damped.

The simplest possibility, suggested by Schelleng, is to attach a mass to the extension of a string between the bridge and tailpiece. This part of

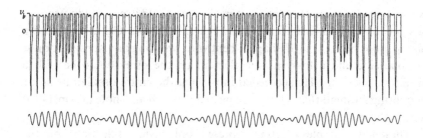

Bow force doubled:

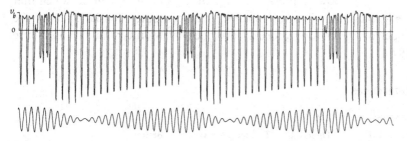

Figure 10.13
Simulation of wolf tones on a computer. For each pair of plots, the upper plot gives the
velocity at the point of bowing, while the lower plot gives the velocity of the bridge. The
bowing pressure for the lower pair of plots was twice that for the top pair (after McIntyre,
Schumacher, and Woodhouse).

the string has the same tension but is much shorter than the part which is
bowed, so its fundamental is normally much higher. Therefore, it is easy
to choose the mass and its position to bring about oscillation at the wolf
tone. If the mass is of a material with large internal losses, for example,
India rubber or, as Carleen Hutchins (quoted at the end of section VI
of Schelleng 1963) suggests, Plasticine, the desired damping can be
achieved.

Eggers (1959), on the other hand, recommends that such a damper be
placed directly on the front of the cello. His damper consists of a 5- to
10-mm sheet of foam rubber with a tuned mass attached. By choosing
its position, the coupling factor can be changed.

If the coupling factor and, especially, the damping, are not chosen
optimally, the addition of another oscillator can have a negative effect,

since it produces another resonant peak. This could itself tend to produce a wolf tone, or to worsen the tone. For this reason, Güth developed a damper consisting of a U-shaped, bent wire, working as a spring. A mass on one side and a lossy plastic element on the other could be slid to different positions. He succeeded in making the entire oscillator and its mounting so small that it could be passed through an f-hole and installed inside the instrument.

These few examples certainly represent only a small selection from the vast array of suggested wolf-tone dampers.

The best solution, however, is to advise the player—without making any changes in the instrument—on how to suppress the wolf tone by choosing the bowing pressure and point of bowing. It can be hoped that applicable rules will be obtained through computer simulation using the model of the bowing process developed by McIntyre and Woodhouse (1979).

References

Backhaus, H., 1930. *Z. Phys.* **62**, 143.

Beldie, J. P., 1975. Dissertation, Technical University of Berlin.

Benade, A. H., 1976. *Fundamentals of Musical Acoustics*. New York.

Cremer, L., and H. A. Müller, 1982. *Principles and Applications of Room Acoustics*. Tr. by T. J. Schultz. London: Applied Science.

Dünnwald, H., 1979. *Acustica* **41**, 238.

Eggers, F., 1959. *Acustica* **9**, 453.

Firth, J. M., 1978. *Acustica* **39**, 252.

Güth, W., 1978. *Acustica* **41**, 177.

Hutchins, C. M., K. A. Stetson, and P. A. Taylor, 1971. *Catgut Acoust. Soc. Newsletter* no. 16, 15.

Itakawa, H., and C. Kumagai, 1952. *Rep. Inst. Industr. Sci. Univ. Tokyo* **3** (1) 5.

Jahnke, E., and F. Emde, 1945. *Tables of Functions*. New York: Dover.

Jansson, E., N. E. Molin, and H. Sundin, 1970. *Phys. Scripta* **2**, 243.

McIntyre, M. E., R. T. Schumacher, and J. Woodhouse, 1983. *J. Acoust. Soc. Amer.* **74**, 1325.

McIntyre, M. E., and J. Woodhouse, 1979. *Acustica* **43**, 93.

Meamari, E., 1978. *Acustica* **41**, 94.

Raman, C. V., 1916. *Phil Mag.*, ser. 6, **32**, 391.

Rayleigh, Lord, 1878. *Theory of Sound*. London: Macmillan. Reprinted 1929.

Reinike, W., and L. Cremer, 1970. *J. Acoust. Soc. Amer.* **48**, 988.

Schelleng, J., 1963. *J. Acoust. Soc. Amer.* **35**, 326.

Wagner, K. W., 1947. *Einführung in die Lehre von den Schwingungen und Wellen* [*Introduction to the Theory of Oscillations and Waves*]. Wiesbaden.

11 The Body of the Instrument as a System of Oscillating Continua

11.1 The three continua

Analysis of the body of the violin as a system of partial masses linked by springs has provided helpful insights and explained certain phenomena; nonetheless, there was no doubt, even from the beginning, that the top and back plates are structures whose mass and bending stiffness are continuously distributed. Furthermore, the top and back, together with the ribs, enclose a cavity in which, as we have already pointed out, the places in which kinetic and potential energy are stored are not completely distinct from one another.

We are therefore faced with the task of treating these structures as continua. At low frequencies, this analysis should confirm the one in the preceding chapter; at high frequencies, this analysis is even simpler than one in terms of a large number of partial masses.

In our analysis in terms of continua, we will again use simplified models. We must ask how well these models approximate reality and at what point they lose their usefulness due to excessive mathematical complexity or insufficient physical clarity.

We might construct a simple model using two beams of constant thickness: 2.5 mm for the top plate and 4 mm for the back plate, connected by 1 mm thick rigid perpendicular beams representing the ribs. Even this simple model will provide a better approximation of high-frequency behavior than one with few degrees of freedom.

If we extend this two-dimensional model infinitely in the third dimension, we then obtain not only the transition to one-dimensional bending waves in plates; we also get coupling to the enclosed air cavity and between the plates via the air cavity. The air cavity, too, is a continuum; but it may be regarded as one-dimensional as long as the wavelength of sound is long compared with its 4-cm height—that is, up to approximately 1,500 Hz. The only waves that occur propagate parallel to the plates; but the air pressure in the cavity and the motions of the plates can nonetheless influence one another.

We may undertake a three-dimensional analysis, incorporating two-dimensional waves in bending and in the air cavity; there is little added mathematical complexity if we regard the plates and the air cavities as rectangular.

The very different bending stiffness of the wood parallel and perpendicular to the grain also adds little mathematical complexity if these directions are taken to be parallel to the edges of rectangular plates.

Strictly speaking, the top and back plates are, however, not plates; they are shells, each of which, in its rest position, has parallel curved surfaces. We can take this complication into account, though we will mostly use a simple cylindrical model.

On the other hand, it is not possible to include more representative boundary shapes in our calculations. No simplified but more representative shape—for example, an ellipse or a trapezoid, as studied by Savart (1819)—leads to results justifying the added mathematical complexity. The waist of the violin body and the f-holes have an influence too great to neglect.

The importance of the theoretical treatment in this chapter, as in previous ones, is that it provides insights into general laws which are also of interest in violin building. These laws lead to test methods that, if used during construction of an instrument, will produce more reliable results than the tapping of the top and back plates before they are glued to the ribs, described in the previous chapter.

An aviation engineer, say, can generate predictive models out of elements such as beams, plates, and shells, with the aid of a computer. No such models will ever be possible for the violin; in any case, predictive models are indispensable only when human lives are at stake. In the realm of art, in which instruments are "played," there is more freedom in design. If an engineering firm equipped with all of today's knowledge and instrumentation had been given the task of developing a string instrument, the resulting design would not be the same as the actual, empirically developed one.

11.2 General laws for plates with bending stiffness

We have already encountered the equation which applies to one-dimensional waves in bending: namely, equation (7.6), where it describes the limiting case of a string with bending stiffness but under no tension. In applying this equation to plates, we replace the mass per unit length with the mass per unit area:

$$m'' = \varrho h, \tag{11.1}$$

where h is the thickness of the plate. We also replace the bending stiffness B with the quantity B', the stiffness per unit width. In B', the area moment of inertia per unit width,

$$I' = \frac{h^3}{12},$$ (11.2)

appears as a geometric factor. The bending stiffness of the plate also differs from that of a rod (or beam) in that the transverse contraction, smaller than the longitudinal extension by a factor μ (μ is the Poisson number), can occur in two dimensions in the rod, but cannot occur in a plate in the lateral direction. This effect leads to an increase in the stiffness in extension from the normal modulus of elasticity E to

$$E' = \frac{E}{1 - \mu^2};$$ (11.3)

B', then, may be written as

$$B' = \frac{Eh^3}{(1 - \mu^2)12}.$$ (11.4)

In terms of velocity v, the one-dimensional equation of waves in bending in the plate is

$$-B'\frac{\partial^4 v}{\partial x^4} = m''\frac{\partial^2 v}{\partial t^2}.$$ (11.5)

As we showed in section 7.1 in connection with a string with bending stiffness, the solutions to this equation include not only wave motions

$$v = \mathrm{Re}\{\underline{\hat{v}}e^{j(\omega t \pm kx)}\},$$ (11.6a)

but also quasistationary near fields

$$v = \mathrm{Re}\{\underline{\hat{v}}_j e^{j\omega t \pm kx}\}.$$ (11.6b)

For a plate supported at $x = 0$ and $x = L$, these near fields go to zero. The velocity function

$$v = \mathrm{Re}\{\underline{\hat{v}}_n \sin k_n x e^{j\omega t}\}$$ (11.7)

with

$$k_n = \frac{2\pi}{\lambda_n} = \frac{n\pi}{L}$$

or

$$L = n\frac{\lambda_n}{2} \tag{11.8}$$

satisfies the boundary conditions requiring velocity to go to zero at both ends:

$$x = 0, L: \quad v = 0. \tag{11.9a}$$

Bending moments must also go to zero at both ends. As already mentioned in connection with (7.2), they are proportional to the second spatial derivative of displacement (curvature), and so to velocity:

$$x = 0, L: \quad \frac{\partial^2 v}{\partial x^2} = 0. \tag{11.9b}$$

Since bending stiffness decreases toward the edges of the back and top plates, this simplified boundary condition is a better approximation than that in which the plates are clamped and near fields occur.

If there are no near fields, however, the modes described in (11.7) and (11.8) are represented by exactly n sinusoidal half-waves extending from $x = 0$ to $x = L$. The pattern is the same as that of a string stretched between rigid supports.

This time, however, the ratios of natural frequencies are different. The equation of waves in bending is of the second order only with respect to time; with respect to position, it is of the fourth order. If we substitute (11.7) into it, we immediately obtain

$$B'k_n^4 = m''\omega_n^2. \tag{11.10}$$

Since

$$\frac{k_{n+1}}{k_n} = \frac{n+1}{n}, \tag{11.11}$$

as with the string, it follows that

$$\frac{\omega_{n+1}}{\omega_n} = \left(\frac{n+1}{n}\right)^2; \tag{11.12}$$

in other words, the ratios of natural frequencies are as $1 : 4 : 9 : 25 \ldots$. This increased spacing of natural frequencies results from the *dispersion* of waves in bending, the increase of their phase speed with frequency:

$$c_B = \frac{\omega}{k} = \sqrt[4]{\frac{B'}{m''}} \sqrt{\omega}. \qquad (11.13)$$

However, this increased spacing is compensated for by the increased number of natural frequencies in a two-dimensional plate.

We will here indicate only that the equation of two-dimensional waves in bending for a plate takes on the form

$$-B'\left(\frac{\partial^4 v}{\partial x^4} + 2\frac{\partial^4 v}{\partial x^2 \partial y^2} + \frac{\partial^4 v}{\partial y^4}\right) = m''\frac{\partial^2 v}{\partial t^2}. \qquad (11.14)$$

(See, e.g., Cremer and Heckl 1973, p. 257.)

It can also be shown that the boundary conditions of a rectangular plate whose length and width are L_x and L_y, supported with rotational freedom all around its edges, are satisfied by the spatial product

$$v = \mathrm{Re}\left\{\hat{v}\sin\frac{n_x\pi x}{L_x}\sin\frac{n_y\pi y}{L_y}\exp\left(j\omega_N t\right)\right\}. \qquad (11.15)$$

(See Rayleigh 1877, vol. 1, sec. 225.) Since this product also satisfies the differential equation (11.14) at certain frequencies, a simple substitution leads to the equation for the natural frequencies:

$$\omega_N^2 = \frac{B'}{m''}\left[\left(\frac{n_x\pi}{L_x}\right)^2 + \left(\frac{n_y\pi}{L_y}\right)^2\right]^2, \qquad (11.16)$$

or

$$f_N = \frac{\pi}{2}\sqrt{\frac{B'}{m''}}\left[\left(\frac{n_x}{L_x}\right)^2 + \left(\frac{n_y}{L_y}\right)^2\right]. \qquad (11.17)$$

If we introduce the speed of propagation of longitudinal waves parallel to the surface of the plate for the material we are using,

$$c_L' = \sqrt{\frac{E}{(1-\mu^2)\varrho}}, \qquad (11.18)$$

we may then write (11.17) as

$$f_N = 0.45c_L' h\left[\left(\frac{n_x}{L_x}\right)^2 + \left(\frac{n_y}{L_y}\right)^2\right]. \qquad (11.19)$$

It is easiest to see the various combinations of $n_x = 1, 2, 3, 4 \ldots$ and $n_y = 1, 2, 3, 4 \ldots$ with the aid of the grid shown in figure 11.1. The spacing of the grid is $\sqrt{0.45c'h/L_x}$ in the x-direction and $\sqrt{0.45c'h/L_y}$ in the

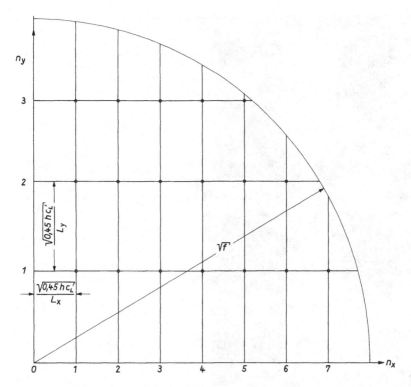

Figure 11.1
The grid of natural frequencies of a rectangular plate supported at its edges.

y-direction. Each intersection corresponds to a natural frequency. The length of the radius from the origin to each intersection is $\sqrt{f_N}$.

The lowest natural frequency, then, is

$$f_{11} = 0.45c'_L h(1/L_x^2 + 1/L_y^2). \tag{11.20}$$

The next higher frequency depends on the ratio of lengths of the edges.

The pair n_x, n_y characterizes not only a particular natural frequency, but also the number of nodes in the x and y directions, given by $(n_x - 1)$ and $(n_y - 1)$, respectively.

In order to allow a better comparison of this grid of nodal lines with the holographic photographs of the top and back plates to be shown later, we here show (figure 11.2) holographic photographs of the natural oscillations of a rectangular plate supported on all sides, for $n_x = 1$, 2 and

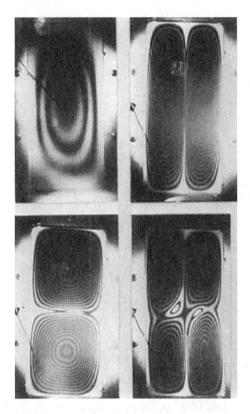

Figure 11.2
Holographic photograph of the natural bending oscillations of a rectangular plate supported at all its edges. Top, $n_x = 1$; bottom, $n_x = 2$; left, $n_y = 1$; right, $n_y = 2$ (after König).

$n_y = 1, 2$. The four photographs are arranged in a matrix, those representing the same values of n_x shown side by side and those representing the same values of n_y shown one above the other. The x-coordinate is vertical and the y-coordinate horizontal in each photograph, and the dimensions correspond to the elongated shape of the violin.[1]

As mentioned before, the existence of nodal lines in two dimensions increases the number of natural frequencies. The amount of the increase can be determined statistically if we consider a frequency so high that very many nodes are enclosed within the radius $\sqrt{f_N}$. The number N of natural frequencies below f_N is given by the area of a quarter-circle, $\pi f_N/4$, divided by the area of one rectangle of the grid, $0.45 c_L' h / L_x L_y$:

$$N = 1.75 \frac{L_x L_y}{c_L' h} f_N = 1.75 \frac{S}{c_L' h} f_N.$$

The number of natural frequencies, then, depends only on the material (c_L'), the thickness of the plate h, and the total area S.

From the corresponding formula for the natural frequencies of a room as derived by Weyl (1912) and Courant and Hilbert (1953), we can conclude that this generalization is not restricted to rectangular surfaces. In its limit form

$$\lim_{t \to \infty} N = 1.75 \frac{S}{c_L' h} f \tag{11.22}$$

it is valid for every boundary shape and boundary condition, including those of the back and top plates of stringed instruments. The only difference is that the frequency at which the generalization is a good approximation may be higher for other shapes than for rectangular ones.

We may even differentiate (11.22) and obtain the simple statement that the number of natural frequencies for a given frequency increment is constant:

$$\lim_{t \to \infty} \frac{\Delta N}{\Delta f} = 1.75 \frac{S}{c_L' h}. \tag{11.23}$$

These formulas are, to be sure, not valid for the lowest natural frequencies; not only because the specific ratio L_x/L_y influences the result,

---

[1] The photographs are taken from student work by O. König in the Institute for Technical Acoustics, Technical University of Berlin, 1970. An orthotropic plate (see n. 2 below) was used, but this makes no difference in the nodal patterns.

but also because our way of dividing the quarter circle into a grid halves the number of nodes along the x and y axes.

It is possible to consider a case in which no nodes occur, corresponding to $n = 0$; in this case, the average distance between nodes is somewhat greater at low frequencies.

The restriction of the number of natural frequencies given by (11.22) holds even when—as is the case—the thickness of the top and back plates vary from one point to another. In figure 11.1, only the location of the intersections would be changed. The presence of the f-holes exerts an effect only by making S smaller.

The grid in figure 11.1 can be taken as an approximation even if we assume a wooden plate of constant thickness whose stiffness in extension is much greater in the direction of the grain (the x-direction) than at right angles to it (the y-direction).[2] Only the spacing of the grid changes. If we assign different velocities of longitudinal waves, c'_x and c'_y to the two directions, then (11.19) need be changed only slightly to arrive at a sufficiently accurate approximation:

$$f_N = 0.45h \left[c'_x \left(\frac{n_x}{L_x} \right)^2 + c'_y \left(\frac{n_y}{L_y} \right)^2 \right];$$ (11.24)

in other words, the spacing of the grid is $\sqrt{0.45hc'_x}/L_x$ in the x-direction and $\sqrt{0.45hc'_y}/L_y$ in the y-direction. The differences in tension stiffness have exactly the same effect as differences in length and width. The two types of differences partially compensate for each other in string instruments, which are much less stiff in the shorter direction.

Measurements by Schelleng (1969) of sample strips give the following average values.

---

[2] Heckl (1960). Heckl first cites the literature to the effect that the equation for waves in bending of such orthotropic plates goes from (11.14) to

$$-\left(B'_x \frac{\partial^4 v}{\partial x^4} + 2B'_{xy} \frac{\partial^4 v}{\partial x^2 \partial y^2} + B'_y \frac{\partial^4 v}{\partial y^4} \right) = m'' \frac{\partial^2 v}{\partial t^2}.$$ (11.14a)

This equation contains three bending stiffnesses; all that can be said about B'_{xy} is that it lies between 0 and $(B'_x B'_y)^{1/2}$. Heckl then shows that the point impedances differ by only 18%, despite these values, which at first appear very different. However, according to Zener (1942), the point impedances are proportional to the densities of those natural frequencies of interest to us here, and so we make no serious mistake in choosing the geometric average of B'_x and B'_y as our value of B'_{xy} in calculating them. This choice leads to (11.24), which differs only slightly from (11.19).

for the top plate (spruce):

$c_x = 5.9 \times 10^5$, $c_y = 1.4 \times 10^5$ cm s^{-1};

for the back plate (maple):

$c_x = 3.6 \times 10^5$, $c_y = 1.8 \times 10^5$ cm s^{-1}.

In terms of the wavelengths in bending, the narrowest part of the top plate and the widest part of the back plate are not smaller than the length of the body of the instrument.

In (11.23), we must replace c'_L with the geometric mean of c_x and c_y:

$$\lim_{t \to \infty} \frac{\Delta N}{\Delta f} = 1.75 \frac{S}{\sqrt{c_x c_y} h}. \tag{11.25}$$

If we put in the values of c_x and c_y given above, then for a violin with $S = 550$ cm^2, we obtain an asymptotic spacing of natural frequencies for the top plate ($\sqrt{c_x c_y} = 2.8 \times 10^5$ cm s^{-1}; $\bar{h} = 0.25$ cm) of

$$\lim_{t \to \infty} \frac{\Delta f}{\Delta N} = 73 \text{ Hz}, \tag{11.26}$$

and for the back plate ($\sqrt{c_x c_y} = 2.6 \times 10^5$ cm s^{-1}; $\bar{h} = 0.4$ cm) of

$$\lim_{t \to \infty} \frac{\Delta f}{\Delta N} = 108 \text{ Hz}. \tag{11.27}$$

At low frequencies, the average spacings are greater, though not by as much as an order of magnitude.

A smaller number of resonance peaks in the frequency functions in figure 10.1 can and must result from the increasing damping at higher frequencies. This causes the peaks to be superimposed on one another; the variations in magnitude then result increasingly from the statistical behavior of the phase angles of the individual natural oscillations.[3]

11.3 Influence of the arching of the top and back plates

The arching of the top and back plates is certainly an essential element in the beauty of the violin's shape. We note, however, that it also serves a technical purpose. The thin top plate, additionally weakened by the

---

[3] This was shown by Schroeder (1954) for a similar problem involving an air cavity.

f-holes, would not be able to support the pressure of the strings transmitted through the bridge, if it were not arched.

Arched plates or shells can support greater loads because—unlike flat plates—they often cannot be deformed in a direction pointing toward or away from their center of curvature without creating tangential compression or tension; much greater forces are needed to produce these deformations. The difference is easiest to see if we examine a cylinder of radius R, displaced radially outward by an amount r due to an internal overpressure. The radius, then, is increased from R to $R + r$, and the circumference from $2\pi R$ to $2\pi(R + r)$. The tangential extension is (r/R), and so the tangential force per unit length is

$$F' = E'h\frac{r}{R}. \tag{11.28}$$

Because of the curvature, this corresponds to a inward-directed force per unit area of

$$F'' = -E'h\frac{r}{R^2}, \tag{11.29}$$

which balances the outward-directed overpressure.

For this reason, pressure vessels are always round, and large spans can be bridged with thin concrete shells, whereas horizontal structures would collapse. For this reason too, the circular walls of the tubes of wind instruments can be regarded as rigid with respect to the sound waves in the air within them.

The requirement for static strength in stringed instruments does not prevent the use of very different curvatures; these differences have, in fact, been characteristic of the instruments of different master builders.

All builders' instruments nonetheless have one feature in common: the curvature changes direction in order that the plates may become plane everywhere at the edges, and meet the ribs perpendicularly. This gradation is necessary because of the complicated shape of the edge. Thus, a region is created in which the lack of curvature reduces stiffness; much of the compliance of the top and back plates is in this region. Though it is an exaggeration, it is helpful to think of the top and back plates as stiff hats with compliant brims. The bass bar in the top plate does not reach all the way to the edges, and adds to this effect. It is precisely this effect that justifies the modeling of a plate as a piston suspended on leaf springs, as

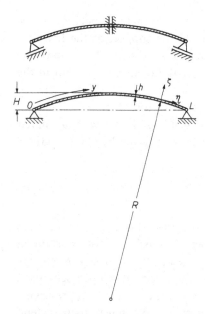

Figure 11.3
Cylindrical shells with tangentially freely movable or rigid supports.

in the previous chapter. The main problem with that model was posed by
the ill-defined "island" in the top plate.

The acoustic behavior of the top and back plates is affected by the way
the added stiffness influences the natural frequencies.

A strict mathematical treatment of the shape of the edge and of the
curvature of actual violin plates is not possible. Consequently, we will
first consider the simple case of a cylindrical plate of uniform radius R.
We will first consider deformations in two dimensions only, in the plane
of curvature; these will reveal that boundary conditions have a strong
influence. The equations are the same as those for a curved beam, except
that we replace m' by m'' and E by E'.

Figure 11.3 shows such a plate, whose rest shape is curved. The dimensions shown correspond approximately to those of a cross section through
a violin top plate between the ribs on either side, with a maximum height
of 1 cm, a width of 9 cm at the waist, and a radius of curvature (assumed
here to be constant) of 11 cm. We take the average shell thickness to be
$h = 2.5$ mm.

As can also be seen in figure 11.3, the coordinate of position is the length of the curved beam with respect to the support at the left side. We call this y, though it no longer corresponds to the y-direction of the plate. We call displacement in the tangential direction η and that in the radial direction ζ.

Radial acceleration, then, depends on the force per unit area, as described in (7.6), which results from bending. Now, with a curved shell, there is an additional force per unit area, given by (11.29). Given these conditions, we would obtain

$$(\varrho_p h)\frac{\partial^2 \zeta}{\partial t^2} = -\frac{E'h^3}{12}\frac{\partial^4 \zeta}{\partial y^4} - \frac{E'h}{R^2}\zeta. \tag{11.30}$$

But the last term, as we see in (11.28), arises from a tangential force per unit width, and this force is, generally speaking, not constant at different points on the curved beam. Consequently, a term $(E'h/R)\partial\zeta/\partial y$ appears in the equation for tangential displacement. Conversely, each change in η leads to a further driving force in the direction of ζ, preserving reciprocity in the terms describing coupling. We must, then, extend (11.30) to

$$(\varrho_p h)\frac{\partial^2 \zeta}{\partial t^2} = -\frac{E'h^3}{12}\frac{\partial^4 \zeta}{\partial y^4} - \frac{E'h}{R^2}\zeta - \frac{E'h}{R}\frac{\partial \eta}{\partial y} \tag{11.31}$$

and we must add

$$(\varrho_p h)\frac{\partial^2 \eta}{\partial t^2} = E'h\frac{\partial^2 \eta}{\partial y^2} + \frac{E'h}{R}\frac{\partial \zeta}{\partial y} \tag{11.32}[4]$$

to account for the tangential motion. The first term on the right side depends on tension in the y-direction. (We have encountered this term before in our study of the behavior of the bow hairs in section 6.3.) If we leave the coupling term out of (11.32), what is left is the ordinary longitudinal wave equation (here again with transverse contraction hindered) with the speed of propagation

$$c_L = \sqrt{\frac{E'}{\varrho_p}}. \tag{11.33}$$

---

[4] See, e.g., Cremer (1955), Kennard (1953). The additional terms Kennard accounts for in our equation (11.31) play an important role for the full cylinder at low frequencies, but not in shallow cylindrical sections such as are of interest to us here.

After we have divided (11.31) and (11.32) by $(\rho_p h)$, E' and ρ_p appear only in this combination. Therefore, we shall rewrite them accordingly.

The wave functions

$$\zeta = \mathrm{Re}\{\hat{\underline{\zeta}}_0 e^{j\omega t \pm jky}\},\tag{11.34a}$$

$$\eta = \mathrm{Re}\{\hat{\underline{\eta}}_0 e^{j\omega t \pm jky}\}\tag{11.34b}$$

embrace the possibility of quasistationary near fields, with an imaginary wave number k. If we substitute (14.34a, b) into our differential equations (11.31) and (11.32), we at first obtain two linear equations for $\hat{\underline{\zeta}}_0$ and $\bar{\eta}_0$, and the vanishing of the determinant of coefficients yields the dispersion equation $\Delta(\omega, k) = 0$:

$$\Delta(\omega, k) = \Delta^4 - \omega^2\left(\frac{h^2}{12}c_L^2 k^4 + c_L^2 k^2 + \frac{c_L^2}{R^2}\right) + \frac{c_L^4 h^2}{12}k^6 = 0.\tag{11.35}$$

If we introduce the wave number in bending from (11.13)

$$k_B^4 = 12\omega^2/(c_L^2 h^2)$$

and the longitudinal wave number from (11.33)

$$k_L^2 = \omega^2/c_L^2$$

which can also be regarded as representing frequency, then (11.35) can be rewritten as

$$(k^4 - k_B^4)(k^2 - k_L^2) - k_B^4/R^2 = 0.\tag{11.36}$$

This form makes it clear that the coupling of bending waves and longitudinal waves becomes weaker and finally goes to zero as $R \to \infty$; in other words, as the shell becomes a flat plate.

In a shell, on the other hand, with its curvature, there are generally always three types of fields: two waves and a near field. The wave numbers k_I, k_{II}, and k_{III} correspond to these three fields.

This situation also corresponds to the existence of three symmetrical pairs of boundary conditions. Two are already familiar, in connection with plates:

$$y = 0, L: \quad \zeta = 0,\tag{11.37a}$$

$$y = 0, L: \quad \frac{\partial^2 \zeta}{\partial y^2} = 0.\tag{11.37b}$$

The third, for a shell with rigid supports, would be:

$$y = 0, L: \quad \eta = 0. \tag{11.37c}$$

It should be noted that L, in accord with the definition of y, represents not the spacing of the ribs but the length of the arc which connects them. In our example, this length is 9.25 cm, while the spacing of the ribs is 9 cm. If we were to compare the resulting natural frequencies with those of a plate of equal length whose radius of curvature $R = \infty$, we would have to substitute this length. In the fundamental mode, this difference amounts to only a semitone, however, so it is of no importance in connection with our present, basic observations.

The inclination at the edges is much more important if we choose to replace the rigid supports with others that are somewhat movable laterally. This change would alter (11.37a), since ζ has a horizontal component.

The mathematically simplest extreme example, in contrast to (11.37c), requires that longitudinal tension (in the y-direction) go to zero at the boundaries:

$$y = 0, L: \quad E'\left(\frac{\partial \eta}{\partial y} + \frac{\zeta}{R}\right) = 0. \tag{11.37d}$$

Since $\zeta = 0$, this leads simply to:

$$y = 0, L: \quad \frac{\partial \eta}{\partial y} = 0. \tag{11.35e}$$

This boundary condition, however, would require the two supports to behave like rollers (i.e., allowing tangential motion), as shown at the top in figure 11.3. This change results in the static position becoming undefined; a vertical guide in the middle remedies this problem. This type of structure would certainly not be useful in civil engineering for a shell, since it would eliminate the stiffening effect of longitudinal tension. In a treatment of the behavior of the violin, it also does not account correctly for the lateral compliance of the ribs, a phenomenon which certainly exists.

We nonetheless discuss this structure first, because it shows the great dependence of the natural frequencies on the choice of boundary conditions, and with minimal mathematical complexity.

In this limiting case, the mathematical expression of the motion becomes greatly simplified, and only one of the possible types of fields is necessary: that which corresponds to a bending oscillation of a beam supported at both ends. It can be seen that the expression

$$\zeta = \mathrm{Re}\left\{\hat{\underline{\zeta}}_0 \sin\frac{n\pi y}{L} e^{j\omega t}\right\}, \tag{11.38a}$$

fulfills the first and second boundary conditions applying to bending waves in plates. If we sustitute (11.38a) into (11.31) or (11.32), we obtain

$$\eta = \mathrm{Re}\left\{\hat{\eta}_0 \cos\frac{n\pi y}{L} e^{j\omega t}\right\}, \tag{11.38b}$$

which fulfills the third boundary condition. The waves we obtain are, then, simple waves with the discrete wave numbers

$$k = \frac{n\pi}{L}, \quad n = 1, 2, 3 \ldots, \tag{11.39}$$

propagating in opposite directions. The characteristic equation of the system results from substitution of (11.39) into the dispersion equation (11.35).

Since the first two boundary conditions are fulfilled, it seems obvious to assume that k differs only slightly from the wave number in bending, k_B, which results when we neglect the curvature.

If we substitute

$$k^2 = k_I^2 = k_B^2 + \Delta(k_I^2) \tag{11.40}$$

into the dispersion equation, the relative deviation may be approximated by

$$\frac{\Delta(k_I^2)}{k_B^2} = \frac{1}{2}\frac{1}{(k_B^2 - k_L^2)R^2}; \tag{11.41a}$$

then, in the frequency range of interest to us, that of the lowest natural frequencies, where $k_B^2 \gg k_L^2$, it follows that

$$\frac{\Delta(k_I^2)}{k_B^2} \approx \frac{1}{2(k_B R)^2} = \frac{1}{2}\left(\frac{\lambda_B}{2\pi R}\right)^2. \tag{11.41b}$$

The longest bending wavelength of interest is $2L$. In the cross section of the top plate of a violin, $L = 9.27$ cm and $r \approx 11$ cm. Consequently

$$\Delta(k_I^2)/k_B^2 = 0.036, \tag{11.41c}$$

confirming that the discrepancy in the fundamental frequency is indeed small. It becomes even smaller for higher-order modes, in proportion to n^{-2}, as long as the relationship $k_B^2 \gg k_L^2$ holds.

We see, then, that the curvature has only a slight effect on the lengths of bending waves of interest to us, as long as the ribs are compliant.

The case in which the ribs are regarded as rigid is much more complicated, since it involves all three fields, each in connection with all three boundary conditions. It is no longer possible to represent even the "bending-wave component" ζ_I by an expression conforming to (11.38a), since the three components of the field do not individually go to zero at $y = 0$ and $y = L$. On the other hand, we can assume that the field quantities are symmetric or antisymmetric with respect to the middle of the arch at $y = L/2$, due to the symmetry of the boundary conditions. It is better, then, for us to use the coordinates

$$u = y - \frac{L}{2} \tag{11.42}$$

in our description. If we do this, it is clear that the vertical displacement of the lowest natural vibration ζ can be described by the expression

$$\underline{\zeta} = \underline{\hat{\zeta}}_I \cos k_I u + \underline{\hat{\zeta}}_{II} \cos k_{II} u + \underline{\hat{\zeta}}_{III} \cos k_{III} u. \tag{11.43}$$

Using (11.40), we can simplify this by replacing k_I by k_B and k_{II} by jk_B. The latter of these two subsitutions is the same as replacing $\cos k_{II} u$ by $\cosh k_B u$.

The wave number k_{III} is certainly determined mainly by longitudinal motions, and so it would seem reasonable to replace it by k_L. However, if we substitute the expression corresponding to (11.40),

$$k_{III}^2 = k_L^2 + \Delta(k_{III}^2), \tag{11.44a}$$

into (11.35) and if we then assume—however k_{III}^2 may be determined exactly—that $k_{III}^2 \ll k_B^2$ in the range of the lower natural frequencies, we obtain the approximation

$$k_{III}^2 \approx k_L^2 - \frac{1}{R^2} = k_L^2 \left(1 - \left(\frac{\lambda_L}{2\pi R} \right)^2 \right)$$

$$= k_L^2 \left(1 - \left(\frac{f_R}{f} \right)^2 \right). \tag{11.44b}$$

While the question of whether or not $k_I^2 \approx k_B^2$ could be answered by determining whether $\frac{1}{2}(\lambda_B/2\pi R)^2 \ll 1$ in (11.41b), it is important in the present case to determine how large $(\lambda_L/2\pi R)^2$ is. A longitudinal wave is much longer than a bending wave in the frequency range of interest. This is easiest to recognize in connection with the last formulation of the "correction term" in (11.44b). This allows us to evaluate its relative size from the ratio of a reference frequency

$$f_R = \frac{c_L}{2\pi R}. \tag{11.45}$$

to the frequency under observation. The frequency f_R represents the inverse of the period corresponding to a full trip of a longitudinal wave around a circle with the given radius of curvature R. We may call f_R the *ring frequency*. In our example, this would be at 2,000 Hz, far above the 195 Hz that we get as the lowest natural frequency in pure bending. The second term in (11.44b), which we first introduced as a corrective term, predominates in the frequency range of interest! For this reason, it is imperative to determine the actual value of the ratio $|k_{III}/k_B|^2$ at 195 Hz, which ratio we earlier assumed to be small. We can calculate its value generally as

$$\left|\frac{k_{III}}{k_B}\right|^2 = \frac{2\pi}{\sqrt{12}}\frac{h}{c_L}\left|f - \frac{f_R^2}{f}\right|; \tag{11.46}$$

in our numerical example, this leads to the just tolerable value 0.066 at the lowest natural frequency in bending. At higher natural frequencies, the quotient becomes smaller and smaller.

In the frequency range of interest, then, k_{III} is an imaginary quantity; we do not need to make formal use of it. We replace (11.42) by:

$$\underline{\zeta} = \hat{\underline{\zeta}}_I \cos k_B u + \hat{\underline{\zeta}}_{II} \cosh k_B u + \hat{\underline{\zeta}}_{III} \cos\left(k_L u \sqrt{1 - (f_R/f)^2}\right). \tag{11.47}$$

Using this expression, we obtain the first and second boundary conditions at $u = \pm L/2$:

$$\zeta = 0:$$

$$\hat{\underline{\zeta}}_I \cos\frac{k_B L}{2} + \hat{\underline{\zeta}}_{II} \cosh\frac{k_B L}{2} + \hat{\underline{\zeta}}_{III}\cos\left(\frac{k_L L}{2}\sqrt{1 - (f_R/f)^2}\right) = 0; \tag{11.48a}$$

$$\frac{\partial^2 \zeta}{\partial u^2} = 0:$$

$$-\underline{\zeta}_{\mathrm{I}} k_{\mathrm{B}}^2 \cos \frac{k_{\mathrm{B}} L}{2} + \underline{\zeta}_{\mathrm{II}} k_{\mathrm{B}}^2 \cosh \frac{k_{\mathrm{B}} L}{2}$$

$$- \hat{\zeta}_{\mathrm{III}} k_{\mathrm{L}}^2 (1 - (f_{\mathrm{R}}/f)^2) \cos \left(\frac{k_{\mathrm{L}} L}{2} \sqrt{1 - (f_{\mathrm{R}}/f)^2} \right) = 0.$$

$$(11.48\mathrm{b})$$

The third boundary condition, $\eta = 0$ at $u = \pm L/2$, requires that the expression for $\eta(u)$ in the coupled wave equations conform to (11.47). This is obtained most simply from (11.32), for each of the components N, as:

$$\eta_N = \frac{1}{R} \frac{d\zeta_N}{du} (k_N^2 - k_{\mathrm{L}}^2); \qquad (11.49)$$

in other words, we obtain from (11.47), neglecting the same terms as before,

$$\eta = -\underline{\zeta}_{\mathrm{I}} (\sin k_{\mathrm{B}} u)/k_{\mathrm{B}} R - \underline{\zeta}_{\mathrm{II}} (\sinh k_{\mathrm{B}} u)/k_{\mathrm{B}} R$$

$$+ \hat{\zeta}_{\mathrm{III}} k_{\mathrm{L}} R \sqrt{1 - (f_{\mathrm{R}}/f)^2} \sin (k_{\mathrm{L}} u \sqrt{1 - (f_{\mathrm{R}}/f)^2}). \qquad (11.50)$$

From this, we obtain the third boundary condition:

$$\eta = 0:$$

$$-\underline{\zeta}_{\mathrm{I}} \sin \left(\frac{k_{\mathrm{B}} L}{2} \right) \Big/ (k_{\mathrm{B}} R) - \underline{\zeta}_{\mathrm{II}} \sinh \left(\frac{k_{\mathrm{B}} L}{2} \right) \Big/ (k_{\mathrm{B}} R)$$

$$+ \hat{\zeta}_{\mathrm{III}} k_{\mathrm{L}} R \sqrt{1 - (f_{\mathrm{R}}/f)^2} \sin \left(\frac{k_{\mathrm{L}} L}{2} \sqrt{1 - (f_{\mathrm{R}}/f)^2} \right) = 0. \qquad (11.51)$$

This time, it cannot be suggested that the three boundary conditions in (11.48a, b) and (11.51) might be fulfilled by an expression for waves in bending alone. The natural frequencies occur where the determinants formed by the coefficients go to zero. This condition leads at first to a difficult transcendental equation in whose arguments the frequency to be found occurs in k_{B}, in k_{L}, and explicitly.

We will not give this equation now, since we will encounter boundary conditions of the same type when we consider the coupling of a plate to an enclosed cushion of air in section 11.6; there the significance of k_L and f_R will be different. In that section, we will examine how the problem can be simplified by introducing an appropriate frequency parameter and by considering which values of the equation's variables are of interest.

We will go no further here than to indicate that calculation using the parameters shown in figure 11.3 results in a natural frequency 7.8 times (!) that of the equivalent flat plate. As we noted earlier, this frequency was almost unchanged in a curved plate free to move tangentially.

This shows the great effect of different boundary conditions, which can hardly be recognized in any actually given mechanical structure. Certainly, the ribs cannot stiffen the edges of the plates so greatly as to force the condition $\eta = 0$. The actual situation lies nearer the limiting case of free tangential motion. Besides, we have pointed out already that the horizontal edge region of the back or top plates, their thinning, and the groove in which the purfling is inlaid concentrate springiness there, so that the middle portion of the plate oscillates as a piston. The additional stiffening due to the curvature has no inertial effect in the middle.

The situation could be different at higher frequencies, at which the middle is divided by nodal lines. Evaluation of (11.48a, b) and (11.51), which embraces all symmetrical cases with an even number of nodal lines, shows that the next corresponding natural frequency is raised by a factor of only 1.19 compared with that of the corresponding flat plate. It is easy to understand how oscillations with areas of opposite displacement ζ generate less tension than unilateral displacements, because the tension and compression partly cancel from side to side. This phenomenon is especially to be expected in the case of all oscillations which are antisymmetric in ζ.

It would, however, be a great mistake to conclude from the one-dimensional case considered up to now that the stiffness due to curvature has a significant effect only under boundary conditions which restrict motion. This conclusion holds only when the displacements change only in the transverse (y) direction.

The other extreme case is that of a purely axial change. (We call the axial direction the x-direction, based on our choice of coordinates for the violin.) In this case, the propagation of waves with radial displacements ζ is already affected by the differential equations.

Using a rectangle cut out of cardboard, it is easy to see that curvature

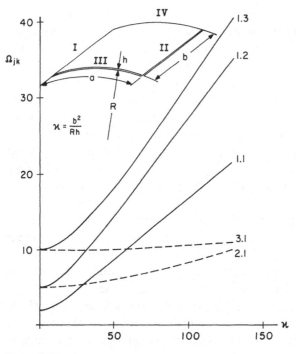

Figure 11.4
Natural frequencies, normalized as in (11.52b), of a cylindrical shell segment free to move
tangentially on all sides, as a function of the curvature parameter as in (11.52a) (after
Emmerling and Kloss).

around the x-axis hardly restricts deformations in the y, z-plane, but
strongly restricts those in the x, z-plane.

It is not within the scope of this book to treat the two-dimensional
theory of cylindrical shells thoroughly: rather, readers are referred to the
works cited above and in what follows.

Emmerling and Kloss (1979) have assembled some of the results for a
cylindrical-shell segment of arc length a (see figure 11.4, top) and longitudinal length b. Various boundary conditions were investigated at boundaries III and IV, but the axial boundaries I and II were free to move
tangentially in all cases. The boundary conditions at III and IV have a
clear effect. But even when these are free to move tangentially, there is a
considerable increase in the natural frequencies, as shown in the curves in
figure 11.4, which are presented as an example of Emmerling's and Kloss's

work. The example shown is for the case $a = b$ and Poisson number $\mu = 0.3$. The abscissa is defined by the parameter

$$\varkappa = b^2/(Rh) \tag{11.52a}$$

This increases with the curvature (the example discussed here corresponds to $\varkappa = 30$). The ordinate is the frequency, also normalized,

$$\Omega_{mn} = \frac{\omega_{mn}}{(\pi/b)^2 \sqrt{E'h^2/12\varrho_p}}. \tag{11.52b}$$

In this case, with the boundary conditions of free tangential motion, Emmerling and Kloss give a formula

$$\omega_{mn}^2 = \frac{E'h^2}{12\varrho_p}\left[\left(\frac{m\pi}{a}\right)^2 + \left(\frac{n\pi}{b}\right)^2\right]^2 + \frac{E}{R^2\varrho_p}\left(\frac{n\pi}{b}\right)^4 \bigg/ \left[\left(\frac{m\pi}{a}\right)^2 + \left(\frac{n\pi}{b}\right)^2\right]^2. \tag{11.53}$$

This formula and figure 11.4 show to what a great degree the arching of the plates can raise certain natural frequencies, even those without nodal lines ($m = n = 1$).

In (11.53), the first term that remains when $R \to \infty$ corresponds to the natural frequencies of a rectangular plate supported rigidly on all sides, as in (11.16). The second term shows the influence of stiffening resulting from the curvature. As we might expect, it becomes greater as R becomes smaller.

Emmerling and Kloss based their calculations on the simplifying assumption that the effect of inertial terms in tangential and axial directions could be neglected. They adopted this simplification from the work of Reissner (1955). He had used it in treating the case in which the shell is curved between all four edges, the radius of curvature between edges I and II being R_b, and that between edges III and IV being R_a. With, again, free tangential motion on all sides, he was able to extend (11.53) to:

$$\omega_{mn}^2 = \frac{E'h^2}{12\varrho_p}\left[\left(\frac{m\pi}{a}\right)^2 + \left(\frac{n\pi}{b}\right)^2\right]^2 + \frac{E}{\varrho_p}\frac{\left[\frac{1}{R_a}\left(\frac{m\pi}{a}\right)^2 + \frac{1}{R_b}\left(\frac{n\pi}{b}\right)^2\right]^2}{\left[\left(\frac{m\pi}{a}\right)^2 + \left(\frac{n\pi}{b}\right)^2\right]^2}. \tag{11.54}$$

Two cases shown in figure 11.4 are of particular relevance to our study of the violin. In case 2.1, with one phase reversal along the arc a, there is only a slight increase in frequency due to the stiffening effect of the arc; in case 3.1, with two phase reversals, there is almost no increase in frequency. This effect is due to the lengths of the arc that oscillate in phase; as the number of nodes increases, these represent ever flatter parts of the shell.

From this and from what we have said earlier about oscillations in the y, z-plane, we can conclude that the points which do not lie near the edges in the grid in figure 11.1, and which represent the natural frequencies of a rectangular plate supported on all sides, hardly change their natural frequencies. Only the points near the edges have significantly raised natural frequencies; some are raised to a great degree.

The number of natural frequencies, however, is not changed. The asymptotic formulas for these, (9.18a) and (9.19), remain valid regardless of curvatures and boundary conditions.

Natural frequencies above f_R, corresponding to the second, largely longitudinal type of wave, are of no interest in connection with the violin, since they scarcely lead to radiation of sound.

11.4 Natural modes of the cavity of the violin

The last of our general observations will deal with some of the basic properties of the cavity of the violin, and of its coupling to the top and back plates. The air in the cavity does not behave simply as a spring, as was assumed in the previous chapter. Rather, mass and stiffness are distributed through it, leading to natural oscillatory modes with nodes and antinodes; this has long been known. The length of the violin's cavity is approximately 35 cm, a half-wavelength at 488 Hz (b above middle c). This is approximately the same length as the column of air in a flute playing the same note.

The only points in dispute have been whether these oscillations have a significant reaction effect on the top and back plates, whether they can be excited by the motions of the plates, and whether the natural modes inside the cavity lead to significant radiation via the f-holes.

Since, once again, the shapes of the boundaries are not amenable to calculation, the exact position of the air cavity resonances can be determined only by experiment.

To this end, Jansson (1973) set up oscillations in the cavity of a violin using an ionophone: a device which produces sound by ionizing air using electrical discharges. He inserted this into the instrument through a small hole drilled near the lower end block, and measured sound pressure at the input of a system of three small tubes each of whose inside diameter was only 0.25 mm, inserted through the same hole and connected to a sound pressure transducer at the outside of the instrument. He coated the body of the instrument thickly with plaster so as to eliminate the excitation of oscillations in the walls of the cavity. The frequency functions of sound pressure level

$$L_p = 20 \log_{10} \frac{p}{p_0} \, dB, \tag{11.55}$$

where p_0 is arbitrary, are shown in figure 11.5: at the top, with the f-holes closed; at the bottom, with them open, and with a surrounding area (reaching as far as the ribs) cleared of plaster to allow free radiation.

In this experiment, every peak in sound pressure level represents a resonance at a natural mode of the air cavity.

With the exception of the Helmholtz resonance at the left end of the lower of the two diagrams, the difference between them in the number and position of the peaks and valleys is slight. In the lower diagram, the peaks are not quite as high and the valleys not as deep; peaks and valleys are somewhat broader. These differences indicate that losses are somewhat greater with open f-holes. (It is not clear whether this is due to radiation or to friction.) Losses are especially evident when the f-holes coincide with areas of high sound pressure, as found in models with flat top and back plates. We can, then, settle the third of our points in dispute: The f-holes do radiate some sound at other air resonances besides the f-hole resonance, but the amount is significant at only a few frequencies.

It is interesting to compare these natural frequencies, which cannot be derived through mathematical calculation, with those we would expect in a rectangular box with approximately the same dimensions as the cavity of the violin (length 34 cm, width 16 cm, and height 4 cm). As mentioned in the discussion of (10.12), sound in air consists of longitudinal waves of the type we examined in section 6.3 for a one-dimensional, solid medium. As the modulus of elasticity here, we substitute the volume stiffness (modulus of compression) K, already used in (10.9), for the tensional stiffness E. In the case of a one-dimensional plane wave, this leads to the relationship

$$p = -K \frac{\partial \xi}{\partial x} \tag{11.56}$$

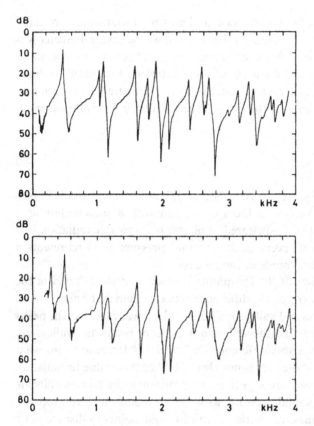

Figure 11.5
Frequency functions of sound pressure level in the cavity of a violin excited with an
ionophone. Top, with the f-holes closed; bottom, with the f-holes open (after Jansson).

between pressure and compression $(-\partial\xi/\partial x)$. If we examine only sinusoidal oscillations and replace displacement with velocity $\underline{v}$ in this vector equation, then

$$\underline{p} = -\frac{K}{j\omega}\frac{\partial\underline{v}_x}{\partial x}. \tag{11.57a}$$

In the three-dimensional case, the gradients of the components of velocity in the y and z directions must be added:

$$\underline{p} = -\frac{K}{j\omega}\left(\frac{\partial\underline{v}_x}{\partial x} + \frac{\partial\underline{v}_y}{\partial y} + \frac{\partial\underline{v}_z}{\partial z}\right). \tag{11.57b}$$

On the other hand, the pressure gradients in these three directions generate the corresponding components of acceleration of a volume element of density ρ:

$$-\frac{\partial\underline{p}}{\partial x} = j\omega\varrho\underline{v}_x, \qquad -\frac{\partial\underline{p}}{\partial y} = j\omega\varrho\underline{v}_y, \qquad -\frac{\partial\underline{p}}{\partial z} = j\omega\varrho\underline{v}_z. \tag{11.58}$$

If we substitute (11.58) into (11.57), we obtain the three-dimensional wave equation for sinusoidal sound waves in air. In Cartesian coordinates, this *Helmholtz equation* is

$$c^2\left[\frac{\partial^2\underline{p}}{\partial x^2} + \frac{\partial^2\underline{p}}{\partial y^2} + \frac{\partial^2\underline{p}}{\partial z^2}\right] + \omega^2\underline{p} = 0 \tag{11.59}$$

The speed of sound is here $c = \sqrt{K/\rho}$, as was already the case in (10.12).[5]

In the present example of a rectangular box, the pressure gradients normal to the walls go to zero at the walls:

$$x = 0, L_x: \quad \frac{\partial p}{\partial x} = 0,$$

$$y = 0, L_y: \quad \frac{\partial p}{\partial y} = 0,$$

$$z = 0, L_z: \quad \frac{\partial p}{\partial z} = 0. \tag{11.60}$$

---

[5] A more thorough derivation which mentions the neglected terms is to be found in, e.g., Cremer and Müller (1982, vol. 2, ch. 1).

It is best to describe the sound field in terms of pressure, because this, unlike velocity, is a scalar; because it is easier to measure; and, last but not least, because it is the field variable which excites our eardrums—the one we hear.

The wave equation and the boundary conditions in our present example are fulfilled by products of cosine functions of the three coordinates:

$$\underline{p}(x, y, z) = \underline{p}_0 \cos\left(\frac{n_x \pi x}{L_x}\right) \cos\left(\frac{n_y \pi y}{L_y}\right) \cos\left(\frac{n_z \pi z}{L_z}\right). \tag{11.61}$$

By substituting this into the wave equation, we obtain the natural frequencies:

$$\omega_N^2 = c^2 \pi^2 \left[\left(\frac{n_x}{L_x}\right)^2 + \left(\frac{n_y}{L_y}\right)^2 + \left(\frac{n_z}{L_z}\right)^2 \right], \tag{11.62}$$

or

$$f_N = \frac{c}{2} \sqrt{\left(\frac{n_x}{L_x}\right)^2 + \left(\frac{n_y}{L_y}\right)^2 + \left(\frac{n_z}{L_z}\right)^2}. \tag{11.63}$$

Once more, the pattern of all possible natural frequencies can be represented by a rectangular grid—this time, a three-dimensional one. The spacings are

$$f_{100} = \frac{c}{2L_x}, \qquad f_{010} = \frac{c}{2L_y}, \qquad f_{001} = \frac{c}{2L_z}. \tag{11.64}$$

This time, the radius to each point in the grid directly represents each natural frequency. The subscripts in (11.64) relate to particular values of n, and n can take on the value zero; in other words, there are one-dimensional wave patterns in all three directions. This was not possible for plates supported on all sides.

In the case of string instruments, however, L_z is so small with respect to the other two dimensions that many resonances in n_x and n_y are encountered before a pattern with $n_z > 0$ falls within the radius $\sqrt{f_N}$ of the sphere.

In a violin—of an average height of 4 cm—this occurs only above

$$f_{001} = 4{,}275 \text{ Hz}, \tag{11.65}$$

in other words, above the range shown in figure 11.5, and above the range in which the question we are discussing is of interest.

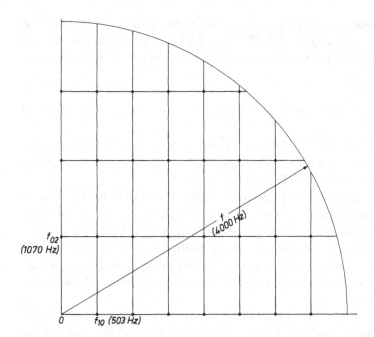

Figure 11.6
The grid of natural frequencies of a shallow, rectangular air cavity.

For this reason, the three-dimensional network of natural frequencies degenerates into a two-dimensional grid; in figure 11.6, it is shown for the chosen dimensions and an upper frequency limit of $f = 4,000$ Hz.

It is not to be expected that these natural oscillations coincide with the peaks measured inside the cavity of an actual violin. Nonetheless, the first three peaks coincide approximately. The second and third peaks are close to one another on the frequency scale because our rectangular model is approximately half as wide as it is long. Closely spaced peaks such as these should also occur in most violins.

While the number of clearly evident measured peaks is only 16 or 17, the number of peaks counted in figure 11.6 is 28. This is also the number to be expected on statistical grounds. We easily obtain this from a quarter circle, $\pi f^2/4$, divided by the area of a two-dimensional element $c^2/(4L_xL_y)$ $= c^2/(4S)$:

$$\lim_{f \to \infty} N_1 = \frac{\pi S f^2}{c^2}. \tag{11.66}$$

The next-smaller integer corresponding to this quantity is 23, based on the values used in figure 11.5. This time, however, we should also account for the axes' halving the values, which this time appear as integers on the axes:

$$\lim_{f \to \infty} N_2 = \frac{1}{2} f \frac{L_x + L_y}{c/2} = \frac{fU}{2c}, \tag{11.67a}$$

where U is the circumference. If we add the next lower integer resulting from such an accounting, which amounts to 5, we arrive at the same value of 28 that we counted in figure 11.6.

The smaller number of peaks seen in figure 11.5 must be based, as in our similar comparison of natural oscillations of plates, on the superposition of increasing numbers of neighboring resonances at higher frequencies. This is especially the case for the length: width ratio of close to 2 : 1, which places many resonances close to one another in any case. Besides, not all natural oscillations can be excited, or sensed equally well, at the positions of the transducers.

In the asymptotic relations (11.66) and (11.67), only the total area and total circumference determine the result, regardless of the shape of the boundary. Here again, it can be established that only the natural-resonance frequencies are shifted. These can fall on top of one another, but their number can be neither increased nor decreased. Even if the cavity is separated into three partial volumes above, below, and at the waist, the total area and the rigid boundaries are not affected.

Whether the 28 natural resonances of the air cavity appear at the outside—whether they are excited by the top and back plates or in turn influence the natural resonances of these plates—can be determined only by a study of the coupling between these media in which bending waves occur, and the one between them in which longitudinal waves occur.

11.5   Coupling between two infinite plates separated by an air cushion

We will first attempt to investigate this question in the most general way possible, in connection with the relationships which are easiest to calculate; so we choose a two-dimensional model consisting of a homogeneous top plate of 2.5 mm thickness, a 4 cm air cushion, and a homogeneous back plate of 4 mm thickness (see figure 11.7).

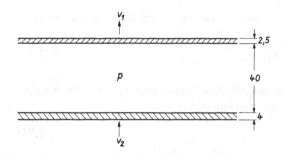

Figure 11.7
A two-dimensional model of the coupling of two plates via an air cavity.

In the first case we examine, these plates are infinitely wide and long; we cannot, then, examine natural oscillations; rather, we examine "natural waves."

In the equations for bending waves, the sound pressure appears as a coupling parameter. It reinforces acceleration in the upper plate and hinders it in the lower plate, in conformity with the sign conventions in figure 10.2. After division by the bending stiffnesses, we obtain two equations for the velocity vectors $\underline{v}_1$, $\underline{v}_2$ and the pressure vector $\underline{p}$:

$$-\frac{d^4\underline{v}_1}{dx^4} + \frac{\omega^2 m_1''}{B_1'}\underline{v}_1 + \frac{j\omega\underline{p}}{B_1'} = 0, \tag{11.68}$$

$$-\frac{d^4\underline{v}_2}{dx^4} + \frac{\omega^2 m_2''}{B_2'}\underline{v}_2 - \frac{j\omega\underline{p}}{B_2'} = 0. \tag{11.69}$$

Similarly, the velocities $\underline{v}_1$ and $\underline{v}_2$ appear as additional field quantities in the equation for waves in air in a cavity with vibrating boundaries; the component of the local decrease in tangential velocity $d\underline{v}_x/dx$ in the equation for compression in the cushion of air (11.57a) has to be supplemented by the perpendicular velocities of the plates divided by the height of the cushion. We must, however, once more note the differing signs:

$$\underline{p} = -\frac{K}{j\omega}\left(\frac{d\underline{v}_x}{dx} - \frac{\underline{v}_1 - \underline{v}_2}{d}\right). \tag{11.70}$$

Combined with the dynamic relationship

$$-\frac{d\underline{p}}{dx} = j\omega\varrho\underline{v}_x, \tag{11.71}$$

this leads to the wave equation for a cushion of air with moving walls

$$\frac{d^2 p}{dx^2} + \frac{\omega^2 \varrho}{K}\underline{p} - \frac{j\omega \varrho}{d}(\underline{v}_1 - \underline{v}_2) = 0. \tag{11.72}$$

We may substitute expressions for plane waves proceeding in the positive and negative x-direction into this equation:

$$\underline{v}_1, \underline{v}_2, \underline{p} = \underline{v}_{10}, \underline{v}_{20}, \underline{p}_0 e^{\pm jkx}, \tag{11.73}$$

and introduce the wave numbers valid when there is no coupling:

$$k_1 = \sqrt[4]{\frac{m_1''}{B_1'}} \sqrt{\omega} = \sqrt[4]{12}\sqrt{\frac{\omega}{c_L h_1}}, \tag{11.74$_1$}$$

$$k_2 = \sqrt[4]{\frac{m_2''}{B_2'}} \sqrt{\omega} = \sqrt[4]{12}\sqrt{\frac{\omega}{c_L h_2}}, \tag{11.74$_2$}$$

$$k_3 = \sqrt{\frac{\varrho}{K}}\omega = \frac{\omega}{c_3}. \tag{11.74$_3$}$$

Although the air is between the two other media, it is designated, as the third medium, by the subscript 3. The three coupled wave equations can now be rewritten as three linear equations for the complex wave amplitudes at the point $x = 0$:

$$[k_1^4 - k^4]\underline{v}_1 \qquad\qquad +\frac{j\omega}{B_1'}\underline{p} = 0, \tag{11.75a}$$

$$[k_2^4 - k^4]\underline{v}_2 \qquad -\frac{j\omega}{B_2'}\underline{p} = 0, \tag{11.75b}$$

$$-\frac{j\omega\varrho}{d}\underline{v}_1 + \frac{j\omega\varrho}{d}\underline{v}_2 + [k_3^2 - k^2]\underline{p} = 0. \tag{11.75c}$$

Where the determinants of the coefficients go to zero, there results a characteristic equation in $k^2(\omega)$, in other words, a dispersion relation:

$$[k_1^4 - k^4][k_2^4 - k^4][k_3^2 - k^2] - \frac{\omega^2 \varrho}{d}\left[\frac{1}{B_2'}(k_1^4 - k^4) + \frac{1}{B_1'}(k_2^4 - k^4)\right] = 0. \tag{11.76}$$

This is a fifth-degree equation in k^2; even without the coupling, there are two waves in bending, each with a boundary field; and a sound wave in air. We should note that the frequency, which we have considered as a given, occurs not only as ω in the coupling terms, but also in k_1, k_2, and k_3.

Despite the relationships in (11.76), which cannot be solved explicitly, there is a relatively simple way to answer the question of interest to us: to what degree the coupling influences the wave numbers, in comparison with those of the uncoupled system. Namely, we can ask whether there are frequency ranges in which there is no influence.

If, for example, we set

$$k^2 = k_1^2 = k_1^2 + \Delta(k_1^2) \qquad (11.77)^6$$

as we already did in (11.40), such a calculation of the disturbance would make sense only when there is only a small amount of "detuning,"

$$\left| \frac{\Delta(k_1^2)}{k_1^2} \right| < 0.1. \qquad (11.78)$$

In this case, we can approximate

$$k_1^4 - k^4 \approx -2k_1^2 \Delta(k_1^2), \qquad (11.79_1)$$

$$k_2^4 - k^4 \approx k_2^4 - k_1^4 = k_1^4 \left(\frac{h_1^2}{h_2^2} - 1 \right), \qquad (11.79_2)$$

$$k_3^2 - k^2 \approx k_3^2 - k_1^2 = \frac{\omega^2}{c_3^2} - \frac{\omega}{c_L h} \sqrt{12}. \qquad (11.79_3)$$

Equation (11.78) then takes on the form

$$\left| \frac{\Delta(k_1^2)}{k_1^2} \right| = \left| \frac{1/2}{(m_1'' d/\varrho)(k_1^2 - k_3^2) - h_1^3/[(h_2^2 - h_1^2)h_2]} \right| < 0.1. \qquad (11.80)$$

The frequency representing the lower limit of validity of the inequality (11.80) is so low (as is known from the study in room acoustics of air cushions fronted with wood panels; see Cremer and Müller 1982, vol. 2, sec. 57) that it cannot possibly fall within the range significant for the violin. This range begins at 200 Hz; using the values in figure 11.7, the correction in (11.80) is only 0.03 at this frequency.

There is, however, a second range in which the inequality does not

---

[6] The following approach is based on the treatment of the problem in Cremer and Müller (1982, vol. 2, ch. 10, sec. 6), in which a rigid back plate is assumed.

hold. The first term in the denominator of (11.80) can go to zero when

$$k_1^2 = \frac{\omega\sqrt{12}}{c_L h_1} = k_3^2 = \frac{\omega^2}{c_3^2}. \tag{11.81a}$$

This condition can be described more clearly by setting the specific wavelengths equal to each other:

$$\lambda_1 = \lambda_3; \tag{11.81b}$$

The wavelength in air, λ_3, decreases as the frequency increases, and that of a natural bending wave, λ_1, decreases as the square root of frequency. The frequency at which the wavelengths are equal is especially important for the radiation of bending waves, as we will show again in section 15.1. This is called the critical frequency f_{cr}. From (11.81a), we obtain the definition:

$$f_{cr} = \frac{\sqrt{12}}{2\pi}\frac{c^2}{c_L h_1}. \tag{11.82}$$

As mentioned in section 11.2, the stiffness of wood parallel and perpendicular to the grain is very different. Consequently, the critical frequencies in the two directions are very different; for a top plate 0.25 cm thick, of average weight, in the direction of the grain with $c_x = 5.3 \times 10^5$ cm s^{-1}, we obtain

$$f_{cr1x} \approx 4{,}870 \text{ Hz}. \tag{11.83a}$$

Perpendicular to the grain, with $c_y = 1.4 \times 10^5$ cm s^{-1}, we find

$$f_{cr1y} \approx 18{,}420 \text{ Hz}. \tag{11.83b}$$

Even in the case represented by the lower of the two frequencies, the range in which the inequality (11.80) no longer holds is above that in which the natural oscillations of the body can be clearly separated from one another.

If we do not consider the actual values we applied to figure 11.7, it is in principle symmetric, and so the characteristic equation (11.76) remains unchanged if we exchange the subscripts 1 and 2. The same is true of the approximate expression

$$k_{II}^2 = k_2^2 + \Delta(k_2^2), \tag{11.84}$$

and for the limit of detuning

$$\left| \frac{\Delta(k_2^2)}{k_2^2} \right| < 0.1. \tag{11.85}$$

Once more, the frequency below which this inequality no longer applies is far below the frequency range of interest. But the inequality also fails to hold as we approach the two critical frequencies. For an 0.4-cm maple plate ($c_x = 3.6 \times 10^5$ cm s^{-1}, $c_y = 1.8 \times 10^5$ cm s^{-1}), representing the back plate of a violin, these critical frequencies are

$$f_{\text{cr}\,2x} = \ \ 4{,}478 \text{ Hz},$$

$$f_{\text{cr}\,2y} = 14{,}330 \text{ Hz}. \tag{11.86}$$

These results are similar to those for the infinite plate representing the top plate.

A change of sign of k_1^2 and k_2^2 is the only difference between the quasistationary exponential near fields and the waves. Consequently, the same restrictions apply at low frequencies. The restrictions at high frequencies are removed, though they could occur even with infinite plates if these were to be excited or disturbed at one point.

In order to derive the last wave number, which is close to k_3, we simply transform (11.76) into

$$k_{\text{III}}^2 = k_3^2 - \frac{\omega^2 \varrho}{d} \left[\frac{1}{B_2'(k_2^4 - k^4)} + \frac{1}{B_1'(k_1^4 - k^4)} \right]. \tag{11.87}$$

To be sure, this transformation is simple only if we can neglect $k^4 \approx k_{\text{III}}^4$ in the two denominators in the range of the lower natural frequencies of interest to us. After undertaking the transformation, we can in fact confirm that it is possible to neglect these quantities.

Equation (11.87) then simplifies to

$$k^2 = k_3^2 \left(1 - \frac{\varrho c_3^2}{d\omega^2} \left(\frac{1}{m_2''} + \frac{1}{m_1''} \right) \right). \tag{11.88}$$

The correction term can be interpreted vividly: It is

$$\sqrt{\frac{\varrho c_3^2}{d} \left(\frac{1}{m_1''} + \frac{1}{m_2''} \right)} = \omega_0^2, \tag{11.89}$$

the natural frequency of the system at which the two plates oscillate in opposite phase around their common center of mass, compressing the cushion of air with its stiffness per unit area

$$s'' = \frac{K}{d} = \frac{\varrho c_3^2}{d}.$$ (11.90)

This is the same as the oscillation of two masses of finite dimensions linked by a spring. Such a structure is called a *Tonpilz* (sound mushroom) in the German literature. In English, confusion may be avoided by designating ω_0 as the mass-spring-mass frequency.

The formula

$$k = \frac{\omega}{c_3}\sqrt{1 - (\omega_0/\omega)^2}$$ (11.91)

characterizes the channel between two inert plates as a high-pass filter, in which wave propagation is possible only above ω_0.

If we substitute the values $h_1 = 2.5$ mm, $h_2 = 4$ mm, and $d = 4$ cm into this formula, we obtain for this additional limiting frequency the value

$$f_0 = 287 \text{ Hz.}$$ (11.92)

This frequency is below the lowest one of the grid in figure 11.6, and below the lowest measured natural frequency of the body of the instrument; but not so low that the dispersion in (11.91) can be neglected.

If this dispersion were not present, the lowest natural frequency of the cushion of air of length L would be

$$f_{1L} = \frac{c_3}{2L}.$$ (11.93)

If, however, we replace the speed of sound c_3 with the value of ω/k derived from (11.91), the actual phase speed in the cushion of air, then f_{1L} is given by

$$f_{1L} = \frac{c_3}{2L}\bigg/\sqrt{1 - (f_0/f_{1L})^2},$$

or

$$f_{1L} = \sqrt{\left(\frac{c_3}{2L}\right)^2 + f_0^2}.$$ (11.94)

The frequency f_{1L}, then, can never fall below f_0. Below f_0, there can be no wave propagation; there are only exponentially decaying boundary fields. When $L = 35$ cm, then, f_{1L} is not 490 Hz, as would be the case for a flute with the distance L between the mouthpiece and the first open hole; rather, it lies at 568 Hz, still within the range of the two lowest natural oscillations of the body (main oscillations).

If we designate the wave numbers of the natural waves generally as k_I, k_{II} and k_{III}, we can conclude our observations with the statement that

$$k_I \approx k_1; \qquad k_{II} \approx k_2 \tag{11.95}$$

at frequencies sufficiently below the critical frequency, but we must use the more complicated relationship for k_{III},

$$k_{III} = k_3\sqrt{1 - (f_0/f)^2}. \tag{11.96}$$

At the very least, this means that it is easy to survey how the natural waves in the cushion of air generate motions in the plates forming its boundaries. Well below the critical frequency, the plates act only as masses with inertia. This leads first to the relationship

$$\underline{v}_1 - \underline{v}_2 = \underline{p}\left(\frac{1}{j\omega m_1''} + \frac{1}{j\omega m_2''}\right), \tag{11.97}$$

which can also be derived by subsituting (11.88) for k^2 in (11.75c). Since, once again,

$$\underline{v}_1 m_1'' + \underline{v}_2 m_2'' = 0 \tag{11.98}$$

must hold in order to conserve the motion of the common center of mass, (11.97) separates into the two equations to be expected on physical grounds:

$$\underline{v}_1 = \frac{p}{j\omega m_1''}, \qquad \underline{v}_2 = \frac{p}{j\omega m_2''}. \tag{11.99}$$

11.6 Coupling between the lowest natural plate vibrations and a finite cavity

The last two equations show that each resonance of the cushion of air must lead to corresponding oscillations of the plates; but it must seem odd that the waves in the plates generate hardly any reaction in the cushion of air. It can be shown, however, that this is the case—again always well below the critical frequency—only when the plates and cushion of air are infinite.

To show the importance of lateral boundaries, we will use a rigid back plate and a top plate which is at first infinite. We will examine a one-dimensional wave in bending, of wave number k_1.

In (11.75c), then, with $v_2 = 0$ and $k = k_1$, the admittance of the cushion of air at the boundary is given by

$$\frac{-v_1}{p} = \frac{-d}{j\omega\varrho}(k_3^2 - k_1^2) = \frac{j\omega d}{\varrho c_3^2} + \frac{(k_1 d)^2}{j\omega\varrho d}. \tag{11.100}$$

The first term represents the inverse of the reactance of the spring stiffness per unit area already mentioned in (11.90). This does not have as strong an effect as in our model in chapter 2 because it is short circuited by the larger, second component. The denominator of this second component would represent the reactance of the inertial mass of the cushion of air if it were moved perpendicularly to the plate. The motions of the cushion of air may, however, consist of incompressible displacements between neighboring narrowed and widened regions (spaced $\lambda_1/2$ apart). In this case, which applied well below the critical frequency, motions normal to the walls distributed over a quarter wavelength laterally are restricted vertically to the height d of the cushion of air; the effective mass is transformed by a factor proportional to $(\lambda_1/d)^2$, and so to $(k_1 d)^{-2}$. It is, so to speak, much easier for the plate to move the air from side to side than to compress and rarefy it.

We discuss this physical interpretation thoroughly because it requires neighboring peaks and valleys. From this requirement we can conclude that a hydrodynamic compensation of the same type occurs to some degree also in the case of oscillations of finite plates and a finite air cavity, and can considerably decrease the coupling of the top and back plates via the cushion of air.

The two lowest modes of the body, to be sure, represent an exception, since—with the exception of that in the island—no displacements increase the volume inside the cavity while others decrease it.

If the situation were otherwise, there would be no coupling of the top and back plates through the spring stiffness of the cushion of air. We recognized and accounted for this spring stiffness as an essential element in our model with separate masses and springs in chapter 10.

When we discuss the bounded system in terms of continua, functions of position must be used to describe velocity in the plates and pressure in the cushion of air; these functions must be of a form which can satisfy the coupled wave equations (11.68)–(3.70).

In order to simplify calculation and to make the results easier to grasp, we will assume that the back plate is rigid ($v_2 = 0$) in the following observations as well, and we will designate v_1 simply as v. The three equations (11.75a–c) simplify to two equations:

$$[k_B^4 - k^4]\underline{v} + \frac{j\omega}{B}\underline{p} = 0, \tag{11.101}$$

$$-\frac{j\omega\varrho}{d}\underline{v} + [k_L^2 - k^2]\underline{p} = 0, \tag{11.102}$$

where we have replaced k_3 by k_L. Where the determinants of the coefficients go to zero, we obtain the dispersion equation

$$[k^4 - k_B^4][k^2 - k_L^2] - \frac{\omega^2\varrho}{Bd} = 0. \tag{11.103a}$$

Even in this form, this equation proves mathematically analogous to the dispersion equation for a curved plate of radius R presented in equation (11.36). It is even possible to rewrite (11.103a) in the form

$$[k^4 - k_B^4][k^2 - k_L^2] - k_B^4\frac{\varrho}{md} = 0. \tag{11.103b}$$

Here, md/ρ appears in place of R^2; this means, however, that the ring frequency $f_R = c_L/2\pi R$ is replaced by the mass-spring-mass frequency, which is now reduced to that of only one mass:

$$f_0 = \frac{1}{2\pi}\sqrt{\frac{\varrho c_L^2}{md}}. \tag{11.104}$$

It here becomes clear that the wave number of the third component of the solution must behave in a way analogous to (11.44a) and so correspond to (11.91), while the wave number of the two other components can be approximated now as it was in connection with those equations, by k_B and jk_B.

Additionally, we can assemble the solutions with respect to the center of the plate ($u = x - \frac{1}{2} = 0$) from symmetrical and antisymmetrical functions of velocity. This is possible, as in our examination of curvature, because the boundary conditions are symmetrical. The lowest natural oscillation, surely the one of most interest to us, is symmetrical. Its solution could be assembled from three components, analogously to (11.47):

$$\underline{v} = \hat{\underline{v}}_\text{I}\cos(k_\text{B}u) + \hat{\underline{v}}_\text{II}\cosh(k_\text{B}u) + \hat{\underline{v}}_\text{III}\cos\left(k_\text{L}u\sqrt{1 - \left(\frac{f_0}{f}\right)^2}\right). \qquad (11.105)$$

This function of velocity must go to zero at the edges:

$$u = \pm\frac{L}{2}, \quad \underline{v} = 0:$$

$$\hat{\underline{v}}_\text{I}\cos\frac{k_\text{B}L}{2} + \hat{\underline{v}}_\text{II}\cosh\frac{k_\text{B}L}{2} + \hat{\underline{v}}_\text{III}\cos\frac{k_\text{L}L}{2}\sqrt{1 - \left(\frac{f_0}{f}\right)^2} = 0. \qquad (11.106)$$

Since the edges are supported, the moments also must go to zero there; so we must also have:

$$u = \pm\frac{L}{2}, \quad \frac{d^2v}{du^2} = 0:$$

$$\hat{\underline{v}}_\text{I}\left(-k_\text{B}^2\cos\frac{k_\text{B}L}{2}\right) + \hat{\underline{v}}_\text{II}\left(k_\text{B}^2\cosh\frac{k_\text{B}L}{2}\right)$$

$$+ \hat{\underline{v}}_\text{III}\left(-k_\text{L}^2\left(1 - \left(\frac{f_0}{f}\right)^2\right)\right)\cos\left(\frac{k_\text{L}L}{2}\sqrt{1 - \left(\frac{f_0}{f}\right)^2}\right) = 0. \qquad (11.107)$$

These boundary conditions, too, are completely analogous to those of the curvature problem we examined previously.

This is true even of the last boundary condition, which relates here, as it did before, to a displacement parallel to the plates. This time, however, the displacement is in the cushion of air instead of in a plate. In both cases, it, and the corresponding velocity v_x, must go to zero:

$$u = \pm\frac{L}{2}: \quad v_x = 0. \qquad (11.108)$$

We easily obtain the three components of the function of pressure from (11.102) by transformation:

$$\underline{p}_N = \frac{-j\omega\varrho}{d}\frac{v_N}{k_N^2 - k_\text{L}^2} \quad (N = \text{I, II, III}). \qquad (11.109)$$

Here, as before, the results are cosine functions with the arguments $k_I u$, $k_{II} u$, and $k_{III} u$. The function of pressure is, then, symmetrical as long as the function of velocity in the plate is also symmetrical.

We obtain the function of the field quantity we seek, v_x, from sound pressure with the aid of the dynamic relationship (11.71):

$$\underline{v}_{xN} = \frac{-1}{j\omega\varrho}\frac{d\underline{p}_N}{du} = \frac{1}{d}\frac{d\underline{v}_N}{dx}\frac{1}{k_N^2 - k_L^2}. \tag{11.110}$$

This relationship is analogous to (11.49), since only factors which depend on N remain when the boundary conditions are fulfilled.

We also note that we can assume that

$$k_B^2 \gg k_L^2 \tag{11.111}$$

sufficiently far below the critical frequency; though this relationship is not valid up to such high frequencies as in the analogous problem in curved plates: c_L represents only the longitudinal speed of propagation in air, not the speed of propagation in a rigid plate, which is greater by an order of magnitude. We obtain, then, the final boundary condition for (11.108):

$$\underline{\hat{v}}_I\left(-\sin\frac{k_B L}{2}\right) + \underline{\hat{v}}_{II}\left(-\sinh\frac{k_B L}{2}\right)$$

$$+ \underline{\hat{v}}_{III}\left(\frac{k_B}{k_L}\frac{f^2}{f_0^2}\sqrt{1 - \left(\frac{f_0}{f}\right)^2}\right)\sin\left(\frac{k_L L}{2}\sqrt{1 - \left(\frac{f_0}{f}\right)^2}\right) = 0. \tag{11.112}$$

It is appropriate to characterize the frequency by the parameter

$$\beta = \frac{k_B L}{2}, \tag{11.113}$$

which varies as the square root of frequency. Correspondingly, we introduce a further parameter to replace f_0:

$$\beta_0 = \frac{\pi}{2}\sqrt{f_0/f_{1B}}, \tag{11.114}$$

in which

$$f_{1B} = 0.45 c_L h/L^2 \tag{11.115}$$

represents the lowest natural frequency in bending, given that the plate need not compress any part of the air cushion.

For f_0, we use the value 287 Hz, which we earlier determined to be appropriate for the violin. In order to retain this value with a rigid back plate and with a top plate of the same average mass as before, we must reduce the distance between the top and back plates from 4 cm to approximately 2.5 cm.

It is also best to reduce L in comparison with the actual length of a violin body, in order to make the lowest natural frequency in pure one-dimensional bending, f_{1B}, correspond to that of the top plate of an actual instrument, which is in fact supported at its lateral edges and at the soundpost and is arched. We use the value $L = 18.4$ cm, obtaining

$$f_{1B} = 177 \text{ Hz} \tag{11.116}$$

and the round number

$$\beta_0 = 2. \tag{11.117}$$

The parameter β_0, however, is not the only one occurring in the boundary conditions; there is additionally the independent relationship k_B/k_L, which appears also in

$$k_L L/2 = \beta k_B/k_L.$$

We therefore introduce another parameter, dependent on f_{cr},

$$\beta_{cr} = \frac{\pi}{2}\sqrt{f_{cr}/f_{1B}}; \tag{11.118}$$

given the calculated value of f_{cr} in the longitudinal direction on the top plate, 4,870 Hz, this takes on the value

$$\beta_{cr} = 8.3. \tag{11.119}$$

Then k_B/k_L can be replaced by β_{cr}/β.

Given these parameters, the characteristic equation that follows from the boundary conditions (11.106), (11.107), and (11.112) takes on the form

$$(\tan \beta)\left[1 + \left(\frac{\beta}{\beta_{cr}}\right)^2 \left(1 - \left(\frac{\beta_0}{\beta}\right)^4\right)\right] + (\tanh \beta)\left[1 - \left(\frac{\beta}{\beta_{cr}}\right)^2 \left(1 - \frac{\beta_0}{\beta}\right)^4\right]$$
$$+ 2\frac{\beta_{cr}}{\beta}\left(\frac{\beta}{\beta_0}\right)^4 \sqrt{1 - \left(\frac{\beta_0}{\beta}\right)^4} \tan\left(\frac{\beta^2}{\beta_{cr}}\sqrt{1 - \left(\frac{\beta_0}{\beta}\right)^4}\right) = 0. \tag{11.120}$$

In calculating the wave numbers and in finding v_x using (11.109), we have already assumed that $(k_L/k_B)^2 = (\beta/\beta_{cr})^2 \ll 1$. Consequently, it would be inconsistent to keep the additional expression $\pm\left(\dfrac{\beta}{\beta_{cr}}\right)^2\left(1 - \left(\dfrac{\beta_0}{\beta}\right)^4\right)$ along with 1. Also, we can set $\tanh\beta \approx 1$ in the frequency range of interest, above $\beta_1 = \pi/2$. Equation (11.120) then simplifies to

$$\tan\beta + 1 + 2\frac{\beta_{cr}}{\beta}\left(\frac{\beta}{\beta_0}\right)^4 \sqrt{1 - \left(\frac{\beta_0}{\beta}\right)^4}\,\tan\!\left(\frac{\beta^2}{\beta_{cr}}\sqrt{1 - \left(\frac{\beta_0}{\beta}\right)^4}\right)$$

$$= \tan\beta - F_1(\beta) = 0. \tag{11.121}$$

As long as the argument of the last tangent is smaller than $\pi/6$, (11.121) simplifies even further to

$$\tan\beta + 1 - 2\beta\left(1 - \left(\frac{\beta}{\beta_0}\right)^4\right) = 0. \tag{11.122}$$

Figure 11.8, top, shows the graphic solution of (11.121). The intersections of $\tan\beta$ with the rest of (11.122), brought to the right side, are what is to be found, with β increasing toward the right. The approximation (11.122) is sufficient to determine the lowest natured value of β, and so the lowest real natural "bending" frequency. The resulting value is $\beta = 2.09$, corresponding to a frequency of 313 Hz. This is lower than the lowest resonance of the body of a violin, yet it is higher than β_{1B} [see (11.116)], clearly showing the influence of the stiffness of the cushion of air.

In order to show that the effect of the cushion of air is smaller at higher natural bending frequencies ($\beta = \pi$, $3\pi/2$, etc.) in figure 11.8, the function $F_1(\beta)$ is plotted out to higher values; however, (11.121) must be used here. (We also enter regions in which $(\beta/\beta_{cr})^2$ can no longer be neglected, but we do not account for this; the resulting corrections would make no basic difference.) Where its argument reaches $\pi/2$, the negative tangent which is part of $F_1(\beta)$ has a pole; $F_1(\beta)$ here goes to $-\infty$ and then jumps to $+\infty$.

This occurs near the lowest natural frequency of the cushion of air (as we shall soon see, when we discuss figure 11.8, bottom), whose value can be approximated by (11.94) as

$$f_{1L} = \sqrt{\left(\frac{34{,}200}{2 \cdot 18.4}\right)^2 + 287^2} \approx 1{,}000 \text{ Hz,} \quad \text{or} \quad \beta_{1L} = 3.73. \tag{11.123}$$

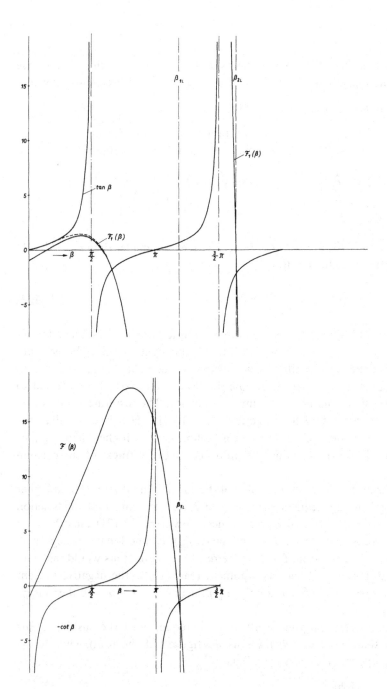

Figure 11.8
Graphic method for determining the natural frequencies of a supported plate of finite length over a closed cushion of air. Top, symmetrical displacement; bottom, antisymmetrical displacement.

The line along which the jump occurs is shown dashed in figure 11.8, top. The curve F_1, now determined mainly by the negative tangent function, now falls rapidly from $+\infty$, intersecting the abscissa near the second natural frequency of the cushion of air, given by

$$f_{2L} = \sqrt{\left(\frac{34{,}200}{18.4}\right)^2 + 287^2} = 1{,}880 \text{ Hz}, \quad \text{or} \quad \beta_{2L} = 5.12 \qquad (11.124)$$

(also shown dashed in figure 11.8). Although the diagram does not show this, it is clear that $F_1(\beta)$ has previously crossed $\tan\beta$ at a high value, just before the pole corresponding to f_{3B}. The position of f_{3B}, then, is not influenced significantly by the cushion of air. The same is also true of f_{2L}, since the value of β at its intersection with the next, rising segment of $\tan\beta$ hardly differs from that where the tangent jumps across the abscissa.

It is also of interest to note to what degree corresponding statements can be made about f_{2B} and f_{1L}. Their modes are antisymmetric, and so can be described by the expression

$$\underline{v} = \hat{\underline{v}}_{\mathrm{I}}\sin(k_B u) + \hat{\underline{v}}_{\mathrm{II}}\sinh(k_B u) + \hat{\underline{v}}_{\mathrm{III}}\sin(k_L u \sqrt{1 - (f_0/f)^2}). \qquad (11.125)$$

If we carry out the same operations on this expression as on the symmetrical expression (11.105), neglecting the same quantities, we obtain the characteristic equation

$$-\cot\beta + 1 - 2\frac{\beta_{cr}}{\beta}\left(\frac{\beta}{\beta_0}\right)^4 \sqrt{1 - \left(\frac{\beta_0}{\beta}\right)^4} \cot\left(\frac{\beta^2}{\beta_{cr}}\sqrt{1 - \left(\frac{\beta_0}{\beta}\right)^4}\right)$$
$$= -\cot\beta - F_2(\beta) = 0. \qquad (11.126)$$

In order to make the presentation in figure 11.8, bottom, conform as well as possible to that in figure 11.8, top, the function $-\cot\beta$ is shown as it intersects $F_2(\beta)$. This time, the substitution of $1/x$ for $\cot x$ in $F_2(\beta)$ is acceptable only in the case of the initial straight line; it no longer holds at the first intersection, very close to $\beta = \pi$. But this is where we would expect the second natural bending frequency β_{2B}, if it were not influenced by the cushion of air. Similarly, the next intersection is very close to β_{1L}—near the resonance of the cushion of air, if we account for dispersion.

Briefly then, the frequency of only the lowest natural bending oscillation is raised significantly by the cushion of air. On the other hand, the natural frequencies of the cushion of air are influenced by the mass of the plate, though only in the dispersion in the speed of propagation, for which the mass of the plate is responsible.

In the chosen example, f_{1L}, at 1,000 Hz, lies, in contrast to the violin, far above the lowest natural frequency of the plate taking air-cushion stiffness into account, viz., 313 Hz. This weakness of the two-dimensional model may seem disappointing. But even if a more rigid plate succeeded in bringing the two frequencies closer to one another, a common effect would not be likely, since the lowest natural oscillation of the cushion of air has an antisymmetrical pressure function—to the first approximation, a sinusoidal function with out-of-phase antinodes at the ends. The lowest natural oscillation of the body of the instrument, on the other hand, has a symmetrical pressure function—to the first approximation, a cosinusoidal function—forced by the symmetrical velocity function. Symmetrical and antisymmetrical solutions cannot excite one another, and so cannot influence one another.

The mismatch between the natural modes has been noted by Jansson and Sundin (1974), who were first to examine the possibility of coupling between oscillations in the body of the instrument and in the cavity. They noted, however, that positive and negative pressure in the cushion of air of a violin are not exerted against the same surfaces; specifically, that differences exist between the volumes displaced by oscillation of the top plate above and below the waist of the instrument. Using holographic photographs, Jansson and Sundin determined that these differences amount to 10% of the total volume displacement.

At least, their experiments suggested that coupling does in fact exist. They placed a normal top plate, with the ribs and soundpost, on a "rigid" back plate, and excited these at the position of the left foot of the bridge. They obtained the sound pressure level as a function of frequency, measured at a point above the top plate, shown in figure 11.9. The first peak at the left is the f-hole resonance; the second is the lowest resonance of the body of the instrument. If they added a 2.8 g mass to the oscillating part of the top plate, they obtained not only a lowering of the resonance of the body, but also a double peak, such as is characteristic of coupled oscillators. Also, by measuring sound pressure inside the cavity, they established that the lowest oscillation of the cavity occurred in the frequency range of the double peak. It is, then, entirely possible that the third peak observed by Beldie in three instruments (see

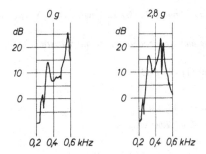

Figure 11.9
Measured frequency function of sound pressure level above a violin top plate supported on a rigid plate through the ribs and soundpost, and excited at the position of the left foot of the bridge. Left, without; right, with an additional mass attached to the plate at an antinode (after Jansson and Sundin).

figure 10.9) resulted from a coupling with the lowest natural oscillation of the cavity.

This explanation is supported by the third peak's occurring in some instruments but not in all, since this coupling depends on details of the distribution of oscillations in the body of the instrument. (If it were not necessary for the end of the body nearer the fingerboard to be narrower in order to facilitiate playing in higher hand positions, it might be assumed that unequal areas above and below the waist might have been developed empirically to serve this acoustical goal.)

Since an increase in the number of peaks leads to a smoother frequency response, probably a desirable property in an instrument, the question arises as to how such coupling can be facilitated. No precise calculation is possible, however; we further note that tapping the top and back plates before assembling the instrument, is also of no use here. It is, however, possible to systematically investigate the frequencies of the modes of the rigidly supported plates of the instrument and of the air cavity; Jansson and Sundin have given one example of how this can be done.

References

Courant, R., and D. Hilbert, 1953. *Methods of Mathematical Physics*, vol. 1. Tr. by the authors. New York: Wiley. Reprinted 1974.

Cremer, L., 1955. *Acustica* 5, 245.

Cremer, L., and M. Heckl, 1973. *Structure-Borne Sound.* Tr. by E. Ungar. New York: Springer.

Cremer, L., and H. A. Müller, 1982. *Principles and Applications of Room Acoustics.* Tr. by T. J. Schultz. London: Applied Science.

Emmerling, E. A. and R. A. Kloss, 1979. *Forsch. Ing.-Wes.* **45**, 6.

Heckl, M., 1960. *Acustica* **10**, 109.

Jansson, E., 1973. *Catgut Acoust. Soc. Newsletter* no. 19, 13.

Jansson, E., and H. Sundin, 1974. *Catgut Acoust. Soc. Newsletter* no. 21, 11.

Kennard, E. H., 1953. *J. Appl. Mech.* **20**, 33.

Rayleigh, Lord, 1877, *Theory of Sound.* London: Macmillan. Reprinted 1929.

Reissner, E., 1955. *Quarterly Appl. Math.* **13**, no. 2.

Savart, F., 1819. *Mémoire sur la constructions des instruments à cordes et à archet.* Paris: Deterville.

Schelleng, J., 1969. *Catgut Acoust. Soc. Newsletter* no. 11, 21.

Schroeder, M., 1954. *Acustica* **4**, supplement 2, 594.

Weyl, H., 1912. *Math. Ann.* **71**, 441.

Zener, C., 1942. *Phys. Rev.* **59**, 669.

12 Observing the Natural Modes and Conclusions

12.1 Making the natural modes visible

The natural frequencies of the body of the instrument, except for the Helmholtz frequency, can be determined only by experiment. An example of this situation is shown in figure 10.1, in which the natural frequencies manifest themselves as peaks of the input admittance of the bridge. Rather than measuring admittance, it is—or at least was until recently—more usual to measure sound pressure in air at a more or less arbitrary point in front of the top plate. With this technique, the level-frequency curve is likely to exhibit a dependence on direction or even on distance in the neighboring field region (see section 14.1). Still, by taking note of the peaks—are not worrying about their heights or widths—we may identify most, if not all, of the natural frequencies. Meinel (1937a), Saunders (1937, 1946, 1953), and Lottermoser and J. Meyer (1957), particularly, have used such "resonance diagrams" of this type to characterize violins.

However, it is very important that the force generator have a very low internal impedance compared to that of the input of the bridge, as the string has. An electrodynamic exciter exemplary in this respect was developed by Dünnwald (1979).

Still, the frequency of a natural oscillation offers no physical insight into how it arises or into its radiational properties. Such insight is provided only by the nodal pattern at that natural frequency. This is almost always represented as a distribution of displacements in the z-direction in the top and back plates.

The first researcher to sense magnitude and phase capacitively—that is, without mechanical loading—was Backhaus (1931).

The same method was used by Meinel (1937b) and later by Eggers, whose results for the cello we have already mentioned in section 10.2, where we described the method in detail. Above all, we should remember that any particular resonant mode is unlikely to occur in isolation, due to the width of the resonance peaks. For this reason, as we will show in chapter 13, a weakly radiated natural mode—of little interest to us—can completely overshadow a well-radiated one.

At the time of Egger's investigations it was a laborious task not only to obtain a nodal pattern, but also to produce a representation (such as figure 10.4) of the results. Today we are able to store displacement amplitudes and phases in a computer; we can program the computer to display the contours of the vibrating object in axonometric relief, and then to rotate this representation around a given axis. Finally, we can allow the displayed object to vibrate in accord with the stored amplitudes and phases.

At the July 1983 Stockholm Acoustic Music Conference, H. A. Müller showed a film of such a computer analysis. The use of a computer also simplified the method of obtaining the amplitudes and phases of the displacements. The violin had only to be struck at the points of interest with a small hammer containing a force sensor, and the resulting displacement at the same point near the bridge recorded. By the principle of reciprocity, one then also knows what the displacement responses at all the points of impact would be if the same impulsive force were to excite the point near the bridge. These responses are then transformed in the frequency domain and there normalized to the spectrum of the specific impulse, so giving the frequency dependence of the corresponding cross admittance, whose magnitude against frequency can be plotted on the computer's display screen; the peaks mark the natural frequencies. At each resonant frequency the complex amplitudes of the stored normalized displacements can be read out and combined, so that the nodal pattern over the entire instrument can also be displayed.

If the goal is only to see into how many neighboring out-of-phase regions the top and back plate are divided, it is sufficient to make the nodal lines visible. It is possible to do this on relatively plane, horizontal surfaces using the method demonstrated by Chladni in 1787. He dusted a surface with fine sand, sawdust, or the like. When the surface was excited at a natural frequency, the grains began to dance around, migrating to parts of the surface that were at rest—namely, to the nodal lines. This procedure caused more wonder in its time than does holography in ours, and is so simple that it is still commonly used. For example, Hutchins used it to demonstrate the ring mode (see section 10.4) of top and back plates not yet incorporated into instruments. It is rather surprising that violin builders make no systematic use of this procedure, at least not to the present author's knowledge. Excitation may be by means of sound transmitted through air, using loudspeakers, or the plate may be directly excited by means of a transducer; if the frequency is swept slowly, one natural mode after another can be seen. By carefully observing how the pattern changes with frequency, it is even possible to deduce whether or not the modes are superposed on one another.

Chladni's "dust figures," however, indicate only the position of the

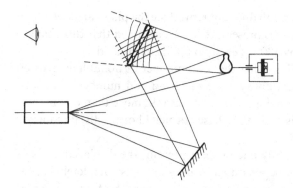

Figure 12.1
Arrangement for time-average holography of an oscillating object (after Reinicke).

nodal lines; they do not show the amplitude of the oscillations in the regions between the lines.

Holography, on the other hand, does show amplitudes, though it is certainly a more difficult technique. We have already described its principles in section 9.3 in connection with the natural oscillations of the bridge. Holograph is a much more necessary aid when studying the many and complicated modes of the body of the instrument than for the few and simple ones of the bridge. For this reason, we only now, in figure 12.1, show our sketch of the experimental apparatus used in holographic interferometry (from Reinicke 1973). Two beams of light proceed from a laser; one, the reference beam, is directed to the holographic plate by a fixed mirror. The plate can be a diffusing disc for direct observation, or a photosensitive medium. The other beam from the laser illuminates the oscillating object; with an additional delay due to the displacement of the surface, this also reaches the holographic plate. In section 9.3, it was shown how a holographic photograph made using a holographic plate illuminated with a reference beam gives the oscillatory displacements of the surface the appearance of the contour lines on a map. The rest position appears as a white line or region. The next white ring occurs where the amplitude of displacement of the oscillating object is about half the wavelength of the laser light; the next after that, where the amplitude is approximately one wavelength, and so forth.

Figure 11.2 gave a simple example of such contour lines: those corresponding to the natural modes 1, 1; 1, 2; 2, 1; and 2, 2 of a rectangular plate supported at its edges.

A holographic photograph does not reproduce the phase angle. As long as only one natural mode is present, it can be assumed that this changes more or less abruptly by 180° as each nodal line is crossed.

From this rule, it follows that the total number of nodal lines crossed along a closed path must always be even. If an odd number of crossed nodal lines appears in a holographic photograph, more than one natural mode is present. In this case, the phases of neighboring displaced areas no longer differ by 180°.

Furthermore, even if only one mode is present, the displaced areas of opposite phase need not be separated by lines or areas completely at rest, which appear as white on the holographic photograph. A shaded nodal line or area indicates that a displacement occurs here as well, as is necessary when damping is present to transfer power from the point of excitation to the furthest regions of the object. Nonetheless, phase angles in the middle of neighboring vibrating areas do differ by 180° in this case.

The relationship between the phase angles of particular top and back plates is defined by the soudpost; with respect to their position in space (and not with respect to the outside of the plates), the displacements ζ_1 and ζ_2 at the ends of the soundpost must be the same.

12.2 Holographic photographs taken during various phases of construction

Among the ever-increasing number of publications of holographic photo-graphs of the modes of the top and back plates, those of Jansson, Molin, and Sundin (1970) deserve particular attention. These are not only of high photographic quality, but also show changes in the nodal patterns at various stages of construction. The back or top plate are shown resting on the ribs, which in turn rest on a rigid plate with a large opening to eliminate the effect of the cushion of air. Then f-holes, bass bar, and, finally, a soundpost resting on the rigid plate are added. It can be seen how the modes depend on each of these stages of construction.

Figure 12.2 shows the four lowest natural modes of the top plate, without f-holes, bass bar, or soundpost. The photographs are arranged the same way as those of a rectangular plate in figure 11.2. Here, too, it seems appropriate to distinguish the natural modes

$$\begin{matrix} 1,1 & 1,2 \\ 2,1 & 2,2 \end{matrix}, \tag{12.1}$$

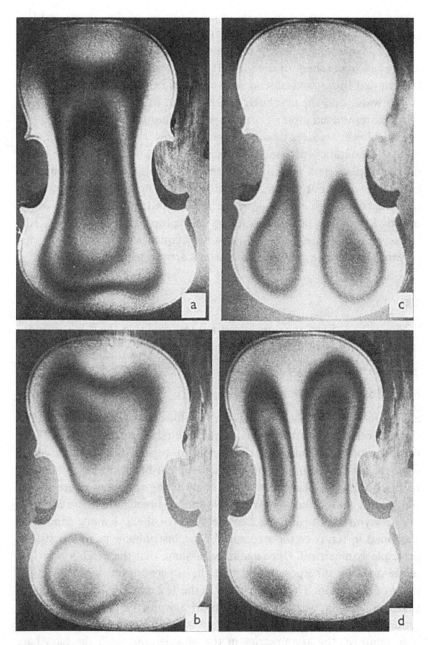

Figure 12.2
Holographic photographs of a violin top plate attached to the ribs, which in turn rest on a rigid plate with a large opening under the top plate. No f-holes, bass bar, or soundpost (after Jansson, Molin, and Sundin).

taking the (vertical) x-axis to be in the direction of the strings and the (horizontal) y-axis to be in the direction of the bow.

The 1, 1 mode (top left) shows how the outer contour lines to some degree follow the shape of the waist. It also proves significant to the nodal pattern that the violin is not so wide above as below the waist.

The waist, and the asymmetry with respect to it, influences the three other patterns even more. The strongest oscillation may be in the upper or in the lower area. As these photographs show, the choice between two points of excitation perpendicular to the plate can have a strong effect on the displacement amplitude. With very little damping, the point of excitation would not matter, unless it were on a nodal line. It is unlikely that the considerable shifting of the maximum to the left in the lower part of 2, 1 reflects strong asymmetry in the top plate; more likely, 1, 2 is unavoidably excited too. The measured frequencies of these two modes lie very close to one another in the matrix arrangement of figure 12.2,

$$\begin{matrix} 550 & 810 \\ 800 & 1{,}000 \end{matrix} \text{ Hz.} \tag{12.2}$$

Figure 12.3 shows the four lowest natural modes in the same arrangement, but with f-holes. The frequencies are

$$\begin{matrix} 420 & 600 \\ 640 & 880 \end{matrix} \text{ Hz,} \tag{12.3}$$

considerably lower than in (12.2). The lowering corresponds to the elimination of a constraint, due to the free edges of the f-holes. These form the outer boundary of the oscillating region in almost all cases, since the white area indicating no displacement begins at the outer edge of the f-holes. Displacement is not equal along the inner edges of the f-holes; the contour lines show this by running directly into these edges.

The asymmetry with respect to the x-axis of the 2, 1 mode cannot be explained in terms of the structure of the instrument, as this is still in principle symmetrical. Once again, we assume that the neighboring 1, 2 mode must be superposed, and that the superposition depends on the point of excitation. The asymmetries in the 1, 2 and 2, 2 modes could be based at least in part on the structural asymmetry of the instrument around the y-axis.

In contrast, the asymmetries in the photographs with the bass bar, shown in figure 12.4, can be explained in terms of the instrument. There

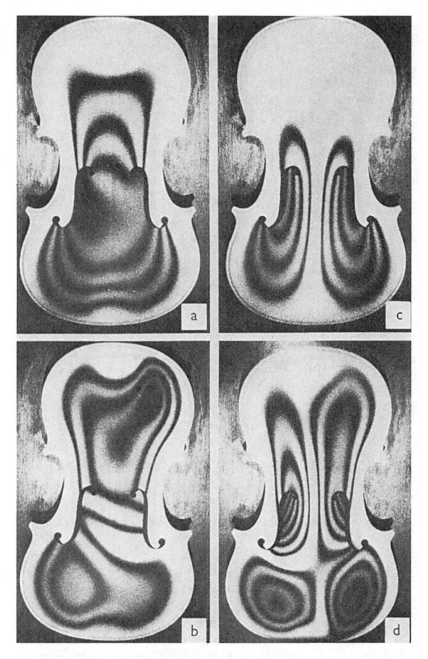

Figure 12.3
As in figure 12.2, but with f-holes (after Jansson, Molin, and Sundin).

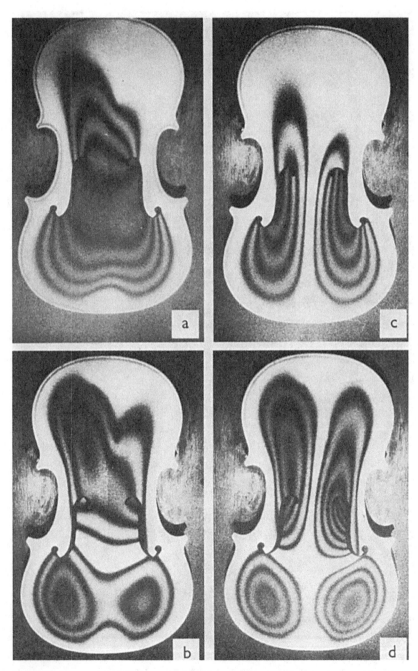

Figure 12.4
As in figure 12.3, but with bass bar (after Jansson, Molin, and Sundin).

is a an enlargement of the oscillating region at the left, though this seems to decrease with increasing frequency. The stiffness of the bass bar has raised most of the frequencies, to give

$$
\begin{matrix} 465 & 600 \\ 820 & 910 \end{matrix} \text{ Hz.} \tag{12.4}
$$

This time, the frequencies corresponding to the 2, 1 and 2, 2 modes are close to one another. In the photograph illustrating the 2, 1 mode, there are clear signs of superposition of the 2, 2 mode.

The soundpost effects even greater asymmetries; here, it generates a point at rest, and so it draws in whichever white area was near it (see figure 12.5). In the 1, 1 mode, it generates a nodal line; a small island consequently appears between the soundpost and the f-hole nearest it. In the 2, 1 mode, the soundpost probably broadens the existing nodal line. In the 1, 2 and 2, 2 modes, the soundpost pushes the oscillating regions somewhat to the side in comparison with figure 12.4.

Additionally, the photographs representing the 1, 2 and 2, 2 modes show typical superpositions of two modes: in both, there occurs an odd number of nodal lines which could be crossed along a closed path. The last three natural frequencies of the group,

$$
\begin{matrix} 540 & 800 \\ 775 & 980 \end{matrix} \text{ Hz,} \tag{12.5}
$$

are close together, leading support to this explanation.

Since the back plate has the same shape and the same anisotropy, holographic photographs of it reveal nothing fundamentally new, as long as it is supported only at the edges by the ribs, which in turn rest on a rigid plate.

On the other hand, the effect of a rigidly supported soundpost is of interest. We must note that the soundpost is at the left in the photographs of the back plate. In all of the individual photographs in figure 12.6, it can be seen that the white areas at the left are larger. There is always a white area next to the lower left corner of the ribs. Once again, the photograph representing the 1, 2 mode has only 3 oscillating areas, as in figure 12.5. Clearly, a superposition occurs; this is understandable in light of the proximity of the middle frequencies of the group

$$
\begin{matrix} 740 & 960 \\ 820 & 1{,}200 \end{matrix} \text{ Hz.} \tag{12.6}
$$

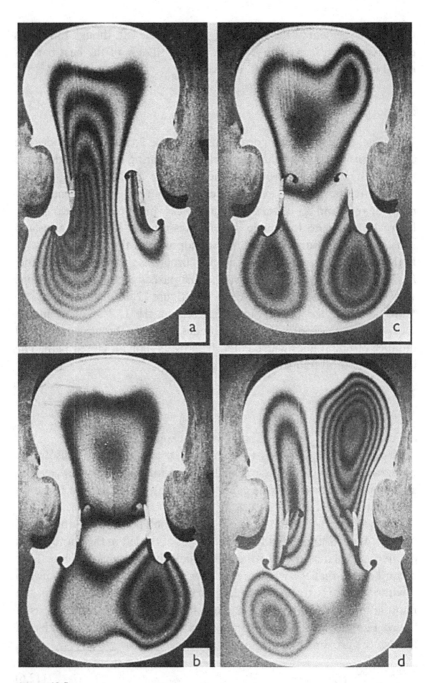

Figure 12.5
As in figure 12.4, but with soundpost between the top plate and the rigid plate (after Jansson, Molin, and Sundin).

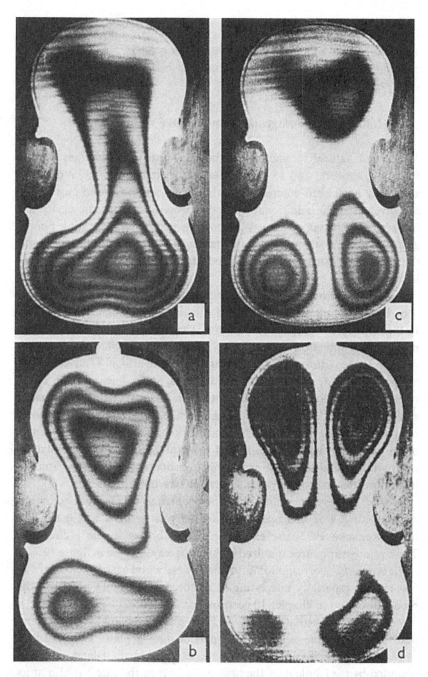

Figure 12.6
Holographic photograph of the back plate of a violin, with ribs and soundpost resting on a rigid plate with a large opening under the back plate (after Jansson, Molin, and Sundin).

In their series for the back plate, the authors include a photograph taken at 1,100 Hz, in which there are 5 oscillating areas, which allows the crossing of closed paths by an odd number of nodal lines, indicating a superposition of at least two modes.

12.3   Holographic photographs of assembled violins

The lowest natural frequencies of the back and top plates are not far from one another, and so we may expect their number to be doubled after they have been connected by the ribs, soundpost, and cushion of air, given that the additional natural frequencies of the air cavity are regarded as of no importance. We would also expect a holographic photograph to show a superposition of these nearby natural frequencies, and the place and direction of excitation should take on an even greater importance.

Figure 12.7 shows two holographic photographs (from figures 6 and 7 of Reinicke and Cremer 1970) of the same instrument at 1,300 Hz. In the upper photographs, excitation was at the upper end of the sound-post in the z-direction; in the lower photographs, excitation was, as in playing, in the y-direction at the upper edge of the bridge. With excitation at the soundpost, the bass bar lies under a nodal line, with an effect opposite that at lower frequencies, at which it is responsible for a parti-cularly large area oscillating in phase. In the photographs at the bottom, with excitation as in playing, the straight nodal line moves to the center of the top plate, between the feet of the bridge. Since the back plate is not partially hidden by the excitation transducer, both photographs of it show particularly clearly that several modes are superposed.

In order to judge the sound-radiating properties, we must use excitation as in playing. For this reason, all of the photographs in Figure 12.8 (from Reinicke 1973) use excitation at the bridge in the y-direction. These photographs are organized in the same way as those in figure 12.7.

Figure 12.8a, corresponding to the lowest wood resonance—here at 550 Hz—is especially interesting to us, because it shows how well the model discussed in chapter 10 corresponds to reality.

We can see immediately that the back plate is entirely in phase, and that the same is true of the greatest part of the top plate. The right foot of the bridge, oscillating out of phase, establishes the island, which is bounded by the f-hole near the bridge not only at the side but also at its

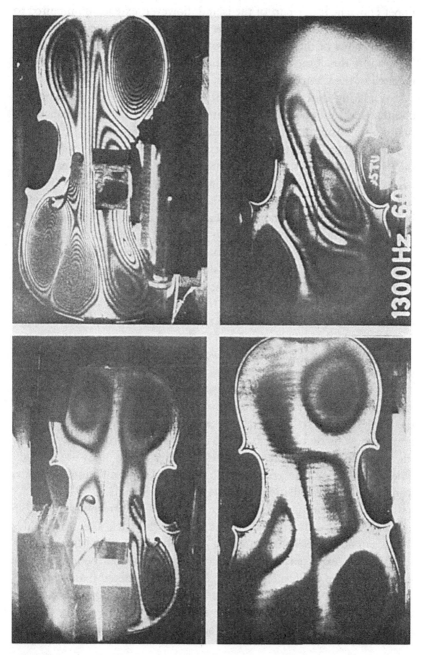

Figure 12.7
Holographic photographs of the top plate (left) and the back plate (right) of a violin
with bridge excited at 1,300 Hz. Top, excitation above the soundpost in the z-direction;
bottom, excitation via the bridge in the y-direction (after Reinicke and Cremer).

upper end. The lower part of the island, however, extends unimpaired to the lower rim of the violin. Still, the island includes only approximately 1/8 of the area of the top plate. Another 1/8 is at rest, and 6/8 oscillates in opposite directions to the back plate—since the soundpost is within the area of the island.

Considering that the peak amplitude of the island shows only four contour lines, and that of the main area shows eleven, it can be concluded that the flux from the main area predominates by far.

The ribs, too, do not move as a unit. This can be determined because they are bounded partly by white areas and partly by gray areas at the edges of the plates, and because contour lines go to the edges between these areas. The uniform motion of the ribs was a problematic assumption in our model. It is clear, however, that the ribs do move somewhat.

Naturally, the back plate and the two areas of the top plate that are out of phase with one another do not move as pistons. Still, it is possible to make use of the model by understanding its velocities as arithmetical averages $\bar{v}$ and by choosing the masses such that their product with $\bar{v}^2$ gives the kinetic energy. Besides, the model was not concerned with constructive calculations, but rather with the simplest possible insight into physical relationships.

Figure 12.8, then, can clearly answer our question: the model is indeed still valid in the range of the first wood resonance.

This is even more to be expected in the range of the f-hole resonance. Holographic photographs cannot prove this, because the air moving in the f-holes cannot reflect the laser beam. Rather, the holographic photographs show us that the body increasingly oscillates rotationally around the x-axis at lower frequencies, as Eggers (see figure 10.3) was already able to determine using electrostatic sensing. This oscillation, however, leads to very little radiation, as we will show in part III.

Clearly, the photographs in figure 12.8b and c, corresponding to 796 and 850 Hz, already show divisions of both the back and front plates which can no longer be described by the model in chapter 10.

It was not expected, however, that the model would be valid in this frequency range. Still, the possibility must be considered that the second wood resonance of the model might fall within this range, and that the model might consequently not represent it correctly.

These two examples also clearly show the effect of several modes at

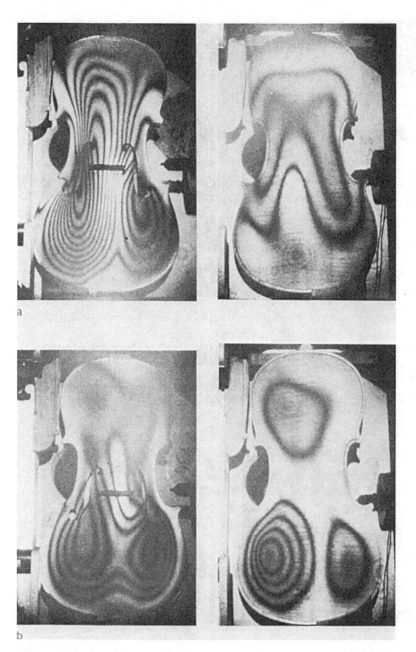

Figure 12.8
Holographic photographs of the top and back plates of a violin excited via the bridge
in the y-direction: a, at 550 Hz; b, at 796 Hz; c, at 850 Hz; d, at 2,666 Hz (after
Reinicke).

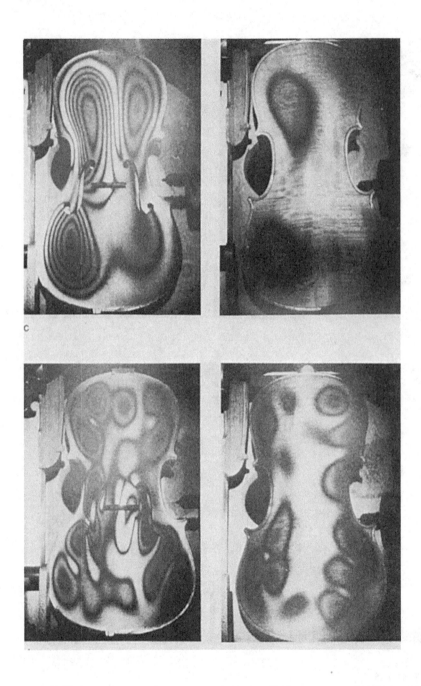

c

once (three oscillating areas separated by three nodal regions in the back plate!)

It is clear that the number of oscillating areas increases with frequency.

For this reason, we go immediately in figure 12.8 to the highest frequency at which Reinecke made holographic photographs using the same experimental arrangement. It is surprising that the back plate exhibits almost as many oscillating areas as the top plate even at this frequency, and that the oscillations are not much weaker. The position of these oscillating areas near the edge suggests that energy transfer is primarily through the ribs. Already, it has been made clear that the ribs cannot be regarded as inertial masses, and that damping is also very high at these frequencies.

Only when we reach the critical frequencies mentioned in section 11.5 could a strong coupling through the air cavity again be possible. In the cello, with its thicker plates, these frequency ranges could be of very considerable interest; but in the violin, even the excitation frequency used in generating figure 12.8d is approximately an octave below the critical frequency for longitudinal waves in the top plate. We can therefore assume—and it is consistent with the appearance of the central white area of the back plate in figure 12.8d—that the peaks and valleys of the front plate cancel each other hydrodynamically in the air cavity. Since such cancellation does not occur at the edges (see figure 15.2 below), a certain energy transfer from the air cavity cannot be excluded from consideration. However, no experiments have yet tested this hypothesis. Here, too, as before, more than one oscillatory mode is clearly at work.

12.4 Possibility of statistical analysis

At higher frequencies the damping increases, and therefore the widths of the peaks become greater; therefore, the peaks of curves of admittance or sound pressure become superposed to a greater and greater degree, and their relation to individual resonances becomes questionable.

We know that the variations in the level curves of sound fields in rooms (similar to that in figure 11.5 for the cavity of the violin) have only a statistical basis at higher frequencies, resulting from statistical phase variations. In the higher frequency range, deterministic statements can no longer be made about the phase angle; they can be made only about the temporal and spatial average values of the squared amplitudes (and therefore the energy). This situation leads to a great simplification of the corresponding mathematical formulas and calculations (see section 16.2).

In air, this transition is inevitable at high frequencies, as the number of additional natural resonances ΔN over a frequency increment Δf increases as f^2. Even in a space which is effectively two-dimensional, such as that between the top and back plates of a violin, this number increases as f, as we have shown in section 11.4.

In problems of noise control, the use of statistical energy analysis (Lyon 1975) has proven successful for plates, for which $\Delta N/\Delta f$ does not increase at all with frequency (see section 11.2). It is in this case appropriate to choose a Δf only great enough that several natural frequencies fall within it.

In all musical problems, it is not only usual, but at higher frequencies also appropriate in terms of auditory psychology to represent all frequency-dependent relationships in terms of $\log f$. Equal ratios, f_1/f_2, are then represented by equal intervals, $\log(f_1/f_2)$, as (nearly) on keyboard musical instruments. Several equally spaced natural frequencies occur even within the interval of a semitone at higher frequencies, and their number doubles with each octave.

The logarithmic scale is the main reason that the peaks representing the natural frequencies occur closer together at higher frequencies even in our analyses of a plate (see, for example, figure 10.1).

It should be possible to make good use of statistical energy analysis at higher frequencies in the study of bowed string instruments. The present author has, however, learned of no attempt to do this, and is himself not in the position to develop such an approach with its necessary precise assumptions, formulas, and conclusions. It is certain that this approach applied to bowed string instruments would have its own particular characteristics. Structures such as the bridge would have to be treated nonstatistically, since they exhibit their lowest natural oscillation in the "deterministic" region. This certainly is evident as a determinable peak, as Moral and Jansson (1982) were able to show. The upper curve in figure 12.9 shows the frequency dependence of the input of the bridge when excited in the direction of bowing. A peak at 3,000 Hz is evident, exactly where Reinicke (see figure 9.6) found an impedance minimum, corresponding to a natural oscillation of the bridge mounted on a rigid support. This evidence already suggests that the admittance peak in figure 12.9, top corresponds to the resonance of the bridge. The lower

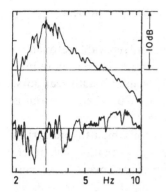

Figure 12.9
Measured admittances at high frequencies. Top, of the bridge input; bottom, of the body input at the left foot of the bridge (after Moral and Jansson).

curve in figure 12.9 shows the input impedance of the body of the instrument under the left foot of the bridge, and is even more persuasive in that this curve does not exhibit the peak at 3,000 Hz. The admittance peak of the bridge resting on the body of the instrument is much broader than the impedance dip of the rigidly supported bridge, but this difference can be explained by the power transferred from the bridge into the body and strings of the instrument. These losses are clearly much greater than the slight internal losses of the bridge.

The air cavity, too, requires some sort of transitional treatment in this frequency range. The ribs, with their restricted deformability, present a similar problem.

The top and back plate are so large with respect to wavelengths in bending that the specific shapes of their boundaries and the details of their thicknesses and arching no longer have an influence on the admittance curve. The only data that have an influence are the total area, the average thickness, and the average properties of the material. The inner damping may be hypothesized to play a large role; this was investigated in experiments on rod-shaped samples undertaken by Ptacnik (1953).

Every builder of string instruments and any person who regards them as artistic masterworks must find it disturbing that they lose their individuality at very high frequencies, according to this analysis, and that, to the degree that the aforementioned averaged values are the same, they should not differ in any way from factory fiddles. It should, however, be noted that this conclusion results only when the instruments are characterized in the frequency domain.

As we have already mentioned, the transition from separable resonances to fluctuations which can only be treated statistically is even more pronounced in rooms, for which it has led to very useful statistical laws. We will show this in detail in sections 16.2 and 16.3. There, a room is characterized only by its volume and decay time.

Nonetheless, it is not at all true that rooms for which the numerical statistical data are the same are regarded as of equal quality. The differences, however, cannot be analyzed in the frequency domain. They must be examined in the time domain by recording impulse responses, as we will see in section 16.4. This is more and more true the larger the room at a given frequency or the higher the frequency in a given room.

Whether a correspondingly high frequency range is of interest in the case of string instruments could be determined only by recording the response at some point on the top plate to an impulse excitation at the bridge, an experiment which has not yet been carried out. The impulse response would certainly be much more complicated than in a room, due to the dispersion of bending waves, the anisotropy of the wood, the differing thickness, and the arching. Furthermore, differing curvatures of the edge could play a role, as is the case in rooms.

This approach is in any case much better justified physically than the idea that the contours of the ribs must stand in simple, rational ratios to specific radii of curvature, ratios like those of the fundamental frequencies of harmonic musical intervals. It should, however, not be excluded that the master builders of the past preferred such ratios in order to keep track of them easily, or because the builders were interested in using tried and true forms; another reason to use these ratios might be to simplify construction. There is, however, no evidence to support the idea that the master builders regarded a numerological mystique as the key to acoustical quality. (Besides, the curvature of the edges is one feature which is easily carried over to factory fiddles.)

As in all problems in the seven-octave range of musical sound, there is a range of low frequencies in which easily separable natural frequencies can be examined deterministically and a higher one at which only statistical statements are possible in the frequency domain; but in between, there is a wide range in which deterministic statements are no longer simple and statistical statements are not yet valid.

12.5 Laws of similarity

Despite the anisotropic plates with uneven thickness and varying arching —above all, despite the boundary shapes which defy any analytical treatment, and the additional complication represented by the f-holes and the soundpost—one very simple general statement is still possible. Namely, if we increase all linear dimensions of an instrument by the same factor μ, and given that the materials are the same, then the wavelengths of all natural oscillatory modes increase in the same ratio, and all natural frequencies decrease as its inverse:

$$\mu = \frac{l_2}{l_1} = \frac{\lambda_2}{\lambda_1} = \frac{f_1}{f_2}. \tag{12.7}$$

We can prove this general law of similarity using the lowest natural modes. We begin with the Helmholtz resonance; i.e., with the lowest natural frequency, without considering the compliance of the top and back plates. In section 10.3, we determined that this natural frequency could be calculated using a formula of the form

$$f = \frac{c}{2\pi} \sqrt{\frac{S}{V(k\sqrt{S} + h)}}, \tag{12.8}$$

in which k is a purely numerical factor. It can be seen that the numerator under the root increases as μ^2, and the denominator as μ^4; as we would expect according to (12.7), f_2/f_1 is then simply proportional to $1/\mu$.

We can also make a similar general statement about the wood resonances. No explicit formulas apply to them; but it is sufficient to examine only the simple, one-dimensional case of a plate of length L and thickness h, supported at its ends, whose lowest natural frequency was derived in section 11.2 as

$$f = C \cdot \frac{h}{L^2}. \tag{12.9}$$

All natural frequencies in bending can be expressed in this form; only the quantity C, which depends on the shape and material, differs in each case; C also depends on the arbitrary choice of L. But if the body of the instrument is everywhere of the same material as before, and the size of all of its parts is altered by the same factor, then C remains constant. In this case, (12.7) takes on the specific form

$$\frac{f_1}{f_2} = \frac{h_1}{h_2} \frac{L_2^2}{L_1^2}. \tag{12.10}$$

In the specific case in which $L_1 = L_2$, this formula also indicates that all natural frequencies are proportional to h; Meinel (1937b) was able to prove this using analysis of the tone color. He concluded in addition that an instrument with too-thick plates would have a bright tone, and one with too-thin plates a dull tone.

If all dimensions of a cello were three times as great as those of a violin, the natural frequencies of the body would be in exactly the same relationship to those of the strings. The Helmholtz frequency—or to be more precise, the f-hole frequency f_0, which according to (10.17) lies somewhat below f_H, due to the effect of the compliance of the plates—would fall at the frequency of the second-lowest open string; and the first wood resonance would fall at the frequency of the second-highest string. Instead, the first wood resonance lies four to five semitones higher on the cello— higher by a factor of $(1.06)^4 = 1.26$ (Hutchins 1967; see also Hutchins 1976). This discrepancy exists because it is technically and also anatomically entirely impossible to make the cello similar to the violin in the sense described above. We need only consider that a cello would then weight $3^3 = 27$ times as much as a violin. And, even though the cello is supported on the floor by the peg, its length cannot be three times as great if an appropriate left-hand technique is to be possible. The actual length of the cello is only a bit more than twice that of the violin. The same is true of its width, in order that the cello fit between the player's knees. Length and width probably were chosen in the same proportion also in order that the top plate would look as much like that of the violin as possible. This similarity was also carried over to the size and shape of the f-holes. Only in the case of the ribs could the factor of 3 be used.

If we consider that the mass element of a Helmholtz resonator is given primarily by the aperture correction, and that this is proportional to $\sqrt{S}$, it follows that the ratio

$$\frac{f_{1H}}{f_{2H}} = \sqrt{\frac{S_1/S_2}{(V_1/V_2)\sqrt{S_1/S_2}}} \quad (= 2.45). \tag{12.11}$$

This is less than the factor of three in the tuning of the strings, by $3/2.45 = 1.22$. (This formula, however, does not account for the differing effect of the compliances of the top and back plates.)

Nor can dimensional similarity be achieved in the structures affecting the oscillations of the body of the instrument, because it is impossible to make the top and back plates of the cello three times as thick as those of the violin. Since, however, the dimensions of the plates are only doubled, lesser thickness is in fact appropriate to compensate for this lack of similarity. If we take $L_2^2 : L_1^2 \approx 4$, we obtain $f_2/f_1 = 1/3$ with

$$\frac{h_2}{h_1} = \frac{1}{3} \cdot 4. \tag{12.12}$$

This would result in plates too thin to support the static loads in the cello. The plates are in fact somewhat thicker (approximately $1.7h_1$), and so the lowest natural oscillation of the body is four or five semitones above the frequency of the second-highest open string.

It is even more difficult to attempt to fulfill the similarity requirements in the case of the viola. Its playing position between the shoulder and the left hand allows only a slight increase in the body length of the instrument; the viola is also grasped between the chin and the shoulder, making it impossible to increase the depth of the ribs much. Both of these requirements of the playing position already impose sufficient difficulties on the viola player.

It could be asked to what degree physical similarity among the instruments of the string quartet is in fact desirable. A difference between the relative spectra could, in fact, be an attractive feature of the tone, as it is for organ stops.

However, Saunders (1953) and Hutchins, later advised by Schelleng, took the opposite point of view. They attempted to fulfill the laws of similarity at least with respect to the lowest natural frequencies. Following the wishes of a composer and a cellist, they took it upon themselves to make additional types of string instruments. Their "family of fiddles" (see figure 12.10) bridged the gap between the viola and the cello with a vertically held instrument which they called the tenor. They also extended the sequence upwards by an instrument called the soprano, tuned an octave above the viola and which had already been used by Bach, and another instrument an octave above the violin, called the treble (referred to in earlier times as a dancing master's violin). Additionally, they redesigned the contrabass as a member of the violin family (the conventional contrabass is a member of the gamba family); and they developed an intermediate instrument in the bass range, the small bass.

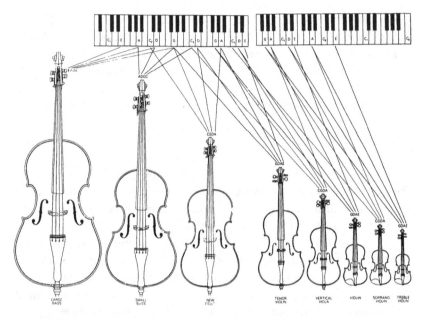

Figure 12.10
Tuning of the open strings and relative sizes of the eight instruments of the "family of fiddles" designed by Saunders and Hutchins and built by Hutchins.

Saunders, Hutchins, and Schelleng were unable to fulfill the requirements for linear dimensions set by the laws of similarity; they used changes in volume and thickness to compensate. Their experiment represented the first dimensioning of string instruments based on physical considerations. Precisely because the usual cello and viola violate the laws of similarity, these researchers developed variants on them in order to place the lowest resonances at the frequencies of the fundamentals of the open middle strings. Hutchins (1967) indicates that musicians were pleased with this change, because the lowest notes on the lowest string sounded fuller. This might seem surprising, as the f-hole resonance is above the fundamental of the second-lowest string. However, the second partial of the lowest string is emphasized by the lowest wood resonance.

This situation is demonstrated by sound pressure measurements in which the string is excited by bowing, and the recorded sound pressure is not filtered. Figure 12.11 shows two such plots of total level as a function of frequency (Hutchins 1967), which Saunders called "loudness curves"

(Saunders 1946). The upper curve represents a normal cello, and the lower one, the newly developed cello called the baritone. In the baritone, the peak of the second overtone lies noticably lower.[1]

As already mentioned, it is difficult in the case of the viola to place the f-hole resonances at the frequency of the fundamental of the second-lowest string, as in the violin. It is not very effective to attempt this by making the f-holes smaller, as the dependence on S in (12.11) goes only as the fourth root, and because decreasing S increases losses due to air friction. Sound flux at the resonance, and so the power radiated from the f-holes, is thereby decreased. Any increase in the volume of the instrument, however, makes playing more difficult. For this reason, Saunders and Hutchins have also developed violas in which shallow ribs and lengths more nearly approximating that of the violin placed the f-hole resonance so high that its lower octave fell on the c string. Figure 12.12 gives an example. Weakness of the lower tones, which had been feared in small violas, did not occur. Loudness was, on the contrary, surprisingly great, and the emphasis on the second partial which brought this about was not found to be in any way unpleasant.

It can be seen that the distribution of the peaks in the loudness curve is most even when the natural frequency of the f-hole resonance is a half-octave below the deepest natural frequency of a wood resonance.

But Saunders and Hutchins could not always find this principle confirmed in precious old violins made by master builders. They even found violins in which the lowest natural wood frequency lay an octave above the natural f-hole frequency, so that the overtone peak of the former coincided with the fundamental peak of the latter.

Saunders (1946) however, points out that most old violins are no longer in the form in which they left their builders' workshops. As concert halls became larger, greater sound power output was required. This could be achieved using heavier strings, which had to be tensioned more highly. The greater force of the strings on the top plate had to be supported

---

[1] The present author thanks Carleen M. Hutchins for correspondence concerning this material. Hutchins (1976) has reported recent investigations in which she states that she must change her opinion that the f-hole resonance would lie a half-octave below the lowest wood resonance in celli; the separation in fact approaches a full octave. Apparently, this lowering of the f-hole resonance is brought about by the relatively great compliance of the top and back plates in the cello. This is also the case for the baritone, which, made like the cello in accord with the rules of similarity, achieves a pleasing sound on the c string.

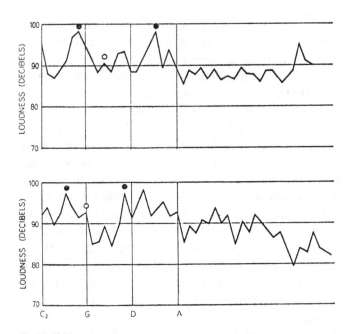

Figure 12.11
Total sound pressure level relative to the pitch of bowed notes. Top, of a normal cello; bottom, of a baritone (after Saunders and Hutchins).

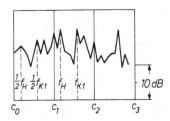

Figure 12.12
Total sound pressure level relative to the pitch of bowed notes of an especially shallow and short viola (after Saunders and Hutchins).

statically with a deeper bass bar. The resulting increase in dynamic stiffness raised the natural frequency of the lowest wood resonance. Despite this physically decisive change, the name of the master builder remained as a symbol of quality.

References

Backhaus, H., 1930. *Z. Phys.* **62**, 143.

Backhaus, H., 1931. *Z. Phys.* **72**, 218.

Dünnwald, H., 1979. *Acustica* **41**, 238.

Hutchins, C. M., 1967. *Physics Today* **20** (2), 28.

Hutchins, C. M., 1976. *Catgut Acoust. Soc. Newsletter* no. 26.

Jansson, E., N. E. Molin, and H. Sundin, 1970. *Phys. Scripta* **2**, 243.

Lottermoser, W., and J. Meyer, 1957. *Instr. Z.* **12** (3), 42.

Lyon, R., 1975. *Statistical Energy Analysis of Dynamical Systems.* Cambridge, Mass.: MIT Press.

Meinel, H., 1937a. *Akust. Z.* **2**, 22.

Meinel, H., 1937b. *Elektr. Nachr. Techn.* **14** (4) 119.

Moral, J. A. and E. V. Jansson, 1982. *Acustica* **50**, 329.

Ptacnik, E., 1953. *Acta Phys. Austria* **8**, 28.

Reinicke, W., 1973. Dissertation, Institute for Technical Acoustics, Technical University of Berlin.

Reinicke, W., and L. Cremer, 1970. *J. Acoust. Soc. Amer.* **48**, 988.

Saunders, F., 1937. *J. Acoust. Soc. Amer.* **9**, 2.

Saunders, F., 1946. *J. Acoust. Soc. Amer.* **17**, 169.

Saunders, F., 1953. *J. Acoust. Soc. Amer.* **25**, 491.

III THE RADIATED SOUND

13 Source Small in Comparison to the Wavelength

13.1 The spherical wave field

The radiation of sound by a violin generally cannot be treated as an exact mathematical problem, due to the violin's complicated shape. There is one exception to this rule, true for all sound sources: the case in which the frequency of the radiated sound is so low that all dimensions of the source may be regarded as small in comparison to the wavelength. But this exception is best understood through an exact treatment of a sound-radiating body chosen so all of its details are amenable to calculation and for which we need set no restrictions on the frequency range.

A suitable sound-radiating body is a sphere of radius a, all points of whose surface oscillate radially and in phase, with the velocity $\underline{v}_a$. This is called a *breathing* (or *pulsating*) *sphere*, or, in the terminology of spherical functions, a *radiator of the zeroth order*. It follows from the symmetry of this source that the field variables must also exhibit spherical symmetry; in other words, they are functions only of distance from the center of the sphere. The vector of particle velocity must, then, always point in a radial direction.

In this sense, the field is one-dimensional, like that of a plane wave. But it does not obey the one-dimensional equation for a plane wave. The radiated power P is distributed over larger and larger spheres as r increases, and the intensity J must decrease as r^{-2}:

$$J = \frac{P}{4\pi r^2}. \tag{13.1}$$

Nonetheless, we may assume that the wavefronts at points a great distance r from the sphere differ little from plane waves. We may call the field at such distances the *far field*; we may regard the relationships that apply to plane waves as approximately valid in the far field, at least within regions no more than half a wavelength in extent. Within such regions, it is valid to say that

$$J = \tilde{p} \cdot \tilde{v} = \varrho c \tilde{v}^2 = \tilde{p}^2/(\varrho c). \tag{13.2}[1]$$

Combined with (13.1), this simplification allows us to describe the dependence of the sound pressure and particle velocity in the far field (i.e., for $r > a \gg \lambda$) by

---

[1] When the quantities representing power are intended to be averaged over time, the field quantities p and v should be replaced by their rms values $\tilde{p}$ and $\tilde{v}$.

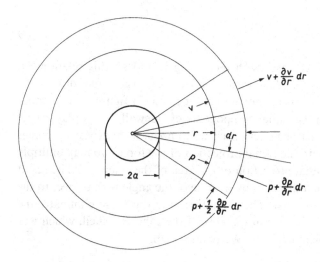

Figure 13.1
Pressure and velocity in a one-dimensional spherical wave.

$$\underline{p} = \underline{p}_a \frac{a}{r} e^{-jk(r-a)}, \tag{13.3a}$$

$$\underline{v} = \underline{v}_a \frac{a}{r} e^{-jk(r-a)}. \tag{13.3b}$$

But the distance r still appears in the denominator, and so these functions do not conform to the one-dimensional plane-wave equation.

There must consequently be differences in the fundamental equations relating p and v. Let us consider small cones within the spherically symmetrical wavefield, as shown in figure 13.1. The apexes of the cones are at the origin, $r = 0$, and each cone spans a solid angle Ω. We describe truncated cones between the radii r and $r + dr$, through whose radial surfaces no sound flux is transmitted. It should be noted, however, that the increase in material, resulting in increased pressure across the truncated cone, depends not only upon the difference in velocities at the inner and outer spherical surfaces, but also upon the areas of these surfaces. The equation

$$\underline{v}(\Omega r^2) - \left(\underline{v} + \frac{\partial v}{\partial r}dr\right)\left(\Omega r^2 + \frac{\partial(\Omega r^2)}{\partial r}dr\right) = (\Omega r^2 dr)\frac{j\omega p}{K} \tag{13.4}$$

holds, and from it we can derive the differential equation

$$-\frac{d\underline{v}}{dr} - \frac{2}{r}\underline{v} = \frac{i\omega\underline{p}}{K}. \tag{13.5}$$

The second term decreases with increasing r, and so this relationship approaches that for the plane wave, (11.57a), for large values of r.

The pressure difference, shown at the bottom in figure 13.1, is more complicated, since the radial components of sidewall forces play a part here. However, a mathematical trick well known in hydrostatics comes to our aid. To obtain the radial projection of the force, we may multiply the force by the radial projection of the area of the sidewalls. The result is the same as if we multiply it by the sine of the angle with respect to the radial direction of interest. The simplified procedure also eliminates the effect of the radial increase in the area of the spherical shell, which was a substantial problem in (13.5). We obtain, then,

$$\underline{p}(\Omega r^2) + \left(\underline{p} + \frac{1}{2}\frac{\partial \underline{p}}{\partial r}dr\right)\frac{\partial(\Omega r^2)}{\partial r}dr - \left(\underline{p} + \frac{\partial \underline{p}}{\partial r}dr\right)\left(\Omega r^2 + \frac{\partial(\Omega r^2)}{\partial r}dr\right)$$

$$= j\omega\varrho(\Omega r^2 dr)\underline{v}, \tag{13.6}$$

from which the same differential relationship follows as for the plane wave:

$$-\frac{d\underline{p}}{dr} = j\omega\varrho\underline{v}. \tag{13.7}$$

By substituting (13.7) into (13.5), we obtain the spherical wave equation

$$\frac{d^2\underline{p}}{dr^2} + \frac{2}{r}\frac{d\underline{p}}{dr} + k^2\underline{p} = 0, \tag{13.8a}$$

which can be written as

$$\frac{d^2(r\underline{p})}{dr^2} + k^2(r\underline{p}) = 0. \tag{13.8b}$$

The corresponding wave number is

$$k = \omega\sqrt{\frac{\varrho}{K}} = \frac{\omega}{c}, \tag{13.9}$$

the same as for the plane wave. The spherical-wave solution involves no dispersion, and the dependence of the sound pressure in the far field on r which we first hypothesized in (13.3a) holds at lesser distances as well, and even at the surface of a sphere of any radius a.

This is not true, however, of the dependence of velocity on r hypothesized in (13.3b). We may calculate the dependence of velocity on r using (13.7), with the function of pressure in (13.3a). Two terms result:

$$\underline{v} = \underline{p}_a a \left[\frac{1}{j\omega\varrho r^2} + \frac{1}{\varrho c r} \right] e^{-jk(r-a)}, \tag{13.10}$$

of which the first falls off as r^{-2}; the second component falls off only as r^{-1}, and conforms to the dependence in (13.3b). The greater the value of r, the more the second component predominates. As indicated above, this component is characteristic of the far field.

Conversely, the first term predominates when r is small; this component is consequently designated the *near-field* term.

The term which governs the field at any given distance depends on the ratio of the two; in other words, on the value of

$$\frac{j\omega\varrho r^2}{\varrho c r} = kr = \frac{2\pi r}{\lambda}, \tag{13.11}$$

a Helmholtz number characteristic of a spherical wave.[2]

If the smallest value of kr—namely, ka—is relatively large, there is no near field: the far field begins at the surface of the pulsating sphere.

Figure 13.2, top, shows the case in which there is no near field, at the moments when there is either a maximum expansion (right) or contraction (left) of the sphere. The motion of the air is indicated by points which would lie at the corners of a square lattice if the air were undisturbed. The lattice is turned 45° to the horizontal. (In the case shown, $ka \approx 3$, so the phase relationships in the diagrams do not exactly correspond to the far-field case discussed above; however, the amplitudes come close to representing this case, since the correction term is $(ka)^{-2} = 1/9$.)

At the bottom of figure 13.2 are shown similar drawings representing the near-field case

$$(ka)^2 \leqq (kr)^2 \ll 1. \tag{13.12}$$

Here we can see the displacement falling off rapidly with increasing r. This corresponds to the more suitable form of (13.10),

---

[2] Each ratio of a characteristic length of a field to the wavelength may be called a "Helmholtz number." Such Helmholtz numbers mean physically similar wave fields, just as do the Reynolds number and Mach number in aerodynamics. The use of Helmholtz's name is based on his being the first to restrict the wave equation to sinusoidal solutions, in which this number appears as a parameter of similarity (Cremer 1971).

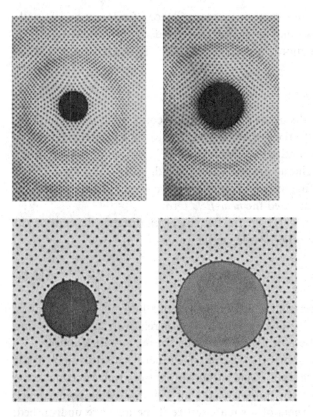

Figure 13.2
The field of a breathing sphere; top: when $\lambda < 2\pi a$ (far field); bottom: when $\lambda \gg 2\pi a$ (near field).

$$\underline{v} = \underline{p}_a a \frac{1}{j\omega \varrho r^2}[1 + jkr]e^{-jk(r-a)}, \tag{13.13a}$$

which includes the special case

$$\underline{v}_a = \underline{p}_a \frac{1}{j\omega \varrho a}[1 + jka]. \tag{13.14b}$$

The quotient of these equations describes the velocity field:

$$\frac{\underline{v}}{\underline{v}_a} = \frac{a^2}{r^2}[1 + jka]e^{-jk(r-a)} \approx \frac{a^2}{r^2}. \tag{13.14}$$

Given (13.12), this results in a *quasistationary field* whose particle velocity falls off as r^{-2}. This corresponds to the hydrodynamic field of an incompressible fluid, in which the same flux crosses every concentric spherical surface:

$$\underline{q}_a = 4\pi a^2 \underline{v}_a = 4\pi r^2 \underline{v}(r). \tag{13.15}$$

We have already examined several cases in which the possibility of compression-free air displacements made a change in density unnecessary. Such is the case here as well. If we set $r = a$ in (13.10), the field admittance at the surface of the radiator is seen to be

$$\frac{\underline{v}_a}{\underline{p}_a} = \frac{1}{j\omega \varrho a} + \frac{1}{\varrho c}. \tag{13.16}$$

This admittance consists of the inertial admittance of a part of the mass of the surrounding air and the admittance of a plane wave—in other words, the reciprocal of the characteristic impedance of air, ϱc. At low frequencies, the first term predominates. The acceleration of air in the near field requires less pressure than does the change in density necessary to bring about radiation.

Nonetheless, some radiation still occurs.

In the limiting case $|ka|^2 \gg 1$, the type of field does not change qualitatively with the distance r; but in the limiting case $|ka|^2 \ll 1$, this is not so. As r increases, we move more and more into a region in which only a propagating wave appears.

We may replace the sound pressure $\underline{p}_a$ in (13.3a) by the velocity $\underline{v}_a$. This may always be regarded as a given at the surface of bodies vibrating in air. We then first obtain, in the general case,

$$\underline{p} = \frac{j\omega\varrho a}{1 + jka}\underline{v}_a\frac{a}{r}e^{-jk(r-a)};$$ (13.17a)

when $|ka|^2 \ll 1$, we obtain

$$\underline{p} = j\omega\varrho(a^2\underline{v}_a)e^{-jkr}/r$$

$$= \frac{j\omega\varrho}{4\pi}\underline{q}_a\frac{e^{-jkr}}{r}.$$ (13.17b)

As we will explain, this fundamental formula incorporates two important statements about breathing spheres for which $(ka)^2 \ll 1$:

1. Sound pressure in the far field increases with sound flux at the source as

$$\underline{q}_a = 4\pi a^2\underline{v}_a.$$ (13.18)

2. Sound pressure increases with frequency, or better stated, it falls off toward low frequencies.

13.2 The point radiator (monopole)

The spherically symmetric radiator for which $(ka)^2 \ll 1$ is also called the monopole. Since this term is also used for static sources, the term "point radiator" is less ambiguous.

The great importance of the monopole is as an elementary radiator; monopoles can be used to build up other types of sources which generate fields which are not spherically symmetric.

Another consequence of the monopole's properties as an elementary radiator is that a point radiator need not be a breathing sphere. Rather, any sound source which creates a sound flux q in the surrounding medium can be regarded as a point radiator if it is small enough with respect to the wavelength.

Only when this is true for each element is the point synthesis mentioned above possible:

$$\underline{p} = \frac{j\omega\varrho}{4\pi}\Sigma\underline{q}_n\frac{e^{-jkr_n}}{r_n},$$ (13.19)

because it is only in this case that a wave from one of the sources can diffract around the others without generating significant secondary waves.

We will now consider the simplest case of point synthesis, that of two sources with the same phase and generating sound fluxes q_1 and q_2. The sources are at positions $+h$ and $-h$ along the z-axis.

We are interested only in large distances r, for which

$$r \gg 2h, \tag{13.20}$$

so that the equation

$$\underline{p} = \frac{j\omega\varrho}{4\pi} \left[q_1 \frac{e^{-jkr_1}}{r_1} + q_2 \frac{e^{-jkr_2}}{r_2} \right] \tag{13.21a}$$

will be satisfied by

$$r_1 \approx r_0 - h\cos\vartheta, \tag{13.22_1}$$

$$r_2 \approx r_0 + h\cos\vartheta, \tag{13.22_2}$$

as shown in figure 13.3. Where (13.20) applies, the difference between $1/r_1$ and $1/r_2$ may be neglected. Consequently, we may write (13.21a) as

$$\underline{p} = \frac{j\omega\varrho}{4\pi r_0} e^{-jkr_0} [q_1 e^{jkh\cos\vartheta} + q_2 e^{-jkh\cos\vartheta}]. \tag{13.21b}$$

If we now assume additionally that the extent of the source is small with respect to the wavelength; that is,

$$2h \ll \lambda, \tag{13.23a}$$

and so

$$kh \ll 1, \tag{13.23b}$$

the phase difference which depends on direction can be neglected; and the smaller kh is, the better the result approximates

$$\underline{p} = \frac{j\omega\varrho}{4\pi} (q_1 + q_2) \frac{e^{-jkr_0}}{r_0}. \tag{13.21c}$$

In this equation, the total flux $(q_1 + q_2)$ appears in place of q_a in (13.17b).

The incompressible, hydrodynamic near field in the immediate vicinity of the two point radiators is, however, certainly very different in the two cases. For this reason, the air masses which oscillate along with the sources and load them are also different.

But with increasing r_0, even this field of total flux $(q_1 + q_2)$ approaches

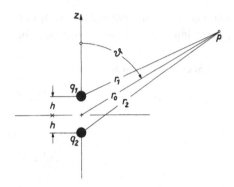

Figure 13.3
Notation for the field of two point radiators.

a spherically symmetric one at distances small with respect to the wavelength; and consequently it still satisfies the definition of a near field. It is therefore not surprising that the spherically symmetric wave field described by (13.21c) develops out of this near field. This is the case at low frequencies with every headphone or loudspeaker diaphragm behind which there is a sealed air cavity.

Consequently, (13.21c) can be extended to any number of sources radiating in phase, as long as they are located within a region small with respect to the wavelength. From

$$\underline{p} = \frac{j\omega\varrho}{4\pi}\Sigma q\frac{e^{-jkr_0}}{r_0},$$ (13.21d)

we obtain the total radiated power,

$$P = 4\pi r_0^2\tilde{p}^2/\varrho c = \frac{1}{4\pi}\varrho c\left(\frac{\omega}{c}\right)^2(\widetilde{\Sigma q})^2$$

$$= \varrho c\frac{\pi}{\lambda^2}(\widetilde{\Sigma q})^2.$$ (13.24a)

If we compare this with the equation defining the radiation flux resistance,

$$P = \mathscr{R}_{rad}(\widetilde{\Sigma q})^2,$$ (13.24b)

we obtain as the value of this resistance:

$$\mathscr{R}_{rad} = \varrho c\frac{\pi}{\lambda^2}.$$ (13.24c)

We have already made use of this formula in section 10.3. We could have obtained this formula from (13.13b) by calculating the real part of $\underline{p}_a/\underline{v}_a$ for $(ka)^2 \ll 1$ and dividing the quotient by the total surface area $4\pi a^2$:

$$\mathscr{R}_{\text{rad}} = \text{Re} \left\{ \frac{\underline{p}_a}{4\pi a^2 \underline{v}_a} \right\}, \tag{13.24d}$$

though in this case only for the spherical source.

The derivation based on the far field is applicable to all point radiators of any shape and flux distribution, including the synthesis in section 10.3 from the fluxes generated by the top and back plates and the f-holes of the violin.

13.3 The dipole

Our second simple example of synthesis using point radiators retains the arrangement in figure 13.3; in this case, the fluxes are equal but opposite:

$$\underline{q}_1 = -\underline{q}_2 = \underline{q}. \tag{13.25}$$

Substituting this into (13.21b), we obtain

$$\underline{p} = \frac{-\omega\varrho}{4\pi} 2\underline{q} \sin(kh\cos\vartheta) \frac{e^{-jkr_0}}{r_0} \tag{13.26a}$$

as the sound pressure in the distant field defined in (13.20). We now also adopt once again the additional restriction of (13.23a); that is, we assume that the separation of the out-of-phase point radiators is small in comparison to the wavelength. It would be tempting to conclude that the radiation would in this case simply cancel out. This is, in fact, the case in the plane $z = 0$, where the components of sound pressure from the two radiators arrive with equal amplitude and in opposite phase. At all other points, especially in the z-direction, the phase difference, though it is small, plays an important role.

In (13.26a), the assumption made in (13.23a) is expressed simply by replacing the sine by its argument; we consequently obtain

$$\underline{p} = \frac{-c\varrho}{4\pi} k^2 (2h\underline{q}) \frac{e^{-jkr_0}}{r_0} \cos\vartheta. \tag{13.26b}$$

There exists, however, a more general way to arrive at this result. Instead of constructing the difference using two point radiators displaced by $2h$ in the z-direction, it is possible to calculate the difference between the sound pressures at two points separated by $-2h$ in the field of a single point radiator. In the limiting case of very small h this leads to

$$\underline{p} = -\frac{j\omega\varrho}{4\pi}(2h\underline{q})\frac{\partial}{\partial z}\left(\frac{e^{-jkr_0}}{r_0}\right); \tag{13.26c}$$

and since r_0 varies according to

$$dr_0 = dz\cos\vartheta, \tag{13.27}$$

if the point of observation is moved in the z-direction, then (13.26c) can be written as

$$\begin{aligned}
\underline{p} &= -\frac{j\omega\varrho}{4\pi}(2h\underline{q})\frac{\partial}{\partial r_0}\left(\frac{e^{-jkr_0}}{r_0}\right)\cos\vartheta \\
&= -\frac{c\varrho}{4\pi}(2h\underline{q})k^2\left(1+\frac{1}{jkr_0}\right)\frac{e^{-jkr_0}}{r_0}\cos\vartheta.
\end{aligned} \tag{13.26d}$$

Equation (13.26a) appears here, in accord with its derivation, as the far-field component of (13.26d).

This case in which there are two point radiators is called the dipole, and the quantity which determines the sound pressure,

$$\underline{M} = 2h\underline{q} \tag{13.28}$$

is called the *dipole moment*. The same words are used to describe stationary point-source pairs, and no distinguishing expression, such as, for example, "double point radiator" has as of yet been applied to dipole radiators.

The most obvious difference in comparison with the simple point radiator is that radiation is not equal in all directions; it still possesses rotational symmetry, but it has a directional dependence on the angle of declination ϑ. If we plot the magnitude of the direction factor given by

$$\Gamma_z(\vartheta) = \cos\vartheta \tag{13.29a}$$

as a polar diagram with respect to ϑ, the resulting directional characteristic is seen to be comprised of two spheres of unit diameter, placed above one another in the z-direction and touching at the origin. Due to the aforementioned rotational symmetry, it is sufficient to express this function as two circles in a vertical plane including the z-axis; we call this, then, the dipole axis. The expression "figure-eight characteristic" is also used, derived from the shape of this function.

The fact that the dipole has an axis has also another consequence. Suppose that the dipole axis does not lie along that of the chosen spherical coordinates, though it may coincide with the x- or the y-axis. Then we must employ the azimuthal angle φ in the description of the dipole. If the dipole lies in the direction $\varphi = 0$, then the direction factor reads

$$\Gamma_x = \sin \vartheta \cos \varphi; \tag{13.29b}$$

if it lies in the direction $\varphi = 90°$, we get

$$\Gamma_y = \sin \vartheta \sin \varphi. \tag{13.29c}$$

Thus any dipole can be decomposed into dipoles directed along the chosen Cartesian axes. This makes it possible to evaluate the magnitude and direction of any dipole by measuring direction factors in the y, z, z, x and x, y planes, as is usual for string instruments. But this description in terms of component dipoles is not as enlightening as one in terms of the actual dipole direction and moment.

The coefficient of $\cos \vartheta$ is retained even after differentiation of (13.26b) with respect to r_0 to derive the radial component of velocity:

$$\underline{v}_r = -\frac{1}{j\omega\varrho}\frac{\partial \underline{p}}{\partial r_0} = \frac{-M}{4\pi}\left(\frac{k^2}{r_0} - \frac{2jk}{r_0^2} - \frac{2}{r_0^3}\right)e^{-jkr_0}\cos\vartheta$$

$$= \underline{v}_r(r_0,0)\cos\vartheta. \tag{13.30}$$

This, however, means that the radial velocities on the surface of a sphere defined by $r_0 = $ const are the same as if a real sphere of radius $r_0 = a$ oscillated in the z-direction with the velocity $\underline{v}_{az}$.

Our initial discussion led us from a source of arbitrary size, the breathing sphere, to a point radiator; now we go in the opposite direction, from the dipole to an oscillating sphere. In accord, once more, with the corresponding first-order spherical function, the cosine, this is called a *radiator of the first order*.

The value of the Helmholtz number ka is, as before, unrestricted for this radiator. Also, the case in which ka is very small makes a generalization possible here, as before. In the present case we need consider only the last term in (13.30), which corresponds to a near field in which v_r falls off as r_0^{-3}. The dipole moment derived from this term is

$$\underline{M} = 2\pi a^3 \underline{v}_{az}. \tag{13.31a}$$

The dipole moment is, then, not simply the product of the volume of the sphere and the velocity v_{az}, but rather $1\frac{1}{2}$ times this quantity. It can be shown that the additional volume multiplied by the density of the medium represents the vibrating mass in the neighborhood of the sphere.

It can be shown that the formula

$$\underline{M} = (V_{\text{body}} + V_{\text{medium}})\underline{v}_{az} \tag{13.31b}$$

is valid for oscillating bodies of arbitrary shape if they are small with respect to the wavelength (Skudrzyk 1954, pt. XI, sec. II). Unfortunately, the calculations are easy in only a few cases. The formula can, however, be determined experimentally: the body to be studied is attached to a thin vertical rod, clamped rigidly at the other end. Then the lowest natural frequency of this vibrating system is measured as it oscillates first in air and then under water.

Figure 13.4 compares displacements once again. At the top is shown the displacement when ka is large. Far-field radiation, clearly dependent on the direction, begins at the surface of the sphere. At the bottom, ka is small, and the sphere pushes the same air mass ahead of it in the $+z$-direction as it draws from the $-z$-region behind it. In this case a velocity in the opposite direction arises at the sides of the sphere. The hydrodynamic short circuiting of the radiation is much more pronounced here than in the case of a zero-order radiator.

This phenomenon is related to a further basic difference between the dipole and the point radiator; the pressure in the far field in (13.26d) increases as k^2, in other words as f^2; or, as it may be better stated, the pressure falls as the frequency decreases.

If an oscillating diaphragm is to radiate low frequencies, its rear must be enclosed by a sealed chamber, or it must be mounted in a large baffle. In this way, a dipole is made into a monopole. For this reason the violin's back and air cavity would be necessary even if they made no direct contribution to the radiated output.

The decrease in the radiation of a dipole at low frequencies has an additional consequence: whenever a radiator is composed of a monopole and a dipole, there is always a frequency below which the sound pressure is governed almost entirely by the monopole. It is for this reason that we neglect the dipole component when we consider the case of two point radiators in phase and very near one another, but of unequal strength.

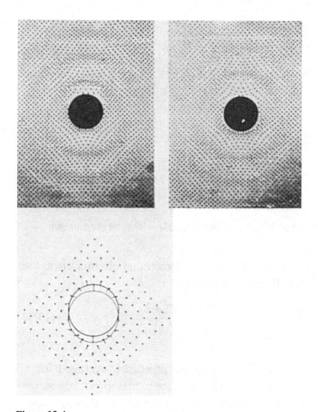

Figure 13.4
The field of an oscillating sphere. Top, when $\lambda < 2\pi a$ (far field); bottom, when $\lambda \gg 2\pi a$ (near field).

If, in that example, we were to account for the dipole component, we would have to replace (13.21c) with

$$\underline{p} = \frac{j\omega\varrho}{4\pi r_0} e^{-jkr_0} \left[(q_1 + q_2) + j(kh\cos\vartheta)(q_1 - q_2)\right] = \underline{p}_I + \underline{p}_{II}\cos\vartheta. \quad (13.32)$$

Even in the direction of maximum radiation ($\vartheta = 0$), the monopole component $\underline{p}_I$ predominates over the dipole component $\underline{p}_{II}$, as soon as $\Sigma q > kh\Delta q$. Since the formulas are valid only for $kh < 0.5$, the monopole component predominates throughout the region in which (13.32) is valid, as long as $\Sigma q > 0.5\Delta q$. For smaller ratios of $\Sigma q/\Delta q$ the transition frequency is lowered.

The ratio of relative output is even smaller for the dipole in comparison

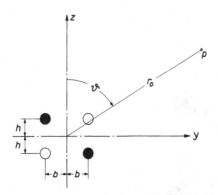

Figure 13.5
Notation for the field of four point radiators in the y, z-plane (tesseral quadrupole).

with the monopole if we take the dipole's directional characteristic into account. We obtain, for the ratio of powers,

$$\frac{P_\mathrm{I}}{P_\mathrm{II}} = \frac{4\pi r_0^2 \tilde{p}_\mathrm{I}^2}{2\pi r_0^2 \displaystyle\int_{-\pi/2}^{+\pi/2} \tilde{p}_\mathrm{II}^2 \cos^2 \vartheta \sin \vartheta \, d\vartheta} = \frac{3p_\mathrm{I}^2}{p_\mathrm{II}^2} = \frac{3(\widetilde{\Sigma q})^2}{(\widetilde{\Delta q})^2}; \qquad (13.33)$$

in other words, the ratio of the rms sound pressures, averaged for all directions, must be multiplied by $\sqrt{3}$. The dipole may produce the predominant component of $\underline{v}$ of a source, and yet a much weaker monopole component may govern the radiated sound.

13.4 Tesseral and axial quadrupoles

Exactly as we constructed a dipole out of two closely spaced out-of-phase point radiators, we can construct a *quadrupole* out of two closely spaced antiphasic dipoles. There are two limiting cases for the quadrupole: in one, the two dipoles lie one behind the other on the same axis; in the other, the dipoles lie side by side.

The tesseral quadrupole

We have already met the case with dipoles lying side by side in section 10.2. This case is represented graphically in figure 13.5; as shown, all four point radiators lie in the y, z-plane. The separation of the radiators in the z-direction is here $2h$, as it was before; separation of the radiators in the y-direction is assigned the value $2b$. This arrangement of four point radiators is called a *tesseral quadrupole* (Morse and Ingard 1968, p. 314).

It is evident that the field no longer possesses rotational symmetry, and that the directional characteristic is much more complicated. We will go only so far as to discuss directionality in the y, z-plane; this plane is in any case where the maxima occur. It is easy to see that the phase differences between the individual point radiators will decrease at angles away from the y, z-plane if the declination ϑ with respect to the z-axis remains constant. Finally, in the x-direction, the effects of the four point radiators cancel each other completely.

We can, however, derive the sound pressure field in the y, z-plane using the same principle that we applied to the dipole in (13.26d). We differentiate the dipole field given in (13.26d) with respect to y and multiply the differentials by $-2b$. In this way, we account for the effects of a change of position dy, which leads to

$$dr_0 = dy \sin \vartheta. \tag{13.34}$$

We consequently obtain

$$\underline{p} = \frac{c\varrho}{4\pi}(4hb\underline{q})k^2 \frac{\partial}{\partial r_0}\left[\left(\frac{1}{r_0} + \frac{1}{jkr_0^2}\right)e^{-jkr_0}\right]\cos\vartheta\sin\vartheta$$

$$= -j\frac{c\varrho}{4\pi}(2hb\underline{q})k^3\left[\frac{1}{r_0} - \frac{2j}{kr_0^2} - \frac{2}{k^2r_0^3}\right]e^{-jkr_0}\sin 2\vartheta. \tag{13.35}$$

The maxima occur at $45°$ angles to the y and z axes (see figure 13.6).

The differentiation of the pressure field in order to determine the radial component of velocity proceeds in the same way as for a single dipole. In the case of the quadrupole, however, sound pressure in the near field falls off as r_0^{-3}; the radial component of velocity, and all other components of velocity as well, fall off as r_0^{-4}. The hydrodynamic short-circuiting is even more pronounced than with the dipole.

The maximum sound pressure in the far field,

$$\underline{p}_{\text{IV}} = -j\frac{c\varrho}{4\pi}[2hb\underline{q}]k^3\frac{e^{-jkr_0}}{r_0} \tag{13.36}$$

varies as f^3; in words, it falls off rapidly toward low frequencies. Given that the flux of all of the point radiators is equal, p_{IV} [see (13.26b)] is smaller than p_{II} by a factor of kb, just as p_{II} was smaller than p_{I} by a

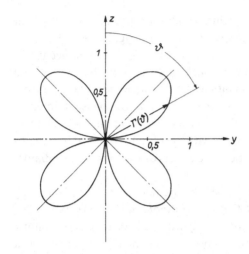

Figure 13.6
Directional characteristic of a tesseral quadrupole in the y, z-plane.

factor of $2kh$. (The factor of 2 arises from the factor $\cos \vartheta \sin \vartheta$, whose maximum is only $1/2$.) The ratio is even greater if, as in (13.33) above, we compare total radiated power.

If we superimpose monopole, dipole, and quadrupole components, as we did earlier for monopole and dipole components only, the quadrupole component may generate a far greater flux; yet the dipole component, and even more so the monopole component, may govern the sound pressure in the far field.

It also holds that the tesseral quadrupole can be oriented in any direction, and can be decomposed into component quadrupoles lying in the three Cartesian planes, with its direction factor given unequivocally in terms of the direction factors in the three planes. Again, the radial velocity on a concentric sphere will have the same directional distribution as the direction factor, and contain four nodal lines, running from pole to pole.

The axial quadrupole

The abovementioned relation between radial velocity on a concentric sphere and directional characteristics holds for all spherical sound fields. In particular, we will here discuss the case where the two out-of-phase dipoles lie one behind the other. We then obtain a third elementary source exhibiting rotational symmetry, called the *axial quadrupole* (Morse and

Ingard 1968, p. 314), which again can be described in terms of the declination angle ϑ when we chose the z-axis to coincide with that of the dipoles. The direction function is given by a spherical function of second order (Morse 1948, p. 315):

$$\Theta_2 = \tfrac{3}{2}(\cos^2 \vartheta - \tfrac{1}{3}).\tag{13.37}$$

This has two in-phase maxima $+1$ at $\vartheta = 0$ and $\vartheta = 180°$ and an out-of-phase maximum -1 at $\vartheta = 90°$, and two zero-crossings at

$$\vartheta = \arccos\left(+(3)^{-1/2}\right) = 55°,\tag{13.38a}$$

$$\vartheta = \arccos\left(-(3)^{-1/2}\right) = 125°.\tag{13.38b}$$

This means that the corresponding vibrating sphere has two nodal lines of radial velocity.

Our remarks concerning decreased tesseral-quadrupole radiation at low frequencies on account of pronounced hydrodynamic shortcircuiting holds as well for the axial quadrupole.

13.5 Application to the behavior of string instruments

In our derivations of the sound pressure in the far field and the corresponding directional characteristics of the point radiator, the dipole, and the quadrupole, we proceeded from the assumption that all dimensions of the sound source are small with respect to the wavelength. If we apply this requirement to the violin, the shortest possible wavelength would be approximately eight times the 35-cm length of the violin body. This would correspond to a frequency of 122 Hz, lower than the lowest fundamental frequency of the violin.

The condition that the wavelength be eight times the instrument's length is somewhat better fulfilled in the case of the cello. Though the tuning is a musical twelfth lower, the cello is not, proportionally, three times as long, having an average length of 75 cm. Still the limiting frequency of 57 Hz, derived in the same way, is lower than the cello's lowest fundamental. So we see that the directional characteristics of elementary radiators hardly apply to string instruments in the x, z-plane.

The situation in the y, z-plane is somewhat better, since the widths of the instruments at the narrowest part (the waist) are only 1/3 as great. The applicable frequencies rise to 366 Hz for the violin and approximately

170 Hz for the cello. In both cases, these frequencies lie above the f-hole resonance, and so near the lowest wood resonances that there is hope for approximate solutions when dealing with the instruments as elementary sound sources.

It therefore seems reasonable to discuss which properties of radiation, particularly directional properties—at least in the y, z-plane—are to be expected on the basis of the laws which are valid for elementary sources.

We begin with low frequencies. As shown in figure 10.4, Eggers established that a sinusoidal input force F_y at the bridge leads to a rotation of the entire body around its x-axis; this motion predominates in the vibration pattern. This pattern corresponds to a tesscral quadrupole in the y, z-plane. Even at the lowest frequencies, then, the instrument does not behave like a simple point radiator.

Nonetheless, F_y could also lead to a translational vibration of the entire body in the y-direction. However, only the ribs, whose area is small, could be regarded as radiating surfaces for this motion.

It is certain, however, that the bridge, functioning as a lever, also leads to point-radiator-like breathing motions of the entire body. Such breathing occurs even at the lowest frequencies, as shown in section 10.2. Only the inevitable island of motion of opposite direction in the top plate leads to a small dipole component.

Statements in the sections just previous to this one, however, clearly suggest that radiation due to the quadrupole and even dipole components is so small at low frequencies that essentially only the circular radiation pattern of the point radiator is of importance, at least in the y, z-plane.

Figure 13.7 shows directional characteristics in the y, z-plane measured by Meinel.[3] That for 290 Hz refers to the f-hole resonance. It has a maximum perpendicular to the top plate, which can be described as the addition of a vertical dipole to a monopole. At the higher frequency of 580 Hz, corresponding to a body resonance, it may be important that the violins no longer small compared to the wavelength even in the y, z-plane. It is astonishing that such uniform sound radiation is found in the region of the lowest body resonances, as the measurements of J. Meyer (1975) reproduced in figure 13.8 demonstrate. The apparent smoothness of these curves is to some degree due to the choice of a logarithmic scale—which

---

[3] Meinel (1937), fig. 16, top left. The diagram has been turned by 180° to bring the bridge to its usual position at the top.

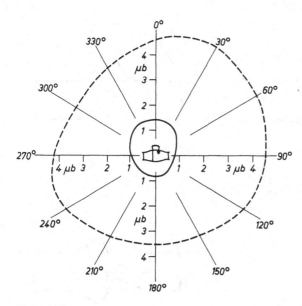

Figure 13.7
Measured directional characteristic of a violin in the y, z-plane (linear scale). Solid line,
at 290 Hz; dashed line, at 517 Hz (after Meinel).

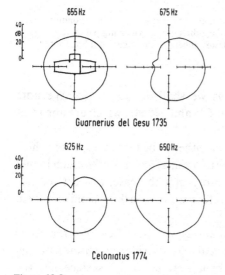

Figure 13.8
Measured directional characteristics of two violins in the y, z-plane (logarithmic scale)
at two frequencies in the range of the lowest wood resonances (after J. Meyer).

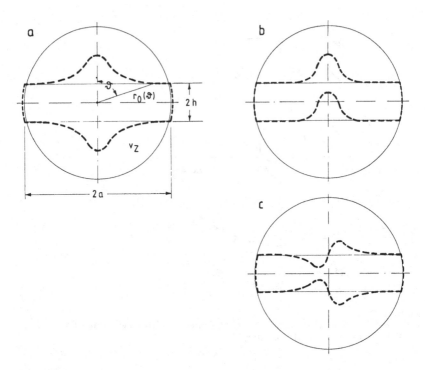

Figure 13.9
a, Oscillation of a breathing sphere of radius a, and (dashed line) the corresponding
distribution of normal velocity restricted to the z-direction by vibrating plates at $z = \pm h$
for $ka \ll 1$. b, the same for a vertically oscillating sphere. c, the same for a horizontally
oscillating sphere.

choice is somewhat problematic, as we shall discuss below. We note
especially the deviations occurring at 625 and 675 Hz, which are apparent
despite the smoothing.

We would expect uniform radiation either to be that of a sphere, which
the violin does not even approach, or to occur at wavelengths much longer
than the source dimensions; this also is not the case, at least for frequencies greater than 500 Hz. Therefore we must ask how this uniform
radiation comes about.

First we can show that the radial stationary velocity field of a monopole
at low frequencies can appear immediately outside the surface of any
finite body if the velocity component normal to the surface has the
appropriate distribution. Figure 13.9a shows a section through a breathing

sphere of radius a, whose Helmholtz number $ka \ll 1$. We now imagine that we are able to put in the inner field, which is that of a monopole, two very light elastic parallel foils of separation $2h$, symmetrical about the equatorial plane. The foils will vibrate with the same normal velocity component v_{z0} as the surrounding oscillating medium:

$$v_{z0} = v_a (a/h)^2 \cos^3 \vartheta, \tag{13.39}$$

shown by the dashed curve in figure 13.9a.

If we now replace the foils by rigid plates, which are forced by some exciting equipment to vibrate with the same v_{z0}, the field outside the plates will remain the same, since the only velocity that could disturb it, v_{z0}, is the same. The pressure will also remain the same, because on account of the impedance mismatch the vibrating body can overcome any reacting pressures, and the tangential velocities will have no influence. But, on the other hand, the field excited by a given velocity of the surface of the body is that of a forced vibration, and its stationary solution, in a linear system, is unique; thus it is the field we wished to produce. In principle our conclusion is the same as that which allowed us to get the $\cos \vartheta$-distribution of the normal velocity, which we may designate as V, for the first-order radiator corresponding to the movement of an oscillating sphere.

Figure 13.9b shows the V-distribution on the plates. Here v_{z0} contains not only a component of radial velocity $v_r \cos \vartheta$, but also one of tangential velocity $v_\vartheta \sin \vartheta$. Since in this case v_r and v_ϑ are given at the surface of the plates by

$$v_{r0} = v_{za} (a/h)^3 \cos^3 \vartheta \cos \vartheta, \tag{13.40a}$$

$$v_{\vartheta 0} = v_{za} (a/h)^3 \cos^3 \vartheta (\sin \vartheta)/2, \tag{13.40b}$$

we get, with due regard to the signs,

$$v_{z0} = v_{za} \left(\frac{a}{h} \right)^3 \cos^3 \vartheta \left(\frac{1}{4} + \frac{3}{4} \cos 2\vartheta \right). \tag{13.41}$$

Although the distributions in figures 13.9a and b are different in detail, both show a peak in the middle. Since the peaks are of equal magnitude, it is easy to see that their superposition would result in a flat circular box, with large velocities only at the middle of one side. This situation recalls that of a violin at its f-hole resonance. On the top plate, the contours of equal V are concentric circles around a center lying between the f-holes, with increasing spacing toward the sides. Only a few such circles, if any, will appear on the back plate. The present superposition can be regarded as a better statement of what we meant when we described the 290-Hz curve in figure 13.7 as a superposition of a monopole and a dipole.

The fact that the lateral boundary of a violin does not lie upon a spherical zone does not prevent us from inscribing the form within a circle. The contours of equal V then end at the ribs and so mean different rib movements in the z-direction. This may not be realistic, but nonetheless can be done, because nothing is contributed to the radiated field. We may also exclude the four corners in order to get the surface as a unique function of the directions ϑ and φ, and we may neglect the arching, which is of little importance in regard to the radiation.

Since the violin also has a dipole component in the y-direction on account of the moment of force that the feet of the bridge exert on the top plate (and which results in the island), we show finally in figure 13.9c the case of a dipole field along the y-axis. Without entering into the details, we again give the v_z-distribution in the y, z-plane:

$$v_{z0} = v_{ya} \left(\frac{a}{h}\right)^3 \cos^3 \vartheta \cdot \frac{3}{4} \sin 2\vartheta. \tag{13.42}$$

With this example we must abandon rotational symmetry. From (13.29b) we know that the azimuth angle φ appears, and the V-contours are no longer circles around the center. This lack of rotational symmetry makes the synthesis much more laborious.

Extension of the foregoing to bodies of dimensions not small in comparison to the wavelength presents great difficulty. In the case of a monopole field the decrease of v_0 as r_0^{-2} changes to one as r_0^{-1}, and we have to take into account a phase delay toward the boundaries. Such a delay even has a physical meaning in the case of the violin, since the power is introduced in the middle, but lost everywhere in the plates.

As long as kh may be regarded as less than 1, i.e., in the region of large V, these changes near the boundaries will not influence our results very much. At least we can answer our question at the beginning of this discussion, by saying that at frequencies in the region of the lowest body resonances V-distributions are possible that would result in an ideal uniform radiation.

13.6 Synthesis of the sound field of any vibrating rigid body by means of spherical fields

The discussion in the preceding section holds in principle for all spherical sound fields that can be regarded as excited by a multipole at the center. We can always calculate a "natural" V-distribution $Q_{mn}(\vartheta, \varphi, f)$, producing at the surface of the vibrating body a spherical sound pressure field characterized by the product of independent functions of the spherical coordinates r, ϑ, and φ:

$$\underline{p}(r, \vartheta, \varphi) = \underline{p}_{mn}\underline{R}_m(r)\Theta_{mn}(\vartheta)\Phi_n(\varphi).\tag{13.43}$$

Here Φ_n means either $\cos(n\varphi)$ or $\sin(n\varphi)$, which may be combined as $\cos(n(\varphi - \varphi_n))$. We also note that Θ is a real function; we are already acquainted with the cases $\Theta_0 = 1$, $\Theta_{10} = \cos\vartheta$, and $\Theta_{20} = 3/2(\cos^2\vartheta - 1/3)$. Equation (13.43) depends on n; but n never surpasses m. Θ_{mn} is always a polynomial in $\cos\vartheta$ and $\sin\vartheta$. The product $\Theta_{mn}\Phi_n$ appears alone if r is constant, i.e., on the surface of each concentric sphere. These functions are therefore called *spherical harmonics*.

In the far region we can define and measure a direction factor $\underline{\Gamma}$ that can be expressed in terms of the spherical harmonics:

$$\underline{\Gamma}(\vartheta, \varphi) = \Sigma\underline{p}_{mn}\Theta_{mn}\Phi_n/\underline{p}_{max}.\tag{13.44}[4]$$

The most complicated multiplier in (13.43) is the complex function $\underline{R}$. It has to change the spherical sound field from a quasistationary motion in the near field to a spherical wave motion in the far field; $\underline{R}$ has been tabulated for rather large m. So we can not only express the direction-factor spherical harmonics, but also the corresponding V_{mn}-distribution can be evaluated for each m, n, and the results can then be superposed, and the sum compared to the measured $V(\vartheta, \varphi)$.

The reciprocal problem can also be solved: A given $V(\vartheta, \varphi)$-distribution can be analyzed in terms of the products $V_{mn}Q_{mn}$, where the Q_{mn} are normalized distribution functions as in figure 13.9, although there are difficulties on account of the complex character of the field at nonspherical surfaces. Nonetheless, the corresponding amplitudes and phases allow us to calculate the direction factor (Cremer 1984). But in both tasks it is evident that the mathematical difficulties increase enormously with m.

---

[4] The reader will find in a paper by Weinreich and Arnold (1980) very careful measurements of the direction factor and numerical tables up to $m = 9$.

Therefore this synthesis or analysis will preferably be applied at distances large in comparison to the wavelength.

References

Cremer, L., 1971. *J. Sound. Vib.* **16** (1), 8.

Cremer, L., 1984. *Acustica* **55**.

Jansson, E., N. E. Molin, and H. Sundin, 1970. *Phys. Scripta* **2**, 243.

Meinel, H., 1937. *Akust. Z.* **2**, 22.

Meyer, J., 1975. *Instrumentenbau* **29**, 2.

Morse, P. M., 1948. *Vibration and Sound*, 2nd edition. New York: McGraw-Hill.

Morse, P. M., and K. U. Ingard, 1968. *Theoretical Acoustics*. New York: McGraw-Hill.

Skudzyrk, E., 1954. *Die Grundlagen der Akustik*. Vienna.

Weinreich, G., and E. B. Arnold, 1980. *J. Acoust. Soc. Amer.* **68**, 404.

14 Wavelength Comparable to the Source Dimensions

14.1 Two point sources at larger distances

Even if the elementary radiators treated in the previous chapter correctly described the behavior of string instruments at low frequencies, we would still have to note that the instruments' dimensions are comparable to the wavelengths at middle frequencies.

We can express some of the essential consequences of this fact simply and vividly by investigating the example of two point radiators. We have already derived the formula for this example in section 13.2:

$$\underline{p} = \frac{j\omega\varrho}{4\pi}(\underline{q}_1 e^{jkh\cos\vartheta} + \underline{q}_2 e^{-jkh\cos\vartheta})\frac{e^{-jkr_0}}{r_0}. \tag{14.1}$$

There, however, we applied this formula only to cases in which the Helmholtz number determining the type of field was small:

$$\mathrm{He} = \frac{2h}{\lambda} \ll 1. \tag{14.2a}$$

This condition led to a quasistationary, incompressible hydrodynamic near field in the region near the source (i.e., for $r \ll \lambda$).

But within this near field, the pattern of the field near the radiators ($r < 2h$) was different from that at greater distances; for instance, the near field of two in-phase point radiators could be expected to be spherically symmetric only where $r \gg 2h$. It would have made sense even in our earlier analysis to have distinguished a neighboring field ($r < 2h$) and a distant field ($r \gg 2h$) as well as a near field ($r \ll \lambda$) and a far field ($r \gg \lambda$). Four combinations of these terms can be defined, as shown in table 14.1. In the earlier synthesis, increasing r led us from the neighboring near field through the distant near field to the distant far field.

If, we now assume that the Helmholtz number of the system of sources is

Table 14.1

	Near Field $r \ll \lambda$	Far Field $r \gg \lambda$
Neighboring Field $r < 2h$	Neighboring Near Field	Neighboring Far Field
Distant Field $r \gg 2h$	Distant Near Field	Distant Far Field

$$\mathrm{He} = \frac{2h}{\lambda} \gtreqqless \frac{1}{2}, \tag{14.2b}$$

then the neighboring near fields separate into two near fields, one for each point radiator, and we enter the far field immediately behind the outer regions of these near fields. The far field's pattern is determined by interference of the waves radiated by both sources. It is impossible in this case to speak of a directional characteristic in the neighborhood of the radiators; the distributions of pressure around spherical shells of increasing size near the center are not similar to one another, and the pressure does not decrease as $1/r$. Only in the distant far field ($r \gg 2h$) can we apply the formula

$$\underline{p} r_0 e^{jkr_0} = \frac{j\omega\varrho}{4\pi} \underline{q}_0 g(\varphi, \vartheta), \tag{14.3}$$

where we can call $g(\varphi, \vartheta)$ the *normalized direction function*. This differs from the direction factor

$$\underline{\Gamma}(\varphi, \vartheta) = \frac{g(\varphi, \vartheta)}{g_{max}} \tag{14.4}$$

in that its greatest value does not have to be 1, but rather depends on the choice of the sound flux $\underline{q}_0$ chosen to characterize the source.

It should be noted that $g(\varphi, \vartheta)$ does not represent the full dependence of the sound pressure on frequency, since ω appears as a multiplicative factor. When different g's are compared, the frequency must be the same in both cases. In what follows, the Helmholtz number will, then, be altered only through changes in the distance $2h$.

As our first example, we will take two point radiators of equal strength radiating in phase. We choose

$$\underline{q}_0 = \tfrac{1}{2}\underline{q}_1 = \tfrac{1}{2}\underline{q}_2. \tag{14.5}$$

From (14.1), which is already adapted to the distant field and independent of the azimuth, we then obtain the normalized direction function

$$g(\vartheta) = \cos(kh\cos\vartheta). \tag{14.6}$$

The magnitude of this function for $2h/\lambda = 1/12$, $1/2$, and $3/4$ is shown in the polar diagrams in figure 14.1, left.

The maximum in this case lies in the plane $z = 0$ ($\vartheta = 90°$). The max-

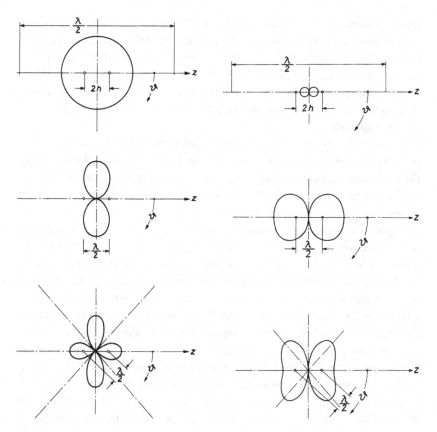

Figure 14.1
Normalized direction functions of two point radiators whose Helmholtz numbers
$2h/\lambda = 1/12$; $1/2$; $3/4$. Left, for in-phase point radiators with $q_1 = q_2$; right, for point
radiators 180° out of phase with $q_1 = -q_2$.

imum is equal to 1, and consequently $g(\vartheta)$ and $\Gamma(\vartheta)$ are identical. In the case of in-phase point radiators, the output is clearly greatest in the limit in which $2h/\lambda$ is small. If $2h/\lambda = 1/2$, the outputs cancel in the z-direction— an example characteristic of the flute and of open organ pipes. When $2h/\lambda = 3/4$, there is another relative maximum; with cancellation occurring in two oblique directions, this forms a conical surface in three-dimensional space.

The case described by (14.5) is exemplified by a single musical instrument at the height h above a more or less totally sound-reflecting stage floor. Though we can exclude any influence of the reflector on the flux at the instrument, we may have to consider the interference between direct and reflected sound waves at the position of a listener. As an example, if a double bass radiates sound from its f-holes at $h = 1$ m, the floor radiation is equivalent to a doubling of q (Lee 1982). This brings about an increase in the sound pressure level of 6 dB in the far field. If the double bass is placed near a hard wall, we may expect another increase by the same amount. It is desirable to place bass instruments to take advantage of this effect, since the bass frequencies are often weakly radiated. For reinforcement to occur, however, the reflecting surface must extend over several wavelengths, and should be rigid so the bass frequencies do not excite it into vibration (see section 14.5). It must also be free of openings through which sound flux can escape. On the other hand, the effect is not disturbed if the surfaces have irregularities (for example the steps of risers) small in comparison with the wavelength.

More important in its immediate application to string instruments is the situation

$$q_0 = \tfrac{1}{2}q_1 = -\tfrac{1}{2}q_2, \tag{14.7}$$

which includes the dipole as a limiting case. Examples of this situation representing the same Helmholtz numbers as before are shown in figure 14.1, right. When the top or back plates of a string instrument are divided into two radiators, they are antiphasic.[1] This case, however, we obtain

$$g(\vartheta) = \sin(kh\cos\vartheta) \tag{14.8}$$

for the normalized direction function. The maximum is no longer 1, but falls off as kh in the range of values of h in which the array may be

[1] In this book, "antiphasic" means $\Delta\varphi = 180°$; "out of phase" means only that $\Delta\varphi \neq 0$.

regarded as a dipole. In this case, $g(\vartheta)$ is not identical to $\Gamma(\vartheta)$, and for this reason $g(\vartheta)$ is better suited to the purposes of our discussion.

We recognize, therefore, that the increase in the Helmholtz number increases radiation. To be sure, the cancellation in the z-direction is retained, but the maximum (in the plane $z = 0$) increases, reaching 1 when $2h/\lambda = 1/2$. The value of the maximum remains 1 at larger Helmholtz numbers, though the direction of the maximum changes (see the example for $2h/\lambda = 3/4$).

In the case of the dipole, increasing the separation of the point radiators up to $2h/\lambda = 1/2$ or, more generally, between the regions oscillating out of phase, is therefore advantageous.

The distance between the antiphasic regions on the body of a string instrument is half the wavelength of the flexural wave. This wavelength decreases only as the inverse square of frequency [see (11.13)]. Since the wavelength of sound in air varies inversely with frequency, the Helmholtz number increases as the square root of frequency. A look at figure 14.1, right, confirms that this condition is favorable to radiation.

It can be stated very generally that dispersion of the flexural waves in the body of the violin lessens the disadvantageous effect of the breakdown of the top and back plates of the violin into antiphasic regions occurring at higher frequencies. More radiation of sound occurs than would seem likely on the basis of the comparisons of the point radiator, dipole, and quadrupole in the previous chapter. We will discuss this effect further in the next chapter, in connection with the limiting case of infinite plates.

We can see from (14.1) the degree to which the normalized direction function depends on the choice of the origin. Only when $|q_1| = |q_2|$ is it self-evident to place the origin halfway between the two point radiators, as we have done previously. If, as a counterexample, $|q_1| \gg |q_2|$, it seems more appropriate to define the position of q_1 as that of the source. But in this case (14.1) would read

$$\underline{p} = \frac{j\omega\varrho}{4\pi}(\underline{q}_1 + \underline{q}_2 e^{-j2kh\cos\vartheta})\frac{e^{-jkr_1}}{r_1}. \tag{14.9a}$$

In deriving this formula, we have once more assumed that the amplitude differences of the two point sources due to difference in their distance from the point of observation can be ignored in the distant field. If this is so, the differences between r_0 and r_1 can also be ignored. But then the phase difference given by $r_0 - r_1 = kh\cos\vartheta$ corresponds to a multiplication by $e^{-jkh\cos\vartheta}$. Given our new choice of an origin and our replacement of r_0 by r_1, we obtain

$$\underline{p} = \frac{j\omega\varrho}{4\pi}(\underline{q}_1 e^{jkh\cos\vartheta} + \underline{q}_2 e^{-jkh\cos\vartheta})\frac{e^{-jkr_1}}{r_1} \cdot e^{-jkh\cos\vartheta}. \qquad (14.9b)$$

This formula shows that only the phase of the direction function is different; the amplitude remains the same. It is the amplitude, however, which is of greatest interest to us.

This result is encouraging, since there is in most cases no way to decide which point should be chosen as the origin. A favorable origin depends in part on the shape of the radiators; but in the case of string instruments even this is not symmetrical in the x-direction. The favorable origin also depends on the flux distribution, and this varies from one frequency to another. Even if it were possible to define an origin, it would be difficult to apply the definition when taking measurements, since the position of the microphone is usually held constant while the instrument is rotated around an arbitrarily chosen axis.

14.2 Shadowing by the body of the instrument

Since we assume that point radiators present no obstruction to the propagation of sound, synthesis using point radiators, no matter how many, cannot account for obstructions presented by rigid bodies whose dimensions are comparable to or greater than the wavelength.

We may assume as a general rule that obstruction increases with the Helmholtz number L/λ. L may represent any dimension characteristic of the obstructing body.

A high frequencies, with plane wave incidence, a shadow zone always occurs behind the obstructing body. This zone becomes narrower with increasing distance, but a pressure transducer immediately behind the obstructing body will always reveal it. If the direction of incidence is changed by 180°, there will be a distinct change in the sound pressure.

The same is true for the reciprocal problem in which we assume that a point sound source is located close to the obstructing body or that a part of the surface of the body is vibrating, and pressure is measured in the distant far field. If we now move our pressure transducer in a full circle around the body or, more simply, turn the body while holding the microphone still, we obtain the same directional function as before. For

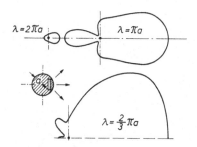

Figure 14.2
Intensity distribution of sound radiated by a sphere if a small part of the surface oscillates
with a given velocity, for Helmholtz numbers $2\pi a/\lambda = 1; 2; 3$ (after Morse).

this reason, it is appropriate in this case as well to speak of "shadowing."

The shadowing is exactly calculable if the obstruction is a sphere, even
if the point radiator is a small part of the sphere's surface.

As already mentioned in section 13.6, each distribution of normal
velocity $V(\vartheta, \varphi)$ at a spherical surface can be synthesized by spherical
harmonics; and, since our shadowing problem can be treated as one of
rotational symmetry, we need only the spherical functions $\Theta_m(\vartheta)$. Thus
we can set

$$v(\vartheta) = \sum_0^\infty V_m \Theta_m(\vartheta). \tag{14.10}$$

Although such a synthesis would also be possible in principle for other
rotationally symmetric bodies, as mentioned in section 13.6, the spherical functions have the great advantage that each V_m can be calculated
independently of the others by an integration over the whole sphere:

$$V_m = (m + 1/2) \int_0^\pi V(\vartheta) \Theta_m(\vartheta) \sin \vartheta \, d\vartheta \tag{14.11}$$

(Morse 1948, p. 319), just as we could calculate the Fourier components
when analysing the displacement of the plucked string and the force on
the pointwise excited string (sections 2.6, 4.2). From each V_m at the
spherical surface we can derive p_m and the direction factor.

Figure 14.2, after Morse (1948, p. 324) shows the directional characteristics for Helmholtz numbers 1, 2, and 3, where we use the same
Helmholtz numbers as for the breathing sphere:

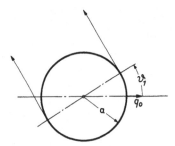

Figure 14.3
Construction to determine the position of the secondary peak in the shadowed region
(see the example in figure 14.2, in which He = 3).

$$He = ka = \frac{2\pi}{\lambda} = 1;2;3. \tag{14.12}$$

Unlike figure 14.1, the directional characteristic is here shown in terms
of intensity,

$$J = \frac{\tilde{p}^2}{\varrho c} = \frac{1}{\varrho c}\left[\frac{\omega \varrho}{4\pi r_0}\tilde{q}_0 g(\varphi, \vartheta)\right]^2 \tag{14.13}$$

$$= \varrho c \left(\frac{\tilde{q}_0}{4\pi r_0 a}\right)^2 [(ka)g(\varphi,\vartheta)]^2. \tag{14.13}$$

(The difference is not only that the plots in figure 14.2 represent squares
of the normalized direction function, but also that the plotted function
is multiplied by the square of the Helmholtz number ka.)

The difference between the front and the back is already clear in the
case in which He = 1; also note that the minimum intensity is, even in
this case, not found precisely at the direction opposite the source.

This effect is much more pronounced when He = 2; a secondary peak
appears at the back. This is because the waves which leave the sphere
around the edges of its equatorial plane add in phase in this direction.
According to Fermat's principle, these waves take the shortest possible
path.

In the case He = 3, another "ghost" peak appears in an oblique
direction. Since the direction function is rotationally symmetric, this
appears as a circluar cup-shaped pattern in three dimensions. This
pattern, too, can be at least made plausible through an examination of
the interference. It can be seen from figure 14.3 that waves leaving both

sides of the sphere at an angle ϑ_1 to the vertical have a path-length difference of $a[(\pi - \vartheta_1) - \vartheta_1]$. If this equals one wavelength, we must expect a maximum. In the present example, the wavelength is $(2/3)\pi a$. Consequently,

$$\vartheta_1 = \pi/6 = 30°, \tag{14.14}$$

which agrees surprisingly well with the results of Morse's synthesis of spherical sound fields (see figure 14.3). Following upon this observation, we will attempt to explain the directional characteristics of the violin in a similar way. The shape of the body of the violin is too complicated to allow exact calculation of the direction factor for pointwise excitation and there is no other choice than to determine the directional properties of the shadowing by measurement.

The first experiments of this type were undertaken by Ising and Nagai (1971). They made use of the principle of reciprocity, by placing microphones at selected positions in front of an instrument and using a plane-wave sound source radiating from different directions in the y, z-plane. The frequency was 100 Hz.

In order to be sure that oscillations of the body of the instrument would not affect their measurements, they filled it with sand. The instrument was, to be sure, not a valuable one.

Lehringer (1972) and Cremer and Lehringer (1973) were later able to show that this precaution was unnecessary. Based on this assumption, Lehringer built small loudspeakers into the body of another cheap violin (see figure 14.4). The loudspeakers were each individually enclosed so their backs did not radiate into the cavity of the instrument. Figure 14.5 shows Lehringer's experimental apparatus, from which we can see that he measured direction functions in the x, z-plane; this plane is of greater practical importance than the y, z-plane in which Ising and Nagai took their measurements. He recorded the amplitudes of the sound pressure as well as its phases. In figure 14.6, we give one example of his diagrams for the top plate of the violin and one for the back plate. The amplitudes are shown radially on a linear scale, and the phase angles are indicated by numbers around the plot of the directional characteristic. In order to make the phase representation easier to understand, the amplitude curve for which phases are between $-90°$ and $+90°$ is drawn as a solid line. The remainder of the curve is drawn as a dashed line; phases here are between $-90°$ and $-270°$. The zero reference for phase is established

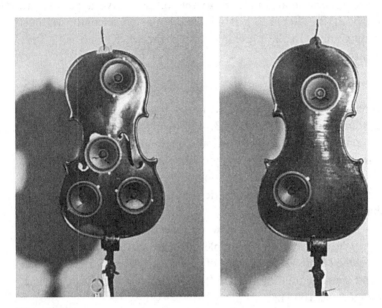

Figure 14.4
Violin with small loudspeakers installed in the body (after Lehringer).

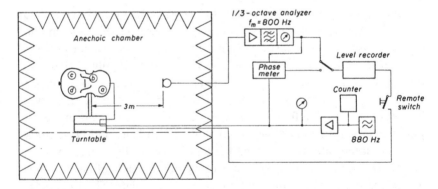

Figure 14.5
Experimental arrangement used in determining direction functions (after Lehringer).

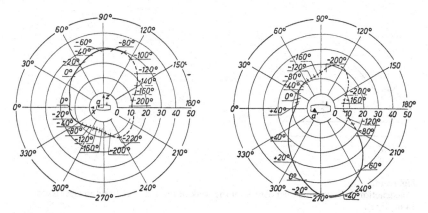

Figure 14.6
Measured direction functions in the x, z-plane with off-center point radiators. Left, in the top plate; right, in the back plate (after Lehringer).

by the position of the source a on the top of the instrument, and by the direction of the neck toward the scroll.

The shadowing and the oblique ghost peaks are easily observed in both diagrams. Once again, we can understand their direction if we apply Fermat's principle and the corresponding coincidence of phases. (See figure 14.7, which corresponds to figure 14.6, left.) If the distances l_1 and l_2 of the point radiator from the two ends of the body are unequal, addition of the waves leaving the instrument can be expected only if the difference $(l_2 - l_1)$ is canceled by a corresponding delay in the phase of the wave proceeding from the edge nearer the source.

This simple consideration already explains why the peak appears in an oblique direction. But even a quantitative evaluation of the sketch in figure 14.7, which represents the situation under measurement, leads to a declination of

$$\vartheta_0 = \arcsin \frac{l_2 - l_1}{l_2 + l_1} = 42° \tag{14.15}$$

with respect to the z-direction. This is in surprisingly good agreement with the measured position of the ghost peak in the shadowed region.

Lehringer's measurements were made at a frequency of 880 Hz, corresponding to a wavelength of approximately 40 cm. Consequently, it is possible to assign a Helmholtz number to the direction function:

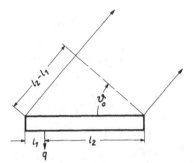

Figure 14.7
Construction to determine the position of the peak in the shadowed region (see figure
14.6, right).

for example, using the circumference of the body in the x, z-plane, 80 cm.
Given this, the Helmholtz number would be 2.

   The shadowing and the ghost peaks are not as pronounced here as
in the example in figure 14.2 with the same Helmholtz number. This
discrepancy is understandable, as the circumference of the body in the
y, z-plane is much smaller. Paths in the y, z-plane certainly also play a
part in the direction function seen in the x, z-plane.

14.3 Synthesis using directional Green's functions

The goal of determining the direction function for various source positions on the violin body could have been attained more simply by placing
microphones near the instrument and invoking reciprocity. The loudspeakers built into the violin in figure 14.4 served the additional purpose
of modeling a directional characteristic, by replacing the positive and
negative regions of a particular distribution of oscillations on the body
of the instrument with point radiators.

   In accord with this goal, it was appropriate to choose an example in
which antiphasic regions occur in the top and back plates of the instrument, but in moderate numbers. For this reason, Lehringer chose an
example from Reinecke's holographic photographs in which four such
regions occur in the top plate of the instrument and two in the back plate.

   Since, however, holographic photography does not show phase angles,
Lehringer went back over the pattern of oscillation point-by-point. Unlike

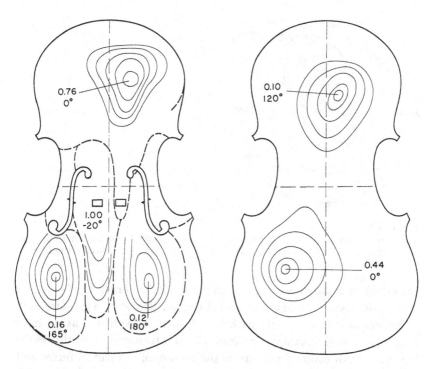

Figure 14.8
Magnetically sensed patterns of oscillation of the top and back plates of a violin at 880 Hz (after Lehringer).

Eggers, he did not use a capacitive probe; rather, he used a magnetic transducer, gluing a small plate of mu-metal to the body of the instrument at 400 separate locations. In this way, he obtained the contour lines in figure 14.8 and the equivalent fluxes and phase angles at the points of peak displacement. Unfortunately, the excitation device covered the region of the front plate above the left foot of the bridge; the relative flux of 1 here must be regarded as arbitrary to some extent.

Lehringer—as earlier, Eggers—observed that the phase angle did not jump 180° at the nodal lines, but rather changed steadily. This was to be expected in a damped system. What was more surprising was that the phase angle at the peaks was not always 0° and 180°, but sometimes took on the distinctly different values of −20°, +120°, and +165°. Lehringer considered the last of these an error of measurement, and replaced it with 180°. He drove the loudspeakers using potentiometers and delay units,

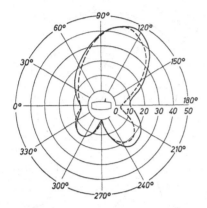

Figure 14.9
The direction function measured when all loudpeakers in figure 14.4 are driven (solid line), and that obtained through synthesis as in (14.13) (dashed line) (after Lehringer).

providing input currents in the correct phase relationships and of sufficient amplitude; the measured directional characteristic is shown in figure 14.9.

He was also able to drive each loudspeaker individually, as we have already seen in the examples in figure 14.6. He then multiplied the resulting complex direction functions by the vectors of the partial fluxes and built up the sum out of these products. The result of this synthesis, which may be represented formally by

$$r_0 \underline{p} e^{jkr_0} = \sum_i \sum_k \Delta \underline{q}_{ik} \underline{G}_{ik}(\varphi, \vartheta) \qquad (14.16)^2$$

is shown as a dashed line in figure 14.9. The slight discrepancy is probably due less to errors in calculation than to inevitable slight differences in the input to the loudspeakers from one measurement to the next.

In theoretical physics, every field generated by a point source is called a Green's function; the functions $\underline{G}_{ik}(\vartheta, \varphi)$, which describe the corresponding direction functions in the distant far field, may be called directional Green's functions.

These differ from the normalized direction function introduced in (14.3) by the factor $j\omega\varrho/4\pi$:

---

[2] Δq here signifies not a difference of fluxes but a specific flux component. The symbol Δ before the q implies only that the method becomes more accurate as the surface elements, and thus their fluxes, become smaller.

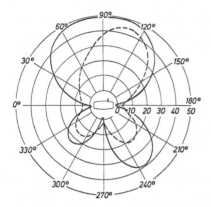

Figure 14.10
The direction function obtained by synthesis (dashed line), and that measured directly
for the violin of figure 14.8, excited at the bridge (solid line) (after Lehringer).

$$\underline{G}(\varphi, \vartheta) = \frac{j\omega\varrho}{4\pi}\underline{g}(\varphi, \vartheta). \tag{14.17}$$

These functions, therefore, account for the frequency dependence as well.
The square of their magnitude is proportional to the intensity as plotted
in figure 14.2. But $\underline{G}$ is a complex quantity; it also includes the phase
angle, which is important for the purpose of synthesis. The double sum
in (14.16) indicates that this formula has a two-dimensional character; it
has to do with the subdivision of an area.

Individual point radiators can no more than approximate a continuous
distribution of velocity, as figure 14.10 shows. Here, the result of the
synthesis (dashed line) is compared with the directional characteristic of
the violin as excited directly at the input of the bridge (solid line). The
quantitative agreement is not very satisfactory, though the positions of
the maxima and minima are reproduced more or less correctly in the
synthesis.

Consequently, synthesis using directional Green's functions has no
advantage over direct measurement in determining any particular directional characteristic. Still, the multiplicative separation of the influence of
the vibration pattern from that of shadowing is of theoretical significance.
In this light, it should be noted that the shapes of individual violins differ
so little that a mean directional Green's function could be derived that

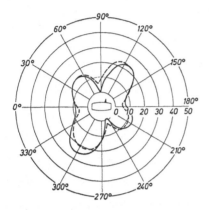

Figure 14.11
As figure 14.9, but without the elementary source b (see figure 14.5).

would apply with little error to any violin. In contrast, the vibration patterns do differ greatly from one instrument to another.

For instance, figure 14.11, like figure 14.9, shows excellent agreement between the measured directional characteristic (solid line), and the synthesis (dashed line) when loudspeaker *b* (see figure 14.5) is switched off. The directional characteristic is then completely different from that in figure 14.9.

The same phenomena explain why the directional characteristics of two instruments shown in figure 13.8 are so very different for slight changes in frequency. The curves are not circular; their dips and the ghost peak indicate that we have already entered a frequency range in which shadowing plays a role.

Figure 14.12 gives a further example of changes in the vibration pattern with slight changes in frequency, and also of the participation of shadowing in establishing the directional characteristic. This example is based on Meinel's earlier measurements, used also in figure 13.7, and taken at frequencies of 922, 950, and 977 Hz; frequencies at which not only the top plate but also the back plate exhibits regions oscillating in opposite phase. The subdivision of the top and back plates is evident in the dipole-like directional characteristic at 950 Hz. This shows greater radiation than at the other two frequencies, as if by resonant reinforcement. The directional characteristics corresponding to the neighboring frequencies, 922 Hz and 977 Hz, deviate from that for 950 Hz and even more from one another.

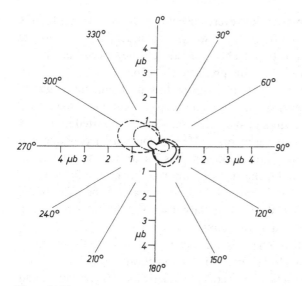

Figure 14.12
Measured directional characteristics of a violin in the y, z-plane (linear scale) at 922 Hz (solid line); 950 Hz (dashed line); and 977 Hz (dot-dash line) (after Meinel).

These characteristics exhibit the typical main and ghost peaks first observed in figure 14.6, and suggest that flux from the top predominates in one case but from the back in the other.

It is, then, possible to explain directional characteristics *a posteriori*; though it is, to be sure, more difficult to predict them *a priori*, even if the vibration pattern is known.

14.4 Shadowing by the player

Through experimentation, we have proven to ourselves that the body of the instrument has a shadowing effect. We next ask ourselves to what extent the parts of the player's body in the neighboring field of the instrument also determine the directional characteristic. The head and the shoulder of a violinist (between which the player clamps the instrument) could have such an effect, as could the entire body of a cellist.

In measurements using artificial excitation mentioned up to this point, it is possible that the experimental apparatus has some shadowing effect; this would be particularly likely with mechanical bowing devices. This possibility is of little interest, however, since it does not correspond to the normal playing situation. If, however, we wish to understand the maxima and minima of the directional characteristic as they occur in an auditorium, we must account for the presence of the player.

Eikholt (1964) was the first to consider this problem, investigating string and wind instruments. He placed a player with an instrument on a chair which stood on a square panel 1.2 m on a side, supported in turn by the wire grid floor of an anechoic room. He suspended a microphone so it could be moved in a horizontal circle of radius 2 m around a center 1.1 m above the floor. Certainly, the microphone was in the far field of the violin; not as certainly, in the distant field, particularly if the effect of the floor panel is considered. We have already discussed this issue in section 14.1 and we shall discuss it again below.

For the violin as well as other instruments, Eikholt defined the reference angle of the directional characteristic as the direction in which the player looks. This probably can be interpreted as a horizontal line perpendicular to the center of a line joining the eyes. Since the position of the instrument may vary, even with respect to a single player, it would perhaps have been more appropriate to refer angles to the instrument.

Since Eikholt needed the presence of a player anyhow, he took it upon himself to play the instruments in some cases. Frequency selection was by means of narrow-band filters in the signal-processing apparatus. The effects of variations in input force at the bridge were compensated for by the use of a reference microphone at a fixed location.

Nonetheless, different players exhibited different directional characteristics; small discrepancies occurred in the nearly circular directional characteristics near 440 Hz (figure 14.13, left); greater ones, at 880 Hz (figure 14.13, right). The influences of nodal lines and shadowing are unmistakable in the latter example. The differing directional characteristics might depend on differences between the players' bodies; more important factors might be differing posture, or ways of playing that excited the instrument differently. The last of these hypotheses seems confirmed by the differences observed when the same player produced the frequency 880 Hz as the second partial of the open a string (figure 14.13, right) or the third partial of the open d string (figure 14.14). Nonetheless, these diagrams show that when complex tones are excited, the result, even with filtering, varies depending on how close other partial tones lie. This effect certainly plays a role for a human listener. In physical investigation,

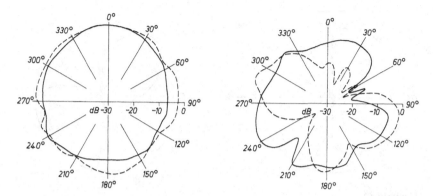

Figure 14.13
Measured directional characteristics (logarithmic scale) in the horizontal plane with two different players i and ii (solid and dashed lines). Left, at 440 Hz (open a string, first partial); right, at 880 Hz (same note, second partial) (after Eikholt).

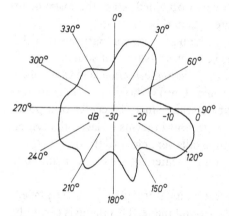

Figure 14.14
Measured directional characteristic in the horizontal plane (logarithmic scale) at 879 Hz, third partial of the open d string (after Eikholt).

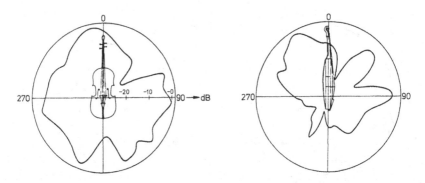

Figure 14.15
Directional characteristics (logarithmic scale) of the violin measured in the horizontal plane. Left, violin held in the x, y-plane; right, in the x, z-plane (after Eikholt).

artificial excitation with pure tones is advantageous because of its unambiguousness. (In general, the differences obtained when the instrument was played normally were not as great as in this example.)

It should be noted in connection with all of these comparisons that the diagrams have a logarithmic scale; i.e., not pressure, but sound pressure level L_p is indicated by radial distance. The highest value is always called 0 dB and is placed 6 cm from the origin. Level is, however, an inherently awkward variable to use in a polar diagram, since the center of a polar diagram should represent zero pressure. But in terms of level, the center represents $-\infty$, and a small amplitude falls very close to the origin. Also, the relative change in a polar diagram is greater the closer the maximum level is to the origin.

The diagrams up to this point represent a natural posture of the player, and microphone positions in the horizontal plane. This plane is certainly of special importance in the concert hall. It does not, however, coincide with any of the principal planes of the violin. On the other hand, figure 14.15, left, represents the case in which the player made a special effort to hold the violin so its x, y-plane was horizontal. Figure 14.15, right, represents the case in which the violin was held so that its y-axis was vertical; the directional characteristic is that of the x, z-plane.

The last of these diagrams, especially, shows a pronounced shadowing with ghost peaks attributable to the upper body of the player. Unfortunately, Eikholt did not undertake comparisons between measurements using artificial excitation with and without the player.

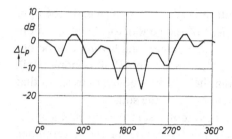

Figure 14.16
Level difference between the directional characteristics of an electroacoustic source with
and without a shadowing player behind it, at 2,000 Hz (after Bergmann and Fischer).

Bergmann and Fischer (1975), however, studied shadowing by placing
the player behind an electroacoustical source at the center of the instrument's position. They measured directional characteristics with and without the player. The player was, in this case, not actually playing an instrument; the curves, once again, have logarithmic scales. Figure 14.16 shows
the results, this time in Cartesian coordinates, at 2 kHz. Shadowing is
naturally more pronounced at this frequency than at 800 Hz.

Bergmann and Fischer kept the microphone in a fixed position and
placed the player and the source on a turntable. They, too, used a reference
microphone, placed above the player on the axis of rotation. They also
attempted to account for the reflection from the stage floor of a concert
hall by placing a 2.85-m × 1.85-m panel under the rotating chair. Since
their microphone was only 1 m away from the source, this reflection clearly
affected the results. However, the microphone was still in the neighboring
field. The mirror-image loudspeaker under the stage floor has to be
regarded as part of the source; the results must have been affected by
interference between the loudspeaker and its image. The distant field, on
the other hand, is of interest in concert halls; and in the distant field, the
stage-floor reflection either is small on account of shadowing by other
players or is in phase with the source. (See our discussion relative to
figure 14.1, left.)

It should be clear from the foregoing that it is very difficult to set
up any generally valid conditions for the measurement of directional
characteristics.

Unfortunately, there are no measurements of directional characteristics for celli with their players. Any concertgoer sitting behind the celli can,

however, observe that they are shadowed by the players; e.g., in the right balcony when the celli are to the right of the conductor. With the more usual placement of the celli facing the audience, shadowing can be observed at positions behind the orchestra. In Berlin's Philharmonic Hall, this problem is mitigated by placing the celli in a quarter circle, so that some face the side and others face the front of the stage.

Bergmann and Fischer also attempted to measure the directional characteristic in two dimensions by varying the height of the microphone between 0 and 2 m over the floor panel. As Eikholt and recently Weinreich and Amold (1980) have shown, however, there are simple means of determining the sound pressure on the surface of a sphere surrounding the source. However, this requires a much greater number of measurements and also of data items for each measurement. If the two-dimensional representation is retained, a second angle must be used as a parameter and a separate curve must be drawn for each significantly different value of this angle. It would, however, be very hard to read such a collection of curves. The data can be better presented by inserting pins into a sphere so that their visible lengths represent values of pressure amplitude (or intensity) at their respective angular positions. This technique was introduced by Thiele (1953) and Meyer and Thiele (1956) in examining reflections in the study of architectural acoustics. The result is a "directional pincushion," called a "hedgehog" (*Igel*) by the authors.

It is doubtful whether such an effort is justified for an instrument whose directionality varies from frequency to frequency and from one individual instrument to another. In fact, this variability and incalculability is characteristic of string instruments and distinguishes them from most wind instruments.

14.5   Efforts to enlarge the radiating area

The radiated sound is reinforced when the bridge connects the thin string to a body whose dimensions are more comparable to the wavelength. Taking this concept further, an increase in the area of the radiating surfaces beyond that of the usual top and back plate might be assumed advantageous.

However, the success of such attempts seems questionable: we already know about the breakdown of the radiating surface into antiphasic regions separated by nodal lines. This holds down its radiation.

Figure 14.17
Violin with enlarged top and back plates (instrument collection, Institute for Musical
Research, Berlin (West)).

Figure 14.17 shows a violin whose inventor, the Rumanian prince
Grigore Stourdza, sought to enlarge the top and back plates wherever
possible. He had to retain the narrowed waist in order to allow bowing on
the outer strings; the upper right quadrant could not be enlarged, because
the left hand has to reach around it in the upper positions on the fingerboard.

The top plate of this enlarged instrument must break down into more
regions; increased radiation could be expected only if the entire back and
top plates (except the island) oscillate so as to expand and contract as one
unit over the same frequency range as with the usual violin. But this
assumption is questionable. It is much more likely that breakdown begins
at a lower frequency; perhaps even below the first wood resonance. Very
generally, it can be assumed that more antiphasic regions occur at all
frequencies and that radiation is consequently worse rather than better.

The enlarged violin rates only as a curiosity; but there is a widespread
speculation among musicians—also regarded as credible by some acousticians—that celli and basses radiate through the peg. Supposedly, the
stage floor or risers are excited into oscillation, increasing the effective
radiating area and the radiated power.

W. Trendelenburg (1925) is certainly correct in noting that celli were
originally held only between the knees of their players: the peg was intro-

duced for nonacoustical reasons, such as lessened fatigue when playing for a long time and greater freedom of movement with respect to the instrument.

The idea that the peg increases radiation is based on a well-known phenomenon: a tuning fork held in the hand is barely audible, but sounds very loud when its foot is rested on a tabletop. The situation with the cello and contrabass is certainly not as favorable. The peg is attached to the ribs and the endblock, parts which receive only a small part of the energy of oscillation. Much energy has already been extracted by internal friction and by radiation from the top and back plates.

Nevertheless, sound transfer through the peg has to be considered as an effect that the player himself may hear and feel. Furthermore, we must consider the possibility of noticeable sound propagation in the walls and floors of concert halls as far as the farthest listeners. For example, it was said that celli and basses sounded "powerful" in the original Leipzig Gewandhaus (unfortunately destroyed in the Second World War) because they directly excited the parqueted floor and the wainscoting of the walls (Bagenal 1929).

One advantage of our time is that such hypotheses can be confirmed or rejected through measurement. E. Meyer and his coworkers took measurements in the Gewandhaus before its destruction (Meyer and Cremer 1933). They were able to record a structure-borne precursor sound wave in the house floor when the stage floor was struck with a hammer; this resulted in hardly any airborne radiation, however, in comparison with the following direct airborne sound. In fact, the airborne sound excited the floor much more strongly than did the structure-borne precursor. When the excitation was a plucked contrabass, the structure-borne precursor could not be detected at the same place.

These experiments led the present author to investigate the question of what fraction of the radiated sound energy is transmitted through the peg. He hypothesized that the increase in radiated sound is so small as to make its accurate measurement difficult in the room with the source; he consequently made use of the ceiling-test arrangement available at that time at the Heinrich Hertz Institute in Charlottenburg (figure 14.18). A test object (in the present case a wooden panel) of 1 m^2 separates two rooms. The upper contains the sound source (in the present case the cello). Both rooms contain microphones. A calibrated attenuator allows presentation of the recorded sound pressures with equal loudness to both ears of the

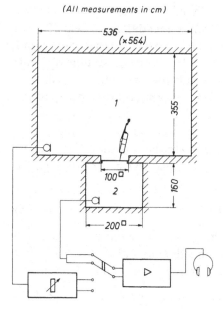

(All measurements in cm)

Figure 14.18
Use of a ceiling test arrangement to investigate the radiation of sound through the peg (after Cremer).

observer. For narrow-band signals the attenuator shows the sound pressure level difference between both rooms in dB (E. Meyer 1931). The evaluation of this difference with respect to the transmission loss of the object under test is based on the rules of statistical room acoustics, as we shall discuss in section 16.2. Today the smaller room shown in figure 14.18 would no longer be regarded as large enough for such treatment. For the low frequencies of interest in the peg problem that room must be regarded as small with respect to the wavelength.

But the use of a cello as sound source without adequate filtering behind the microphones, not available at the time, presents two further problems: First, the differing sensitivities of the loudness-comparing ears at different frequencies and levels; second, the comparison is based on some averaging of all frequencies produced by the cello.

Today it would not be difficult to repeat these first gross evaluations of the power radiated through the peg in rooms of more realistic size, and to obtain the frequency dependence by means of 1/3-octave filters.

Then the following formulas, already used for those early observations, would be correct: An attenuator showing a level difference D in dB gives the ratio of the mean energy densities of room 2 (below) to that of room 1 (above). According to the equations we shall derive in section 16.2, this ratio is equal to the ratio of the powers radiated into both rooms divided by the ratio of the so-called equivalent absorption areas A_2/A_1 of the rooms.

Before the peg is placed on the wooden panel, the power radiated in room 1 is only that radiated immediately by the body of the cello in air. We call this P_{air}. The power entering room 2 is that transmitted through the panel: P_{trans}. So we get

$$10^{-D/10dB} = E_2/E_1 = (P_{trans}/P_{air})(A_1/A_2).\qquad(14.18)$$

After the peg has been placed on the wooden panel in both rooms the power is increased by the equal amounts that are radiated over peg and panel. So we get, if we designate the corresponding D and E by primes:

$$10^{-D'/10dB} = E_2'/E_1' = ((P_{trans} + P_{peg})/(P_{air} + P_{peg}))(A_1/A_2).\qquad(14.19)$$

Now, as a matter of fact, the arrangement in figure 14.18 already demonstrated that the effect of the peg was scarcely to be heard in room 1, but was very noticeable in room 2. (This indicates that persons living downstairs in an appartment house would notice the additional radiation through the peg much more than listeners in the room with the instrument.) But if P_{peg} can be regarded as small compared to P_{air}, we can eliminate P_{trans} by substracting (14.18) from (14.19) and so get the formula for the ratio P_{peg}/P_{air} that we are interested in:

$$10^{-D'/10dB} - 10^{-D/10dB} = (P_{peg}/P_{air})(A_1/A_2).\qquad(14.20)$$

The ratio (A_1/A_2) is evaluated from measurements of the ratio of the energy densities, when a cello tone of the same power is radiated by loudspeakers into both rooms.

Using these relations it was found that P_{peg}/P_{air} was only 5% for the bass and 1% for the cello. But this rough estimate is for all partials taken together. It says nothing about much larger increases in particular frequency regions, which, indeed, might be very noticeable in the room in which the instrument is played.

Fuchs (1964) undertook the investigation of real risers. It was not possible for him to use a two-room test stand, so he had to make com-

Figure 14.19
Apparatus for determining the sound radiated from a cello via the peg resting on a riser, in an anechoic chamber (after Fuchs).

parisons in the room containing the instrument. He preferred an anechoic chamber to a reverberant room. Figure 14.19 shows a riser lent by the Berlin Philharmonic Orchestra, consisting of panels of 1-cm thickness nailed to a 130 × 80 × 40-cm frame. This riser was placed over a larger wooden panel, and the combination was suspended in the anechoic chamber. The cello was mounted on a trestle and bowed mechanically; open strings were played, and also a major scale, fingered by hand. It was established that the static force of the peg on the floor does not affect the sound transfer. Figure 14.20 shows the increase in sound pressure level as a function of the (filtered) frequency at 4 measuring points. Three of these were 1.4 m above the stage floor; the remaining one, 1.2 m. Though these points were chosen somewhat arbitrarily and are all in the neighboring field, the discrepancies between the recorded levels were small.

If the results at all frequencies are averaged, the principal result is in agreement with that of the previously described experiment: namely, the

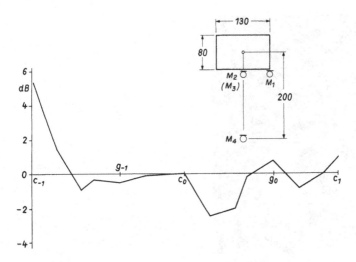

Figure 14.20
The increase of sound pressure level vs. frequency, when the peg is placed in contact with the riser as in figure 14.19 (after Fuchs).

average change in the radiation is very slight. Only in the range of the lowest fundamental frequency (64 Hz.) is there a significant increase (approximately 6 dB). If we assume that the body of the instrument and the riser both set air into motion in phase at this low frequency, this must mean that the total flux is doubled by a resonance of the riser panel with the cushion of air underneath it. This is a physically surprising, possibly desirable result; the body of the instrument exhibits no resonance in this frequency range, and not much radiation, since its dimensions are small compared with the wavelength.

It should also be noted, however, that radiation is decreased in other ranges, possibly due to interference resulting from the average distance between the body of the instrument and the floor. Another possibility is that the riser no longer reflects sound, but rather absorbs it through vibration—an effect which can happen with any type of paneling (see section 16.3). Due to this effect, it is clearly better for the stage floor under instruments without a peg to be as rigid as possible.

Beranek et al. (1964) also found ranges in which radiation was decreased. They had musicians play their instruments with the peg resting on concrete, and then on risers with 12.5-mm thick plywood floor panels 10 or 20 cm over the concrete. Figure 14.21 shows the increase in level as a

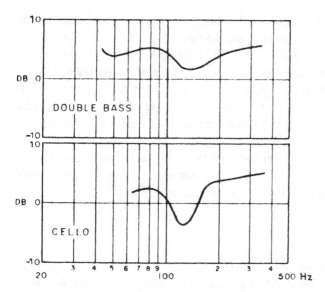

Figure 14.21
Increase in level measured in a concert hall when the peg is placed on a hollow wooden riser after being placed on a concrete floor. Top, contrabass; bottom, cello (after Beranek et al.).

function of frequency. In this case, too, a decrease in radiation was registered for the cello in approximately the same range.

On the other hand, Beranek et al. recorded a greater increase in other frequency regions for the contrabass. Generally, they found a much greater influence of the peg than the previous researchers, particularly in the range above 200 Hz. Here, more power seems to be radiated from the riser than from the instrument. This result is physically improbable, and also poses the question whether the quality of the sound might not be reduced if the component radiated by the valuable, carefully built instrument is smaller than that radiated by a floor representing ordinary carpentry work. Beranek et al. emphasize that the musicians were capable of keeping sound pressure level constant within 1.5 dB; but on the other hand, the musicians maintained that it was easier to play on the risers and that the tone quality was better. We cannot exclude the possibility that the musicians might have been induced to play louder on the risers. Certainly, the musical result can benefit from such a reaction; but it becomes very difficult to determine the physical facts of the matter.

We must also consider that the musician might be influenced by the oscillation of the riser, felt through the feet. This effect can differ according to the construction of the riser and to the position of the peg and the player's feet. It is understandable, then, that great artists take not only their own precious instruments but also their own risers with them on concert tours. Whether there is an advantage in making holes in such risers in order to produce an additional Helmholtz resonance has not yet been proven.

But if the artist believes that he plays better because of such refinements, the acoustician should not discourage him unless it can be proven that there is an actual detrimental effect on the radiation.

References

Bagenal, H., 1929. *J. Royal Inst. Brit. Arch.* III **36**, 756.

Beranek, L. L., F. R. Johnson, T. J. Schultz, and G. B. Watters, 1964. *J. Acoust. Soc. Amer.* **36**, 1247.

Bergmann, M., H. M. Fischer, 1975. Student work, Institute for Technical Acoustics, Technical University of Berlin.

Cremer, L., and F. Lehringer, 1973. *Acustica* **29**, 137.

Eikholt, J., 1964. Thesis work, Institute for Technical Acoustics, Technical University of Berlin.

Fuchs, H., 1964. Student work, Institute for Technical Acoustics, Technical University of Berlin.

Ising, H., and Y. Nagai, 1971. Unpublished measurements, Institute for Technical Acoustics, Technical University of Berlin.

Lee, J. B., 1982. *J. Acoust. Soc. Amer.* **71**, 1610.

Lehringer, F., 1972. *Berichte der DAGA-Tagung* [*Reports of the DAGA Convention*]. Stuttgart.

Meyer, E., 1931. *Sitzungsberichte Preuss. Akad. Wiss.* Math. Phys. Kl. p. 166.

Meyer, E., and L. Cremer, 1933. *Z. techn. Phys.* **14**, 500.

Meyer, E., and R. Thiele, 1956. *Acustica* **6**, 425.

Morse, P. M., 1948. *Vibration and Sound,* 2nd edition. New York: McGraw-Hill.

Thiele, R., 1953. *Acustica* **3**, 291.

Trendelenburg, W., 1925. *Die Natürlichen Grundlagen der Kunst des Streichinstrumentenspiels.* [*The Natural Basis of the Art of String-Instrument Playing*]. Berlin.

Weinreich, G., and E. B. Arnold, 1980. *J. Acoust. Soc. Amer.* **68**, 404.

15 Wavelength Small in Comparison to the Source Dimensions

15.1 The critical frequency

The effect of neighboring bodies on directionality is a difficult problem at wavelengths comparable to the dimensions of the instrument. When wavelengths are shorter than the dimensions of the source, the problem becomes much simpler. For example, when a trumpeter holds his instrument above the head of the player in front of him, the "brilliant" high frequencies are not affected by objects behind or even somewhat to the side of the trumpet's bell.

Similar statements may be made about string instruments. At very high frequencies, sound radiated from the top plate of the instrument is independent of that radiated from the back plate; and the high-frequency top-plate radiation is unaffected shadowing by the back plate and by the player's body.

We need only note that the number of antiphasic regions becomes greater in both the transverse and longitudinal directions as frequency increases; however, we pointed out in section 14.1 that the radiation is not decreased much, since the dimensions of the regions are determined by the length of flexural waves. As frequency is increased, this wavelength becomes longer in comparison to the wavelength in air.

This improvement in radiation is most pronounced for an infinite plate set into vibration by a sinusoidal wave propagating in the x-direction. The velocity vector for such a wave is

$$\underline{v} = \underline{v}_0 e^{-jk_B x}. \tag{15.1}$$

This wave generates a sound pressure field with the same dependence on x. We will examine only the half-space $z > 0$ above the plate. We may then write:

$$\underline{p} = \underline{p}_0(z) e^{-jk_B x}. \tag{15.2}$$

This sound pressure field must fulfill the conditions for the wave equation (11.59) that we derived in two-dimensional form as we examined phenomena in the cavity of the instrument:

$$\frac{\partial^2 \underline{p}}{\partial x^2} + \frac{\partial^2 \underline{p}}{\partial z^2} + k^2 \underline{p} = 0. \tag{15.3}$$

After substituting (15.2) into (15.3), we obtain the ordinary differential equation

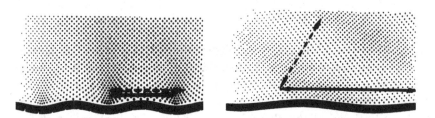

Figure 15.1
Motion of particles near a plate excited so as to produce a flexural wave propagating in
the x-direction. Left, when the flexural wavelength is shorter than the wavelength in air;
right, when it is longer.

$$\frac{d^2 \underline{p}_0(z)}{dz^2} + (k^2 - k_B^2)\underline{p}_0(z) = 0. \tag{15.4}$$

Two entirely different types of fields are obtained as solutions to this
equation, depending on whether

$$k < k_B \tag{15.5a}$$

or

$$k > k_B. \tag{15.5b}$$

Since λ is proportional to f^{-1}, while λ_B is proportional to $f^{-1/2}$, the limiting case

$$k = k_B \tag{15.6a}$$

or

$$\lambda = \lambda_B \tag{15.6b}$$

corresponds to a definite frequency. We have already encountered this
critical frequency in section 11.5.
    As we approach the critical frequency from below ($k < k_B$), the solution
takes the form of a surface wave; this has the same form as a surface wave
in water. In the present case, sound pressure falls exponentially with
normal distance from the oscillating plate:

$$\underline{p} = \underline{p}_{00} e^{-Jk_B x - \sqrt{k_B^2 - k^2}\, z}. \tag{15.7a}$$

The corresponding displacements are illustrated in figure 15.1, left, by
means of the visual device already familiar to us from figures 13.2 and

Figure 15.2
Hydrodynamic short circuit and sound radiation below the critical frequency. Top, in an infinite plate; bottom, in a finite plate (after Heckl).

13.4. In the present case, there is no radiation below the critical frequency.

Only above the critical frequency is the surface wave of (15.7a) replaced by an obliquely radiated plane wave. Its pressure is given by

$$\underline{p} = p_{00} e^{-jk_B x - j\sqrt{k^2 - k_B^2}z},$$ (15.7b)

as shown in figure 15.1, right.

The *trace wave number*,

$$k_B = k \sin \vartheta,$$ (15.8)

describes the propagation direction ϑ, defined with reference to the z-direcion:

$$\vartheta = \arcsin \frac{k_B}{k} = \arcsin \frac{\lambda}{\lambda_B} = \arcsin \frac{c}{c_B}.$$ (15.9)

This last relationship allows us to interpret ϑ as a *Mach angle*. Here, the supersonic phase speed of the flexural wave takes the place of velocity of the front of an object traveling at a supersonic speed. Since c_B increases as $\sqrt{f}$, the direction in which sound is radiated begins parallel to the plate in the limiting case $f = f_{cr}$, then approaches the z-direction without ever reaching it.

This complete separation between a frequency range with radiation and one without it vanishes if we consider a finite plate. At the boundaries, as illustrated in figure 15.2, regions occur whose flux is no longer compensated for by that of neighboring antiphasic regions (Cremer and Heckl

1973, p. 469). This phenomenon also explains the increasing discrepancy when comparing results for an infinite plate to those for one with a decreasing ratio of the length of the plate L to that of the flexural wave λ_B.

The problem of radiation from finite plates was given an exact mathematical treatment in three dimensions by Gösele (1953) in connection with questions of sound isolation by suspended ceilings in buildings.

We are interested only in outlining the behavior here, so we will examine the problem only in two dimensions. We will assume that two plates oscillate one above the other, as is the case for string instruments. We will, however, assume that these plates vibrate at all points in opposite directions so that the sound pressure field is symmetrical about a plane halfway between and parallel to them. This plane may be conceived of as a rigid wall (see figure 15.3). Given this assumption, we may approximate the regions of the waves in opposite phase by point radiators with the same flux. We may then calculate the directional characteristics in the x, z-plane using point-radiator synthesis as in (13.19). When there are n equidistant point radiators of alternating phase, the synthesis takes the form

$$\underline{p} = \frac{j\omega\varrho}{4\pi r_0}\underline{q}e^{-jkr_0} \sum_{m=1}^{n} e^{j\frac{m}{n}kL\cos\vartheta}(-1)^m. \tag{15.10}$$

(The origin lies at a distance L/n outside the plates, where L/n is the separation between two of the sources.)

If we set $L/n = \lambda_B/2$ and assume that

$$\frac{kL}{n} = \pi\frac{\lambda_B}{\lambda} = \pi\frac{\lambda_{cr}}{\lambda_B} = \frac{\pi}{4}n, \tag{15.11}$$

and consequently that $\lambda = \lambda_B$ with $n = 4$, then (15.10) becomes

$$\underline{p} = \frac{j\omega\varrho}{4\pi r_0}\underline{q}e^{-jkr_0} \sum_{m=1}^{n} e^{jmn\frac{\pi}{4}\cos\vartheta}(-1)^m, \tag{15.12}$$

as illustrated in figure 15.3.

Although the formula was derived under the assumption that there are antiphasic point radiators, it is formally valid even when $n = 1$ and $\lambda_B/\lambda = 1/4$. The directional characteristic is nearly circular in this case. The next case, in which $n = 2$ and $\lambda_B/\lambda = 2/4$, has a dipole-like directional characteristic. In this case, there is no radiation in the z-direction; but this is no longer so when $n = 3$ and $\lambda_B/\lambda = 3/4$; the three sources do not cancel one another out in the z-direction. The lobe in the z-direction is,

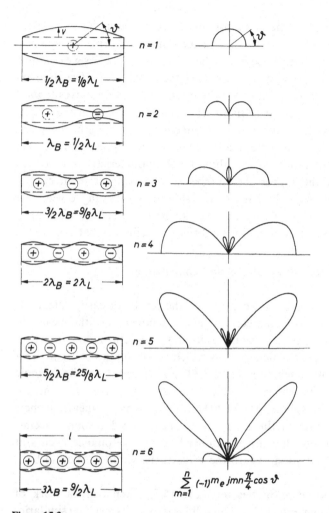

Figure 15.3
Directional characteristics of a symmetrically oscillating two-plate system calculated using point-radiator synthesis. The sources are separated by half-wavelengths of the flexural waves.

however, small; also, the radiation in the x-direction has increased. Radiation in the x-direction reaches a maximum at the critical frequency, corresponding here to the case in which $n = 4$ and $\lambda_B = \lambda$. As we continue the progression ($n = 5, 6$; $\lambda_B/\lambda = 5/4, 6/4$), the radiation pattern approximates that of an infinite plate; the direction of radiation moves toward longer values of ϑ, defined in figure 15.3 as measured from the x-axis. The finite length of the plate produces a finite maximum, but the sharpness of the maximum increases, and small secondary peaks appear in increasing numbers. The derivation of this directional characteristic is based on symmetry; still, the higher the number n, the better the directional characteristic agrees with that on one side of a radiating object (for example, the top plate of a violin) even if the other side (for example, the back plate) is at rest or exhibits a different oscillatory pattern.

15.2 Experimental observations and conclusions

Now, as before, we must not expect that the results of easily calculable cases will indicate more than certain general tendencies of the measured directional characteristics. Having said this, we return once more to Meinel's (1937) measurements, from which the examples in figure 15.4 are taken. The solid line representing 2,323 Hz reminds us of the example $n = 5$ figure 15.3, in which the peaks lie at oblique angles. In contrast, the dashed line corresponding to 2,630 Hz, only a slightly higher frequency, exhibits a maximum in the z-direction. The reduced response seems, however, to indicate that no pronounced natural resonance occurs at this frequency. Radiation toward the back of the violin is small in both cases.

We find the same properties in figure 15.5, taken from a work of Backhaus and Weymann (1939). The solid line exhibits a strong similarity to that of figure 15.4, made at nearly the same frequency. However, the comparison shows once again that directional characteristics vary from one instrument to another. This varation can be due only to differences in the nodal patterns. For the same reason, directional characteristics of the same instrument in figure 15.4 at frequencies only one or two whole-tones apart prove here to be very different.

In order to arrive at general conclusions—for example, with regard to the placement of instruments in the orchestra—it is necessary to simplify the representation of the results. J. Meyer (1972) sought to do

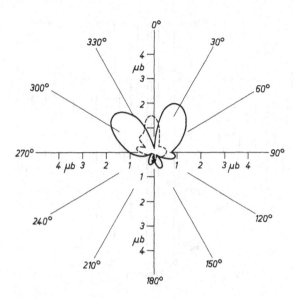

Figure 15.4
Measured directional characteristic of a violin in the y, z-plane at 2,323 Hz (solid line)
and at 2,630 Hz (dashed line) (after Meinel).

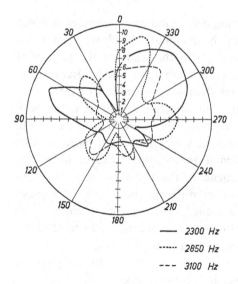

—— 2300 Hz

······ 2850 Hz

– – – 3100 Hz

Figure 15.5
Measured directional characteristic of a violin in the y, z-plane at 2,300 Hz (solid line);
2,850 Hz (dotted line); and 3,100 Hz (dashed line) (after Backhaus and Weymann).

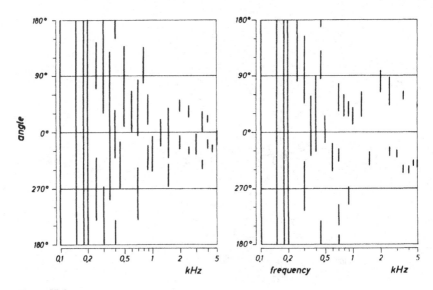

Figure 15.6
Measured ranges of angles in which the intensity of sound radiated by a cello does not
fall below half the maximum value. Left, in the y, z-plane; right, in the x, z-plane (after
J. Meyer).

this as shown in figure 15.6, by using bars to indicate the angles at which
intensity sinks to no less then half the maximum. (Such representations
of half-power bandwidth are common in the study of the physics of
vibration, especially in connection with resonance peaks.) Representations such as Meyer's serve better than a compendium of directional
characteristics to show the change in directional characteristics with
frequency. Sound is radiated in all directions at low frequencies; then,
with rising frequency, the characteristics vary more and more, so that
radiation in directions perpendicular to the plates can even become
minimal; finally, at higher frequencies, the radiation separates into
maxima in an increasing number of different directions. (We must,
however, note that the absolute value of the maximum intensity as
shown in the diagram can be very different from one frequency to another.)
At very high frequencies, radiation as seen from the y, z-plane is only in
the z-direction: in other words, it is in the x, z-plane. In this plane, it
separates into two maxima symmetrical to the z-axis.

Figure 15.7, also from Meyer's work, shows that this pattern represents

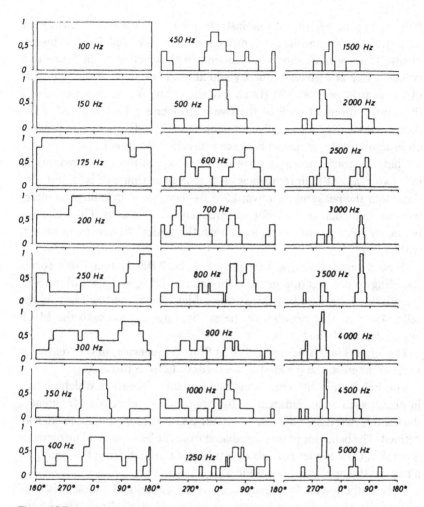

Figure 15.7
Probability distributions of preferred directions for five different celli showing angular regions in the *x, z*-plane in which intensity does not fall below half the maximum value, at various frequencies (after J. Meyer).

more than a peculiarity of one instrument. Here, representations of the same frequencies and in the same planes are assembled for five instruments. The diagrams show the frequency with which a bar in figure 15.6 will be found at a given frequency and angle. (The slope at the right side of the central peak at 350 Hz clearly shows the five steps representing the contributions of each of the five instruments.) The position of the peaks in these diagrams is of general relevance; the diagrams also provide an insight into the variation between individual instruments.

These diagrams not only prove that the separation of the maximum into two peaks at high frequencies is a general attribute of celli, but also show that the peaks move toward $\vartheta = 0$—here again meaning the direction perpendicular to the front—as frequency is increased. This behavior is exactly what we would expect above the critical frequency in accord with (15.9).

If we assume the critical frequency to be 2,800 Hz—a value corresponding to propagation in the x-direction, with $c_x = 5.3 \times 10^5$ cm s^{-1}, longitudinally in wood of the average thickness of the top plate of the cello, 4.4 mm—the position of the maxima agrees well with the Mach angle obtained from (15.9).

The angle $\vartheta = 0$ is never reached at finite frequencies; in the frequency range of interest, the peaks still clearly deviate from this angle.

This behavior is different from that of a plate all parts of which vibrate in phase, such as a loudspeaker diaphragm. With increasing frequency, the main forward lobe of such a radiator becomes more and more sharply defined. The behavior of sound radiated from the bell of a wind instrument is similar. (The higher partials of a tuba are reflected from the ceiling in a hall; in the open air, they are lost.)

Only in the lowest frequency range is the directional characteristic of a violin simple: radiation is more or less equal in all directions. Above this range, there is no way to predict the directional characteristic. It can be said only that a maximum points first in one direction, then in another, but that on the average the radiation fills the half-space into which the top plate faces. Definite directions, such as given by the Mach angle, occur only at very high frequencies.

For this reason, it is also very doubtful whether a soloist's turning the front of an instrument toward the audience has any significant effect; in fact, a compensation of the partial fluxes frequently occurs precisely in the z-direction as parts of the front of the instrument oscillate antiphasically.

Wavelength Small in Comparison to the Source Dimensions

To the contrary, it has been confirmed over and over that the first violins of an orchestra can be heard well from seating positions at the rear of the stage, as long as these are above the instruments' x, y-plane. Listeners are at a significant disadvantage only where the head and upper body of the violinist or the entire body of the cellist shadow the instrument; we discussed this problem in the previous chapter. Complicated directional characteristics and variability from one instrument to another are, after all, characteristic of string instruments.

References

Backhaus, H., and G. Weymann, 1939. *Akust. Z.* **4**, 302.

Cremer, L., and M. Heckl, 1973. *Structure-Borne Sound.* Translated by E. Ungar. New York: Springer.

Gösele, K., 1953. *Acustica* **3**, 243.

Meinel, H., 1937. *Akust. Z.* **2**, 62.

Meyer, J., 1972. *J. Acoust. Soc. Amer.* **51**, 1994.

16 The Influence of the Room

16.1 Room dimensions comparable to the wavelength

In the introduction to this book, we have already noted that no room in which a string instrument is played is small enough to have a physical influence on its vibrations. In other words, if we excite the instrument using a mechanical or electroacoustical apparatus, the same oscillations occur in the string and the body of the instrument whether the room is large or small, reverberant or dead.

The room, however, does make a great difference to the player; and this difference can have a profound influence on the quality of playing. The player needs to feel that the "tone" he generates is transmitted into the room; he can only judge this if he can hear the room responding to his playing.

Since the player is very near the sound source, he needs a louder room response than a listener. For a listener, spatial acoustic cues are audible which are masked for the player by the direct sound. To this extent, the room is even more important to the listener than to the player.

We begin our observations with small rooms. As always, we define the term "small" in comparison to the wavelength.

All players, even performing artists, must practice or warm up for performances in small rooms; most home music practice and performance is in rooms with rather low ceilings.

Most small rooms are rectangular. Consequently, their lowest natural frequencies, which are largely unaffected by the presence of people or furniture, can be calculated using the formulas (11.62) and (11.63). These formulas are valid notwithstanding that we originally introduced them in connection with the natural frequencies of the much smaller cavity of the violin body.

The lowest natural frequencies in the grid in figure 11.6 are far apart; the individual resonance peaks can be detected in the plot of sound pressure against frequency. This is even truer if some of the peaks are absent. Peaks corresponding to waves traveling along the length or width of the room may be damped by curtains or upholstered furniture along one of the walls; the room may be very long or wide, or perhaps it may open into a neighboring room. The remaining natural ocillations between the floor and ceiling then stand out even more.

When the ceiling height is 2.4 m, the lowest of the vertical natural frequencies is approximately 70 Hz, the d ($\equiv d_{-1}$) on the c string of the cello. The next resonance is at 140 Hz, the d ($\equiv d_0$) on the c string of the

viola, and the third, at 210 Hz, is already within the frequency range of
the g string of the violin.

These resonances can have exactly the same effect on the player and the
listener as if they were due to the instrument itself.

Artists do not attribute any negative effects to the additional resonance
peaks which a cello or double bass can produce in a riser by means of
energy transmitted through the peg (see section 14.5). But room resonances
are different, because the room in which the artist warms up is different
from the one in which he or she performs.

For this reason, it is a good idea to damp out resonances in warm-up
and practice rooms (for example in conservatories), using tuned absorbers.

The formula (11.62) used in calculating natural frequencies assumes
that the walls of the room are parallel. It is therefore often assumed
incorrectly that resonances can be prevented by placing the walls out of
parallel or by using a corrugated ceiling. These measures are not effective
in preventing resonances whose wavelengths are long—the ones with
which we are here concerned. The only effect is to make calculation of the
natural frequencies more difficult or impossible. The number and approximate values of the natural frequencies are not affected. We examined the
number of natural frequencies in section 11.4 using the asymptotic
formula (11.66) for the number of natural resonances below a given
frequency in a shallow, rectangular cavity. Extended to three dimensions,
this leads to

$$\lim_{f \to \infty} N = \frac{4}{3}\pi \frac{Vf^3}{c^3}. \tag{16.1}$$

The only way to quell a room resonance is a sufficient and specially
tuned absorbing treatment on the walls and ceiling. A carpet would not
be effective at these low frequencies.

In section 16.3, we will briefly describe why absorption at low frequencies is attainable only using vibrating elements. These must be tuned
to the natural resonances which are to be damped. Details of the construction of such devices, however, are beyond the scope of this book.

16.2 Statistical treatment of rooms of moderate size

The number of resonances below a given frequency increases with the
volume, as (16.1) shows. Differentiating this formula, we obtain the

number of resonances within a frequency band:

$$\lim_{f \to \infty} \frac{\Delta N}{\Delta f} = 4\pi V \frac{f^2}{c^3}. \tag{16.2}$$

Since a musical semitone is defined by

$$\frac{\Delta f}{f} = 0.06, \tag{16.3}$$

we obtain

$$\Delta N_{1/2} = 2 \times 10^{-8} V f^3 \tag{16.4}$$

as the number of room resonances per semitone, with V in m^3. In a room whose volume is only 200 m^3, the size of an ordinary classroom, this, even in the range of the lowest fundamental of the cello ($f = 64$ Hz), leads to at least one resonance at every semitone. This may still be a relatively small number of resonances; but there are already eight times as many at the lowest fundamental of the viola. If we increase the volume of the room to 1,000 m^3, the number of resonances is quintupled.

Briefly, then, the number of resonances increases rapidly with increasing frequency and room size; due to damping, each resonance peak has a certain finite width, so it can respond to excitation over a range of frequencies. Therefore, even a sinusoidal tone will excite several resonances.

We may, then, employ statistical methods. Through these, it is possible to determine only the amplitude of sound pressure, or its square; the statistical equations express only average energy or power.

If the excitation is constant—for example, a sustained violin tone—the radiated power P_0 must equal the sum of all power losses in sound-absorbing elements of the room. If we assume that these are distributed over surfaces S_i with different absorption coefficients, we obtain the power balance

$$P_0 = \sum P_i. \tag{16.5}$$

The values of P_i depend on the angles of incidence ϑ. The absorption coefficient α_i (the ratio of absorbed to incident sound intensity) may depend on ϑ; but also, the surface S_i, as figure 16.1 shows, intercepts only a projection, $S_i \cos \vartheta$, of a wave arriving at an angle ϑ. Assuming that the wave is of intensity J, we obtain P_i by means of the relationship

$$P_i(\vartheta) = J S_i \cos \vartheta \, \alpha_i(\vartheta). \tag{16.6}$$

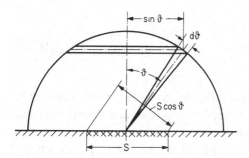

Figure 16.1
Sketch illustrating (16.6) and (16.7).

In averaging over all possible angles of incidence, statistical room acoustics assumes that all are equally probable; in order to achieve this, the absorbing surfaces must be rather equally distributed in the room. Granted this assumption, the component of intensity falling within the range of angles of incidence between ϑ and $\vartheta + d\vartheta$ is proportional to the ratio of the corresponding zone of a sphere to its total area:

$$dJ = \frac{2\pi \sin \vartheta d\vartheta}{4\pi} Ec = \frac{Ec}{2} \sin \vartheta d\vartheta. \tag{16.7}$$

In this equation, c represents the speed of sound and E the energy density. E is measurable, as it is proportional to the square of pressure:

$$E = \frac{\tilde{p}^2}{\varrho c^2}. \tag{16.8}$$

This relationship is easiest to understand for a plane wave, from the relationship between intensity and energy density,

$$J = \frac{\tilde{p}^2}{\varrho c} = Ec. \tag{16.9}$$

In integrating over all possible directions of incidence, it must also be noted that only waves moving toward the wall contribute to P_i; the range of integration of ϑ is, then, only from 0 to $\pi/2$. Consequently,

$$P_i = \frac{Ec}{2} S_i \int_0^{\pi/2} \alpha_i(\vartheta) \cos \vartheta \sin \vartheta d\vartheta = \frac{Ec}{4} \alpha_{im} S_i, \tag{16.10}$$

where

$$a_{im} = \frac{\int_0^{\pi/2} \alpha_i \cos \vartheta \sin \vartheta d\vartheta}{\int_0^{\pi/2} \cos \vartheta \sin \vartheta d\vartheta} \qquad (16.11)$$

represents an average accounting for the weighting factors $\cos \vartheta$ and $\sin \vartheta$. The first of these represents the decrease in projected area, and the second, the greater probability of oblique incidence.

Substitution of (16.10) into (16.5) gives the relationship between the power P_0 of the source and the energy density:

$$P_0 = \frac{Ec}{4} \sum \alpha_{im} S_i. \qquad (16.12)$$

The sum characterizing the room,

$$\sum \alpha_{im} S_i = A, \qquad (16.13)$$

is called the *equivalent absorption area*, in keeping with its dimension. Absorbing objects within the room such as chairs and persons may each contribute to this sum by increments δA_k, so we expand it to two terms:

$$A = \sum \alpha_{im} S_i + \sum n_k \delta A_k. \qquad (16.14)$$

If the same amount of power is radiated into two different rooms, the resulting energy densities are inversely proportional to the equivalent absorptive areas:

$$\frac{E_1}{E_2} = \frac{A_2}{A_1}. \qquad (16.15)$$

We have already made use of this equation in section 14.5, in connection with a measurement of the sound transmitted through the peg of a cello.

In a room whose equivalent absorptive area is known, (16.12) allows us to measure the sound power radiated into the room, since this is proportional to the square of sound pressure in the room. This is the most important application of (16.12); a typical example is the testing of a noise source in a room designed to be as reverberant as possible.

Beldie (1975) applied this method to the measurement of curves of total radiated power of violins. The upper curve in figure 16.2 shows the recorded sound pressure level as a function of frequency; the lower curve shows the input admittance of the bridge of the same instrument, repeated

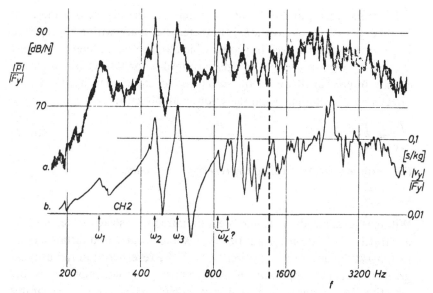

Figure 16.2
Sound pressure level measured in a reverberant room (top) and a logarithmic recording of bridge input admittance (bottom) as functions of frequency (after Beldie).

from figure 10.1. The frequency and admittance scales are logarithmic. The use of a reverberant room in measuring the upper curve is apparent in the superimposed small fluctuations. The coil of the electrodynamic transducer delivered a constant-force frequency-modulated signal to the input of the violin bridge; he used this narrow-band noise signal in order to excite several of the room's resonances at all times. Nonetheless, the lowest resonant peaks show clearly; they are also apparent in the admittance curve. As expected, the Helmholtz resonance is much more pronounced in the radiated sound than in the admittance curve; higher resonances, however, are less well radiated, due to the subdivision of the top and back plates into antiphasic regions.

In recording a resonance curve in a reverberant room, it is expedient to average the results at several microphone positions. This procedure is less arbitrary than using a single microphone location in an anechoic chamber. A method intermediate between these two extremes was used by J. Meyer (1975). He measured sound pressures in an anechoic room, with six microphones in different directions in the plane of the bridge; the microphone outputs were summed after rectification.

Formula (16.12) allows us also to determine the degree of validity of frequency curves and directional characteristics made with a microphone close to the sound source in rooms which are not completely anechoic. Reflected sound is overwhelmingly responsible for the statistical energy density in (16.12); if we set this equal to the energy density due to the direct sound, assuming equal radiation in all directions,

$$\frac{4P_0}{cA} = \frac{P_0}{4\pi r^2},$$
(16.16)

the result is a distance called the *reverberation radius* r_H:

$$r_H = \sqrt{A}/(4\sqrt{\pi}).$$
(16.17a)

Within this radius, direct sound predominates in the steady state. We may arbitrarily say that the distant field of a source begins at a distance twice its longest dimension (see section 14.1); if the reverberation radius is far greater than this, approximate measurements of frequency curves and directional characteristics of violins are possible even if there is measurable reverberation in a room.

The measurement error is reduced at a pronounced maximum of a directional characteristic. There, r_H increases to

$$r_H = \sqrt{A\gamma}/(4\sqrt{\pi}),$$
(16.17b)

where γ is the ratio of maximum to average intensity.

Assuming that the average absorption coefficients are nearly the same, then r_H must increase with the room's length, width, and height. In general, then, if fields in which direct sound predominates are to be measured in a nonanechoic room, the room must be rather large.

Musicians' comparisons of instruments in a large concert hall, then, may not be affected by particular characteristics of the hall, as long as the listener stays within $r_H/2$ of the player. Certainly, the hall must also be quiet.

16.3 Reverberation

At the beginning or end of an interval during which power is introduced into a room, the energy content EV of the room changes according to the difference between the power input and the power loss:

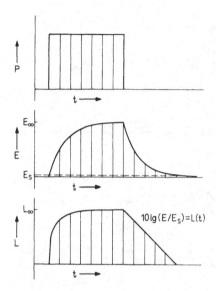

Figure 16.3
Time functions showing growth and decay. Top, power input; middle, energy density; bottom, corresponding level.

$$\frac{d(EV)}{dt} = P - \frac{Ac}{4}E. \tag{16.18}$$

It follows that the energy decays exponentially after the power input has stopped ($P = 0$):

$$E = E(0)e^{-Act/(4V)}. \tag{16.19}$$

Conversely, when the power input begins, the room's energy content is calculated as the final, steady-state value minus the same exponential function:

$$E = E(\infty)(1 - e^{-Act/(4V)}). \tag{16.20}$$

In figure 16.3, the time function of input power is shown at the top; the energy density in the room, with its growth and decay, is shown in the middle. The dashed line near the bottom of the middle drawing represents the energy density E_s corresponding to the threshold of audibility. The time interval from the end of power input until the energy density reaches this value is called the *duration of reverberation*. The energy

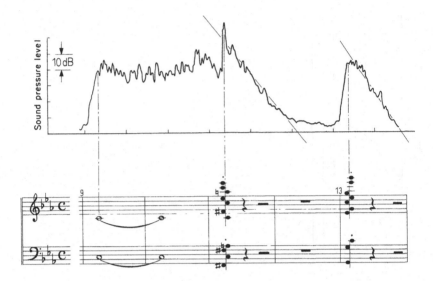

Figure 16.4
Level recordings showing reverberation during a concert (Beethoven, Coriolanus Overture op. 62, measures 9–13) (after E. Meyer and Jordan).

density level corresponding to this value is shown at the bottom. In the level diagram, the reverberation appears as a falling straight line, reaching the abscissa at the threshold of audibility.

The level diagram is better adapted to a quantitative evaluation of reverberation. Level recorders, which register the level directly, have consequently been developed. Figure 16.4 shows a diagram produced by such a recorder during the playing of the first measures of Beethoven's Coriolanus Overture (Meyer and Jordan 1935). Straight lines have been fitted to the decay slopes; the slope of these straight lines characterizes the decay constants in (16.19).

The first researcher to investigate room reverberation quantitatively was Sabine (1900, 1923). He introduced the concept of the *reverberation time*, the time necessary for the level to fall 60 dB or for the energy density to fall to 10^{-6} of its steady-state value. He set

$$e^{-\mathrm{Act}/(4V)} = 10^{-6t/T},$$
(16.21)

from which it follows that

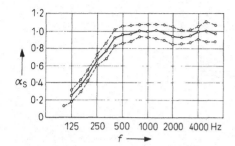

Figure 16.5
Frequency curve of the absorption coefficient of a 5-cm layer of rock wool, with
confidence intervals, as measured in various laboratories (after Kosten).

$$T = \frac{6 \ln 10}{4c} \frac{V}{A},$$ (16.22a)

and with V in m³ and A in m²,

$$T = 0.163 V/A.$$ (16.22b)

The reverberation time is of the same order of magnitude as the duration
of reverberation Sabine determined experimentally, though the two
should not be confused with one another. The reverberation time depends
only on the room, while the duration of reverberation depends also on
the steady-state initial value E_0 and the threshold value E_s determined by
the background noise.

The absorption coefficients are dependent on frequency, and so, then,
are the values of the equivalent absorption area A, and consequently of
the reverberation time. Conversely, measuring the reverberation time at
different frequencies is a simple way to determine the frequency dependence of the absorption coefficients.

Porous materials generally absorb sound more completely with increasing frequency. Figure 16.5 shows a typical dependence of the absorption
coefficient on frequency (Kosten 1960). Such materials are always found
in a music room, in upholstery and in the clothing of persons in the
room. Since, then, a significant part of the value of A is set by the number
of persons in the room, the quotient V/A governing the reverberation time
is largely determined by the volume per seat.

Rooms in which high frequencies are absorbed disproportionately

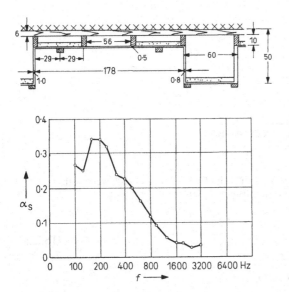

Figure 16.6
Absorption coefficient vs. frequency for a coffered wood ceiling measured in a reverberation room.

sound dull. Special absorbers effective at lower frequencies are necessary in order to deal with this problem. These almost always consist of elements excited into vibration at their natural frequencies. Wooden panels suspended in front of walls and below ceilings are the most common type of such absorbers. (We have already noted their effects in our discussion of risers in section 14.5.) Figure 16.6 shows the frequency dependence of the absorption coefficient of a coffered wooden ceiling, measured in a reverberation room.

Helmholtz resonators, too, have been used to absorb low frequencies. It is interesting that the same means used to increase the radiation of the violin are used in room acoustics for absorption. To be sure, the damping is greater in the absorptive devices.

For a long time after Sabine's pioneering work, the reverberation time and its frequency dependence were regarded as the sole criterion of the quality of the acoustics of a room. They remain important today; there is an appropriate range of them of rooms of every size, and for every use to which rooms may be put. The room is too "dry" if the reverberation time is too short. Frequency output curves with many peaks, complicated

directional characteristics, and strong individual variation, are characteristic of string instruments; a mixture of many long-lasting reflections is
best for them. But even for string instruments, the reverberation time can
become so long that the listener can no longer distinguish individual notes
in rapid passages.

There are limits to the desirable frequency dependence of reverberation,
too. In larger rooms, it is good if the reverberation time rises toward the
bass; in smaller rooms, such a characteristic can lead to an audible
"overhang" in the bass. The question is still open, however, as to the best
reverberation characteristics for a good string sound.

16.4 The room impulse response

If only the reverberation time were important, the particular shape of the
room would not matter. Such is however, not the case, even with smaller
rooms. A room with a circular floor plan or a cupola is poorly suited for
listening to music or for taking acoustical measurements, due to acoustical
focusing effects.

Faults in the shape of larger rooms are more noticeable yet. This
becomes evident if we record pressure versus time following the radiation
of an impulse in two rooms of the same shape but different sizes in such a
way that we can distinguish the direct sound and individual reflections.
The larger the room, the more these will be separated in time.

We have already encountered impulse responses in section 8.4, in
connection with the problem of the excitation of the string. There, they
occurred in a one-dimensional space. In room acoustics, such diagrams
are called *echograms*, since they were originally used in research on the
presence and origin of echoes. An echo is a reflection for which the delay
is most evident. In order for the "auditory event" of an echo to appear,
the minimum delay between the direct sound and the reflection must be
approximately 100 ms. If the delay is less than 50 ms, our sense of hearing
perceives the repetition of a signal only as a welcome increase in its
loudness and definition.

Although the only difference between the impulse responses of our two
rooms of the same shape is in the time scale, the subjective evaluation is
entirely different. The limits mentioned above are exceeded more easily
in large rooms than in small ones.

The six echograms, recorded as $p^2(t)$, at the top in figure 16.7, recorded

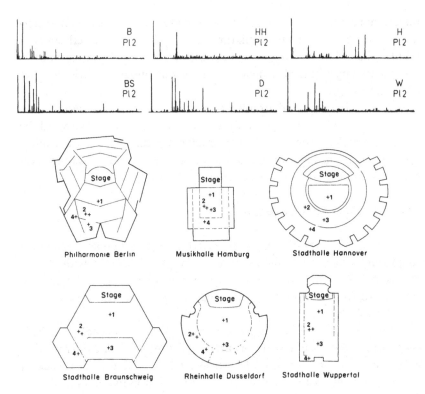

Figure 16.7
Echograms ($p^2(t)$) recorded in the six concert halls whose floor plans are shown at the bottom. Source: a pistol on the stage. Pressure-sensitive microphones at position 2 in each case (after Lehmann).

by Lehmann (1976), show how greatly the impulse responses of rooms can differ. The examples record the response to a pistol shot on the stage at the comparable listener positions number 2 in each of the six concert halls whose floor plans are shown at the bottom in figure 16.7.

Widely differing impulse responses occur not only in rooms whose shapes differ as much as those in figure 16.7, but even in the same room when the sending and receiving positions are different.

The same is true of all criteria based on the evaluation of echograms. Such criteria mostly lead to a comparison of the "early" energy component with the rest, the "late" energy, or the total energy. Thiele (1953) determined the ratio

$$\vartheta = \int_0^{50\,ms} p^2\,dt \Bigg/ \int_0^{\infty} p^2\,dt;\qquad\qquad\qquad (16.23)$$

this is called the *distinctness coefficient*, since the upper limit of the integral in the numerator corresponds to the "limit of perceptibility" at 50 ms for speech. On the basis of investigations using musical signals, Reichardt et al. (1975) suggested that the boundary between early and late be 80 ms for music, instead of 50 ms. Reichardt suggested using the logarithmic expression

$$C = \log_{10}\left[\int_0^{80\,ms} p^2\,dt \Bigg/ \int_{80\,ms}^{\infty} p^2\,dt\right]dB \qquad\qquad (16.24)$$

as a measure of the clearness of musical performances. Although musical information is of a different character than that of speech, it is certain that some degree of distinctness is desirable there as well, and it is appropriate to give this quality another name. On account of its logarithmic scale, C may be called the *clearness index* (Cremer and Müller 1982, vol. 2, p. 432).

Another criterion that emphasizes the importance of early reflections, without introducing a more or less doubtful subjective boundary, is the *center-of-gravity time*, or for short, the *center time* t_c, defined by

$$t_c = \int_0^x tp^2\,dt \Bigg/ \int_0^{\infty} p^2\,dt. \qquad\qquad\qquad (16.25)$$

There are optimal ranges for the criteria ϑ, C, and t_c. Musical presentations in anechoic chambers are unsatisfactory, and so are those in reverberant rooms.

Certainly, however, an acoustical consultant must take care that the largest possible number of early reflections, those in the first 80 ms after the direct sound, reach the player and listeners. Musicians need early reflections also, in order to hear one another.

The direction from which reflections arrive is also important. Since we can distinguish directions of arriving sound much better in the horizontal plane than in the median plane, we are more aware of reflections from the sides than of those from overhead. This does not, however, decrease the importance of early reflections from the ceiling to loudness and distinctness (Barron 1971). But the "spatial impression," to use Barron's term, is based on early reflections from the sides (Kuhl 1976). Recent experiments have attempted to correlate this impression with an objec-

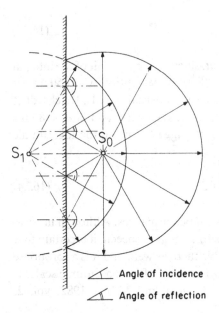

Angle of incidence

Angle of reflection

Figure 16.8
Reflection from a flat wall.

tively measurable criterion based on the signals received by the microphones at the left and right ear positions of a dummy head. However, the
present author believes that it would be premature to include these
methods, still under discussion, in this brief survey.

16.5 Additional geometric considerations in large rooms

In order to design the walls and ceiling of a room so that early reflections
reach the player and listeners, it is helpful to trace sound rays. Fermat's
principle states that sound follows the path along which the travel time is
shortest; consequently, sound rays are straight lines, as long as they do
not encounter obstacles.

When the obstacles are smooth, large surfaces, and at high frequencies,
it is also appropriate to regard reflection as geometric: the incident and
reflected rays lie in the same plane as the perpendicular to the reflecting
surface; and the angles of incidence and reflection with respect to this
perpendicular are equal. Reflection from a plane wall may, then, for
example, be modeled by a mirror-image source at the same distance from
the wall as the actual source, but behind it (see figure 16.8).

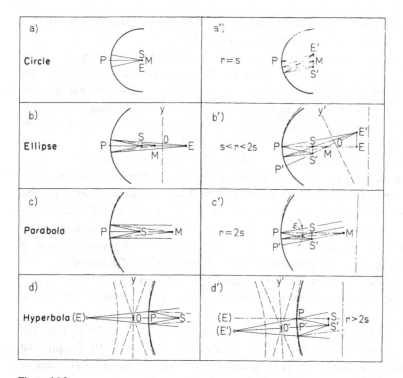

Figure 16.9
Effect of a concave mirror of increasing radius MP, for a fixed source distance SP and receivers at E. Left, with the source on the axis; right, off-axis.

Due to their simplicity, geometric constructions using rays have been applied even when their validity is questionable; still, this approach has proven useful in arriving at simple approximations.

Geometric room acoustics is the oldest type of room acoustics, first applied only to curiosities such as whispering galleries, multiple echoes, and the like, most of which involved concave reflecting surfaces.

Most such phenomena could be explained simply, using laws for concave mirrors borrowed from optics. Figure 16.9 surveys the characteristic cases of these laws: the distance between the source S and the vertex P is kept the same. The curvature of the wall, given by the center M, varies. Reflected rays are concentrated at a receiving point E which is a) at the source; b) behind it; or c) at an infinite distance. In d), the sound appears to originate at a mirror-image source behind the wall and more distant from it than the actual source; since *EP* is larger than *SP* in this case, the divergence of the rays is reduced.

The application of geometric laws in directing sound began in this century.[1] At first, it was thought that concave surfaces behind an orchestra would be advantageous. Generally, however, it has proven better to use reflection to increase the divergence of sound rays rather than to diminish it or even concentrate the rays at a focus. Convex surfaces increase divergence, as is clear from case d) in figure 16.9 if the source (in this case a mirror-image source) and receiver are exchanged.

An example of the use of convex surfaces is shown in the longitudinal section of the Berlin Philharmonic Hall shown in figure 16.10, top. The ceiling is suspended in convex sections like those of a tent. The reflectors hanging above the orchestra, shown in black in the floor plan, serve to assure that the musicians can hear one another, and project the string sound into the seating close to the orchestra. These reflectors are convex too, toward the sides as well as the front and back. Also, the reflectors shadow the unavoidable peak angle of the ceiling, preventing its generating a strong echo.

The effort to generate early reflections, particularly those from the sides, is also evident in the floor plan shown in figure 16.10. Despite the desire of the architect, Hans Scharoun, to place listeners on all sides of the sound source, reflections were achieved by interrupting the ascending rows of seating by vertical walls; the overall appearance is similar to that of a terraced vineyard.

The excellent concert halls built in the nineteenth century were narrow, and had balconies with horizontal soffits along their side walls; this combination generated lateral "cue-ball" reflections and directed sound down to the floor. Today, on the other hand, it is considered important to provide a good view of the stage from all seating positions, even with a larger number of listeners.

The efficacy of correctly placed and angled reflecting surfaces can be tested at the drawing board; the task is simple if the surfaces are perpendicular to the plane of the drawing. Alternatively, optical modeling techniques may be used: the sound ray is replaced with a light ray, and optical mirrors are used to model the walls. Today, the paths of rays

---

[1] Readers interested in pursing this subject further are referred to Cremer and Müller (1982, vol. 1, ch. 5).

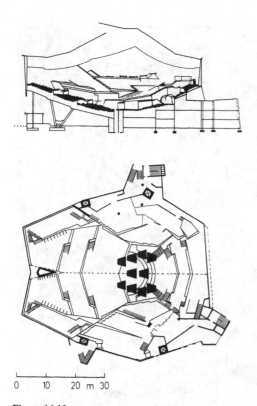

0   10   20 m 30

Figure 16.10
Longitudinal section and floor plan of Philharmonic Hall, (West) Berlin (architect, Hans
Scharoun).

can also be traced using computers (Krokstadt et al. 1968). It is more
effective yet to build real acoustical models and to conduct tests using
sound whose frequency is raised in inverse proportion to the scale of
the models. The wavelengths are reduced in proportion to the scale, and
so the Helmholtz numbers are not affected. Diffraction and diffuse
reflections, which occur when the dimensions of obstacles or surface
irregularities are comparable to the wavelength, can then be modeled
accurately.

It can be seen that such diffuse reflections (in which sound is spread
out more or less in all directions) are especially appropriate in achieving
an even distribution of sound from all directions as the number of
reflections increases. The acoustical quality of the baroque theater was

Figure 16.11
Loge wall of the Cuvilliés Theater in Munich.

largely due to this phenomenon. Figure 16.11 shows the loge wall of the Cuvilliés Theater in Munich, a particularly beautiful example.

The presence of many diffuse reflections, particularly early reflections, is especially important for string instruments, due to their complicated frequency dependences and directional characteristics.

References

Barron, M. J., 1971. *J. Sound. Vib.* **15** (4), 475.

Beldie, J. P., 1975. Dissertation, Technical University of Berlin.

Cremer, L., and H. A. Müller, 1982. *Principles and Applications of Room Acoustics.* Tr. by T. J. Schultz. London: Applied Science.

Kosten, C. W., 1960. *Acustica* **10**, 400.

Krokstadt, A., S. Strom, and S. Sorsdal, 1968. *J. Sound Vib.* **8** (1), 118.

Kuhl, W., 1976. Lecture before the DAGA, Heidelberg.

Lehmann, P., 1976. Dissertation, Technical University of Berlin.

Meyer, E., and W. Jordan, 1935. *Elektr. Nachr. Techn.* **12**, 217.

Meyer, J., 1975. *Instrumentenbau* **29**, 2.

Reichardt, W., O. Abdel Alim, and W. Schmidt, 1975. *Acustica* **32**, 126.

Sabine, W. C., 1900. *The American Architect.*

Sabine, W. C., 1923. *Collected Papers*, No. 1. Cambridge, Mass.: Harvard University Press.

Thiele, R., 1953. *Acustica* **3**, 291.

Index

Printed in the United States
By Bookmasters